Ancient & Historic
METALS

CONSERVATION AND SCIENTIFIC RESEARCH

Ancient & Historic
METALS

CONSERVATION AND SCIENTIFIC RESEARCH

Proceedings of a Symposium
Organized by the J. Paul Getty Museum
and the Getty Conservation Institute
November 1991

Edited by
DAVID A. SCOTT, JERRY PODANY, BRIAN B. CONSIDINE

THE GETTY CONSERVATION INSTITUTE

Symposium editors: David A. Scott, the Getty Conservation Institute; Jerry Podany and Brian B. Considine, the J. Paul Getty Museum
Publications coordination: Irina Averkieff, Dinah Berland
Editing: Dinah Berland
Art director: Jacki Gallagher
Design: Hespenheide Design, Marilyn Babcock / Julian Hills Design
Cover design: Marilyn Babcock / Julian Hills Design
Production coordination: Anita Keys

Printed in Singapore

Library of Congress Cataloging-in-Publication Data
Ancient & historic metals : conservation and scientific research :
proceedings of a symposium organized by the J. Paul Getty Museum and the Getty Conservation Institute, November 1991 / David A. Scott, Jerry Podany, Brian B. Considine, editors.
p. cm.
Includes bibliographical references.
ISBN 0-89236-231-6 (pbk.)
1. Art metal-work—Conservation and restoration—Congresses.
I. Scott, David A. II. Podany, Jerry. III. Considine, Brian B.
IV. J. Paul Getty Museum. V. Getty Conservation Institute.
VI. Title: Ancient and historic metals.
NK6404.5.A53 1995
730'.028—dc20 92-28095
CIP

Cover photograph: Bronze sheathing tacks from the HMS Sirius. Courtesy of the Australian Bicentennial Authority. Photography: Pat Baker.

PICTURE CREDITS

Bassett and Chase *Considerations in the Cleaning of Ancient Chinese Bronze Vessels.* Figures 1–4: Courtesy of the Honolulu Academy of the Arts, Honolulu. Photography: J. Bassett; Figure 5: Courtesy of the Arthur M. Sackler Gallery, Smithsonian Institution, Washington, D.C.; Figures 6–8: Courtesy of the Freer Gallery of Art, Smithsonian Institution, Washington, D.C.
Bonadies *Tomography of Ancient Bronzes.* Figure 1: Courtesy of Jason Franz; Figures 2–10: Collection of the Cincinnati Art Museum. Photography: Steve Beasley.
Chapman *Techniques of Mercury Gilding.* Figures 1–3: Courtesy of Maison Mahieu, Paris; Figures 4–5: Courtesy of the Trustees of the Victoria and Albert Museum, London.
Chase *Chinese Bronzes.* Figures 1–2: Courtesy of China Institute in America; Figure 3: Courtesy of C. S. Smith. Photography: Betty Nielson, University of Chicago; Figures 4, 9–15, 17–21. Courtesy of the Freer Gallery of Art, Smithsonian Institution, Washington, D.C.; Figures 5–6: Data from Johnston-Feller 1991; Figure 7: Courtesy of Chen Yuyan, University of Science and Technology of China, from her work with Mike Notis, Lehigh University, Bethlehem, PA. Samples made for the Freer Gallery of Art by Rob Pond, Baltimore, MD; Figure 8: Study Collection, Freer Gallery of Art, SC-B-2. Photography: E. W. Fitzhugh; Figure 16: University of Michigan Museum of Art, Ann Arbor. Estate of Oliver J. Todd, no. 1974/1.180.
Grissom *The Conservation of Outdoor Zinc Sculpture.* Figure 2: Courtesy of the Missouri Historical Society, St. Louis; Figure 3: Courtesy, The Winterthur Library, Printed Book and Periodical Collection. Figure 7: Courtesy of John L. Brown Photo.
Matero *Conservation of Architectural Metalwork.* Figures 1, 3: Courtesy of Ohio State University Archives.
MacLeod *Conservation of Corroded Metals.* Figures 1–3: Courtesy of the Australian Bicentennial Authority. Photography: Pat Baker; Figure 4: Courtesy of the British Museum (Natural History), London.
Marabelli *The Monument of Marcus Aurelius.* Figure 1: Courtesy of Accardo, Amodio, et al. (1989); Figure 2: Courtesy of Accardo et al. (1985); Figures 5a–b and 6: Courtesy of Accardo et al. (1983).
Ogden *The Technology of Medieval Jewelry.* Figure 1: Courtesy of the Yorkshire Museum, York; Figure 2: Courtesy of the Trustees of the British Museum. Photography: N. Whitfield and K. East; Figure 6: Courtesy of W. Duckzo; Figures 8–13, 16, 17, 19, 20: Courtesy of the Trustees of the British Museum; Figures 18, 22: Courtesy Fitzwilliam Museum; Figure 23: Courtesy Cambridge Centre for Precious Metal Research archive.
Schrenk *The Royal Art of Benin.* Figures 1, 2: Gift of Joseph H. Hirshhorn to the Smithsonian Institution in 1966. Photography: Jeffrey Ploskonka, National Museum of African Art; Figures 3, 4, 6, 8: Gift of Joseph H. Hirshhorn to the Smithsonian Institution in 1966. Photography: the author; Figure 5: Purchased with funds provided by the Smithsonian Institution Collections Acquisition Program in 1982. Photography: the author; Figure 7: Gift of Joseph H. Hirshhorn to the Smithsonian Institution in 1979. Photography: the author; Figure 9: Gift of Joseph H. Hirshhorn to the Smithsonian Institution in 1977. Photography: the author.

THE GETTY
CONSERVATION INSTITUTE

The Getty Conservation Institute, an operating organization of the J. Paul Getty Trust, was created in 1982 to address the conservation needs of our cultural heritage. The Institute conducts worldwide, interdisciplinary, professional programs in scientific research, training, and documentation. This is accomplished through a combination of in-house projects and collaborative ventures with other organizations in the United States and abroad. Special activities such as field projects, international conferences, and publications strengthen the role of the Institute.

Contents

Preface

The articles contained in this publication represent the proceedings of a three-day Symposium on Ancient and Historic Metals held at the J. Paul Getty Museum in November 1991. The conference was produced through the collaborative efforts of the Getty Museum and the Getty Conservation Institute with special funding provided by Harold Williams, chief executive officer of the Trust. The broad range of time periods, geography, and technologies discussed here reflects an important shared goal of the Getty Museum and the Getty Conservation Institute: to encourage the dissemination of knowledge that supports and furthers the conservation of cultural heritage throughout the world.

In planning the conference, the organizers sought to bring together conservators, conservation scientists, curators, and museum staff with an interest in the technology, history, structure, and corrosion of ancient and historic metalwork. They invited papers on subjects that not only spanned different time periods, but also reflected a wide range of subject matter. As the diversity of articles in this volume clearly shows, their efforts were amply rewarded. The objects studied range from Nigerian to Chinese bronzes, Zimbabwean to British gold, from the fittings of ships wrecked on the shores of Australia to pots buried for centuries beneath inland lakes, and from architectural iron to historical monuments.

To each of the authors we offer our warm gratitude for their work. We look forward to further collaborative conferences addressing topics that reflect important issues in the field of conservation. We would also like to extend our thanks to all those who made the symposium possible, particularly the staff of the J. Paul Getty Museum, who made most of the practical arrangements for the participants, designed and printed the program, and arranged for the speakers' travel and accommodations in Los Angeles.

In preparation of these proceedings for publication, we wish to thank the book's editors David A. Scott, Jerry Podany, and Brian B. Considine of the Getty Museum; as well as Irina Averkieff and Jacki Gallagher of the Getty Conservation Institute publications department; independent editorial consultants Dinah Berland and Dianne Woo; and everyone else who participated in bringing the valuable knowledge shared at the symposium to a larger audience. We hope the work presented here will serve to stimulate further investigations in the conservation of ancient and historic metals now and into the future.

Miguel Angel Corzo, Director
The Getty Conservation Institute

John Walsh, Director
The J. Paul Getty Museum

Foreword

Relatively few papers have been published in the conservation literature in recent years dealing specifically with new conservation treatments for metals. This reflects the fact that a certain degree of homeostasis has been reached on the subject. As conservators, however, we are all aware of the continuing difficulties posed by the treatment of outdoor statuary and the preservation of archaeological ironwork, areas in which continued research is still required. The Symposium on Ancient and Historic Metals, held at the J. Paul Getty Museum in November 1991, was organized for the purpose of reflecting current views on methods now in use for metals conservation, particularly in respect to ancient and historic objects.

The intention of the symposium was to focus on objects rather than archaeometallurgical aspects of smelting, extraction, or refining of metals. Conservation treatments for metal objects are subject to continued reevaluation by the profession, and the relation between treatment and technology of the metalwork is an important one. Without an appreciation of how a metal object was made and finished, it is difficult to imagine applying a conservation treatment with any justification or control.

Some of the issues concerning conservation treatments currently being reevaluated are those relating to the cleaning of patinated ancient bronzes and the corrosion of outdoor bronzes. As the sophistication of analytical and technical studies increases, it is becoming increasingly apparent that the cleaning of ancient bronze surfaces can remove evidence of association and burial context, even when careful mechanical cleaning is undertaken. These concerns are addressed in articles by Chase and Bassett. A considerable amount of work has also been published recently that discusses the etiology of basic copper sulfates and their relationship to the corrosion process of statuary exposed outdoors. Lins reassesses the evidence for the formation of some of these corrosion products based on new research reported here.

Looking at the corrosion of archaeological bronzes, Schweizer discusses the identification and investigation of patina in the classification of bronze surfaces from different land and lake environments, and Schrenk presents a detailed examination of the bronze surfaces of sixteenth- to seventeenth-century objects from the Benin Kingdom, Nigeria. In considering marine corrosion of bronze and other metals, MacLeod describes the examination of objects recovered from shipwreck sites in Australia.

The important restoration which has been carried out on the equestrian monument of Marcus Aurelius in Rome has not been previously well described or available in English. The work of Marabelli and Fiorentino included here provides a very interesting example of a detailed conservation and restoration project. The subject of outdoor statuary is further considered in articles by Grissom and Holm, each of whom discuss the often neglected subject of the numerous historic cast-zinc sculptures in Europe and the United States that are becoming an increasing cause for concern as they deteriorate.

The corrosion of archaeological iron and the methods of treatment for more recent architectural ironwork also pose considerable difficulties for the conservators charged with their care. Keene reviews the survival rates for treatments carried out on archaeological iron from the Museum of London, while Matero examines historic American architectural ironwork finished by surface treatment. Radiography has long been accepted as very important in the examination of metals, and more recent industrial developments have led to the application of radiographic tomography. Bonadies offers an account of tomographic studies of ancient bronzes using industrial imaging systems.

Studies of gold objects tend to reveal a great deal about the technology of the society in which a given piece was produced. Oddy, Ogden, and Chapman examine decorative goldwork and manufacturing techniques in early African, medieval European, and eighteenth-century European precious metalworking, respectively.

The symposium from which this volume was compiled would not have been possible without the support of Harold Williams, chief executive officer of the J. Paul Getty Trust, as well as the encouragement of John Walsh, director of the J. Paul Getty Museum, and Miguel Angel Corzo, director of the Getty Conservation Institute. In conclusion, we wish to extend special appreciation to Frank Preusser, former associate director for programs at the Getty Conservation Institute, for supporting the idea of the conference and for guidance throughout the planning process.

David A. Scott
Jerry Podany
Brian B. Considine

The Monument of Marcus Aurelius: Research and Conservation

MAURIZIO MARABELLI

The equestrian monument of Marcus Aurelius, the most famous bronze monument of antiquity, is all that remains of the twenty-two *Equi Magni* that once adorned Late Imperial Rome. It was created according to the characteristic iconography of the so-called Type III style of the period following 161 C.E. and is thought to be connected with the celebration of a military victory of the emperor, perhaps in 173 C.E. (Fittschen 1989; Torelli 1989). The statue represents Marcus Aurelius with his right arm and hand in a relaxed pose, while his left hand is positioned as if holding the horse's reins, which are missing. The horse, of Nordic breed, is represented in the act of drawing up from a trot.

The gilt equestrian statue was probably erected in the area of the Fori and later moved to the Lateran Plaza, presumably in the eighth century following the political decline of the Imperial Fori. In the tenth century, according to the *Liber Pontificalis,* the *Caballus Constantini,* as the monument was then known, was visible in the Campus Lateranensis near the basilica of the same name and the patriarch's residence. This position corresponded to the new religious and political center of medieval Rome (De Lachenal 1989).

After the historical memory of Emperor Marcus Aurelius had been expunged, the monument first became a symbol of Constantine and papal authority. Then, in the twelfth century, according to the *Mirabilia Urbis Romae*, the statue was considered an effigy of a knight defending Rome against the barbarians. At the end of the twelfth century the statue probably underwent its first crude restoration. A further restoration certainly took place from 1466 to 1475 in at least two stages when the monument was placed on a new stone base, as shown in Filippino Lippi's fresco in the church of Santa Maria sopra Minerva (De Lachenal 1989). This restoration, carried out by the medalist Cristoforo Geremia da Mantova and the goldsmiths Corbolini and Guidocci, cost a total of 970 gold florins. About fifty years later, in January 1538, Paul III Farnese had the monument transferred to Capitoline Hill.

A new pedestal, commissioned from Michelangelo in 1539, was finally constructed in 1561 and is still visible today.

Two subsequent restorations took place, one in 1834–36 and another in 1912. The first was principally concerned with the static condition of the monument, while the second was an unscientific restoration of the surface with the addition of new dowels and the consolidation of preexistent patches and dowels (De Lachenal 1989).

In 1980 preliminary analyses of surface-corrosion products and an acoustic-emission and ultrasonics survey of the monument were carried out. The results of these tests revealed a defective structure and an extensive sulfur-dioxide attack on the surface (Marabelli 1979). In January 1981 the equestrian statue was moved to the Istituto Centrale per il Restauro (ICR) in San Michele, where it remained until the completion of the restoration in 1988. In December 1984 the results of the research were summarized in an exhibition and a catalogue (Aurelio 1984); other important results on casting and assembly techniques (Micheli 1989) and on gilding (Fiorentino 1989) were published subsequently.

The major investigations of the ICR laboratories preceding and accompanying the monument's most recent restoration included the following:

1. Static condition and structure of the monument
2. Nondestructive testing: fabrication and repair techniques
3. Analysis of the alloys
4. Thermal behavior of the monument
5. Climate and pollution: time of wetness and damage function
6. Patinas and types of corrosion
7. Process and condition of the gilding

Static Condition and Structure

Evaluation of the static condition showed that the monument rests essentially on two of the horse's legs, the left-front and the right-back, while the left-back leg acts as a balance to the oscillations of the structure caused by wind, among other disturbances. The right-front leg is raised.

Structural examination of the monument and its tensile state was carried out or coordinated by the ICR Physics Laboratory, primarily using two different techniques: finite element mathematical (FEM) model and speckle interferometry. The purpose of these measurements was to assess the limits of stability of the bronze structure under the stress of its own weight.

Initially, the weights of the horse and horseman were calculated experimentally. The distribution of thickness was measured in each case, paying particular attention to the horse and what came to be considered its critical points (bearing legs and belly). Using a steel hook equipped with a strain-gauge element, the weight of the horseman was determined with reasonable accuracy to be 620 kg ± 6 kg (Accardo et al. 1984). The same technique was used to calculate the weight of the horse at approximately 1,300 kg.

Ultrasonics were used to determine the thicknesses of the metal. For example, the average thickness of the four legs was calculated as follows: left-front, 5.9 mm; right-front, 5.4 mm; left-back, 5.8 mm; and right-back, 5.8 mm. The average thickness of the belly measured 5.5 mm and 5.6 mm. Variations in thickness (standard deviations) were found to be fairly restricted (Table 1).

In order to develop a method for structural calculation of the finished elements, the form of the horse was reproduced on a computer by transferring the coordinates of the surface from photogrammetric images. The surface of the horse was subdivided into a grid structure corresponding to 365 shell elements, 406 nodes, and 36 high-stiffness beams. The schematic structure was then simulated for conditions of stress. The movements of the horse as a rigid body were calculated at considerable loads—in particular, under the weight of the horseman. The area that showed the most stress turned out to be the juncture of the left-front leg (Accardo, Amodio, et al. 1989). Figure 1 shows the movement of the mathematical model, magnified 109 times, as the horse moves forward and to the right under the weight of the horseman.

FEM model calculations were integrated with repeated linear measurements of displacement, using linear variable differential transformers (LVDT), of the raised front leg in all three directions. Calculations were also made with the application of

TABLE 1. *Statistical elaboration of the ultrasonic measurements of thickness (mm). (For symbols, see page 6.)*

Area	*X*	*S*	*τ1*	*τ2*
Left-front leg	5.9	1.3	0.3	1.8
Right-front leg	5.4	1.5	2.1	4.6
Left-back leg	5.8	1.0	0.7	5.7
Right-back leg	5.8	1.5	1.1	2.5
Left side, repair	5.2	1.2	1.0	1.9
Left side, repair	6.6	1.6	−0.2	−0.2
Right side, repair	5.1	1.2	0.8	2.6
Right side, repair	7.1	1.0	1.4	−0.7
Belly, left side, V1	5.4	1.1	1.5	9.3
Belly, left side, V2	6.1	1.1	−0.1	6.3
Belly, left side, V3	6.6	1.5	−2.4	2.5
Belly, left side, V4	4.5	1.8	1.5	−0.2
Belly, left side, V5	4.9	1.2	1.1	−0.1
Belly, left side, Vt6, repair	5.3	1.6	1.0	−0.3
Belly, right side, V7	6.1	1.4	−0.05	−0.6
Belly, right side, V8	5.3	1.9	−0.05	−1.5
Belly, right side, V9	5.3	2.0	−0.1	−1.5
Belly, right side, V10	5.2	1.2	0.6	2.1
Belly, right side, Vt11, repair	6.7	2.2	−0.1	−0.1
V1 + V2 + V3 + V4 + V5 =	5.6	1.7	0.1	−0.7
V7 + V8 + V9 + V10 =	5.5	1.7	−0.2	−0.8

FIGURE 1. *Displacement of the horse under the weight of the horseman (×109).*

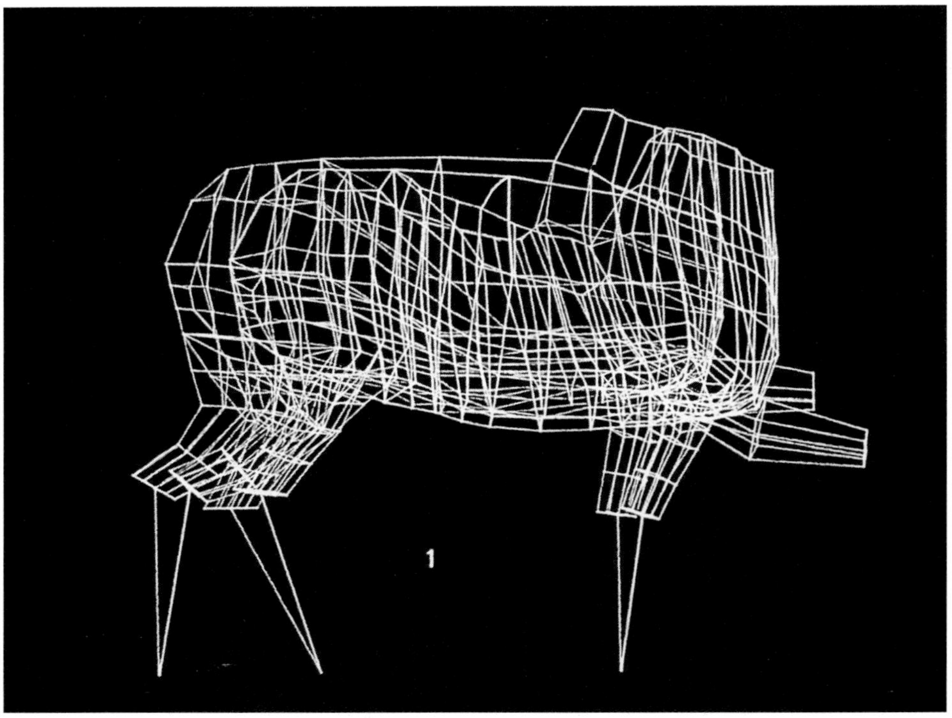

strain gauges (twenty-one groups of three elements), mostly attached to the inside of the left-front leg, and with the figure of the horseman placed on the horse in every experiment (Accardo, Bennici, et al. 1989). The greatest displacement of the left-front leg was concluded to be approximately 3 mm.

Among the possible hypotheses of attachment of the monument to its base, the one that corresponds to the minimum tension, according to the FEM model, presupposes a rigid fastening of the legs to the stone, with a forward displacement of the tip of the hoof of the left-front leg of 0.1% of the distance between this point and the corresponding back leg. This method of attachment would have been much easier to achieve than an internal framework of light, stiff metallic elements, which would have presented some difficulties in execution and maintenance (Accardo, Amodio, et al. 1989).

At the same time, the structural deformations of the horse were determined optically under a stress equal to approximately one-fourth the weight of the horseman. The structure was photographed with laser illumination (514.5 nm), first under the deformations caused by the added weight of the Marcus Aurelius, and later under normal conditions. This resulted in a kind of double exposure (Accardo et al. 1985). The photographic representation of a small area of the surface under laser illumination shows up on the film as an initial series of light and dark spots (speckles). A second series of spots corresponds to the first but is slightly displaced as a result of the deformations, producing a typical interference pattern (Young fringes).

The measurement of these displacements can be obtained by illuminating the photographic film with the same coherent light and measuring on a magnifying screen the period of the interference fringes that corresponds to the small selected area (in effect, measuring the distance between each successive fringe). From these data it is possible to determine the distance between two coupled speckles on the

FIGURE 2. *Speckle image of the neck and muzzle of the horse.*

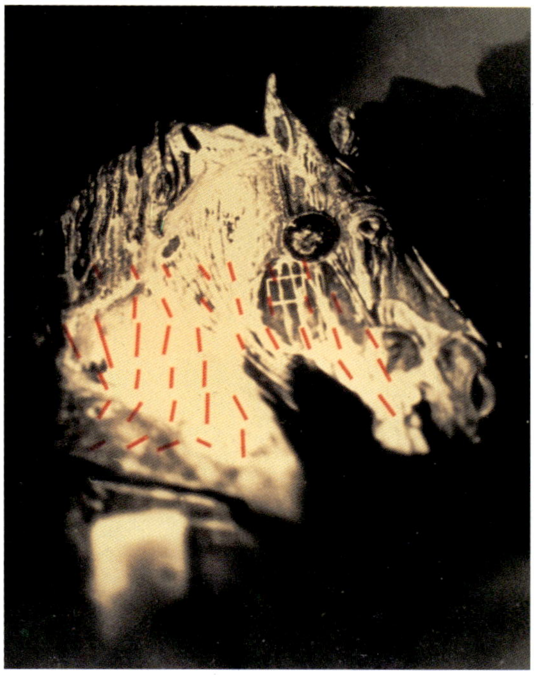

film (d) and thereby the real displacement (L) of the structural deformation in the small area. Given the enlargement factor of the camera (M), d = ML.

Figure 2 shows the speckle image of the horse's neck and end muzzle, superimposed on the image of the surface illuminated with incoherent light; a series of segments corresponding to the displacements caused by elastic deformation of various microareas is visible. The length of the segments is proportional to the extent of the linear deformations (3 mm maximum) and their orientation to the direction of the displacements (Accardo et al. 1985).

One can deduce from these experiments that the structure of the monument, particularly that of the horse, undergoes a certain modest deformation in the elastic range when submitted to a force equal to the weight of the horseman. This is especially the case at the juncture of the left-front leg. Nevertheless the bearing legs easily withstand the weight of both statues, exhibiting a rather skillful casting under ultrasonics, showing uniform thickness reinforced with a tin-lead alloy filling.

The forces and subsequent deformations (elastic, for the most part) caused by weight, even when considered in the general context of other stresses to which the structure was submitted—such as thermal stress (discussed herein) primarily, and wind pressure (which can reach maximum values of about 57 kg/m^2) secondarily—never reach levels great enough to compromise the conservation of the monument.

Nevertheless, the numerous gaps, disjunctions, and irregularities of the structure, as well as the serious damage caused by relocations of the monument in past centuries, worried medieval conservators. These early restorers attempted, therefore, to displace some of the weight of the horseman onto two small stone columns that functioned in compression. The columns are visible in Pisanello's early fifteenth-century drawing of the left side of the monument (De Lachenal 1989). This drawing also shows a small column supporting the belly of the horse, perhaps intended to consolidate the structure at what was perceived to be the point of greatest stress.

Nondestructive Testing

Nondestructive testing played a fundamental role in the preliminary phase of study. In addition, the data obtained were essential in determining the process by which the monument was fabricated.

The ICR Chemistry Laboratory examined the major sections of the two statues at more than 10,000 measurement points using ultrasonics. Researchers divided the surface into areas of smaller dimensions, subdivided each area into a grid of 2 cm squares, then transferred each value onto a flexible acetate sheet laid out along the curvatures of the surface.

Table 1 shows the thickness values of some areas with statistical values calculated, such as the standard deviation S, the curtosis $\tau 2$, and the skewness $\tau 1$, or asymmetry coefficient. The horse's four legs indicate remarkable homogeneity of casting, probably achieved by rotating the clay forms containing the molten wax. The overall average value of the thicknesses (x) of the entire bronze ranges from 5 mm to 6 mm, with minimums of 3 mm and maximums of 8 mm (Canella et al. 1985).

A radiographic survey (with 300 radiograms) by Micheli, together with endoscopic examination and direct observation, permitted the identification of the constituent sections. The statue of the horseman is made up of seventeen parts, separately cast and then joined together; the individual parts (head, arms, legs, and sections of drapery) were cast by the indirect, lost-wax method.

The horse is made up of fifteen sections (muzzle and neck, body in eight parts, legs, and tail), also cast separately by the same technique and then assembled (Micheli 1989). This was the most logical and simple process for casting bronzes of large dimensions, for which a single casting would have presented unmanageable difficulties. Not only the legs of the horse but also the other self-contained parts (the head, arms, and legs of the horseman) were obtained by pouring molten wax into a negative mold and distributing it by rotating the mold.

Radiograms have shown that the original sections underwent a slow process of cooling that, on one hand, prevented large cracks and cavities and, on the other hand, contributed to the separation of lead and slag into stratified bands in a frontal direction away from the solidification of the metal (Micheli 1989). The original solderings were made by pouring the molten metal directly and often discontinuously along the edges of the sections using, where possible, preexistent mechanical junctures.

The classification of the repairs to the monument proved rather complex. The first type of treatment, contemporaneous with the fabrication, was the filling in of missing parts, pores, and spongy areas in the cast with small (a few centimeters in diameter at most) rectangular dowels. Polygonal dowels of various sizes were also used in the same situations to repair either defects in casting or imperfections in the junctures between sections. A later type of repair, difficult to date, was used to fix extensive damage or large holes in the cast. In this method, cordlike strips of metal were used to join the cast with plates made to size, slightly smaller than the lacunae. The soldering was accomplished by pouring molten metal into the interior of the lost-wax casting. The molten-metal solder covered the edges of the juncture, forming a cordlike strip that penetrated the interconnecting spaces between the cast walls and the repair plates laid against them. The same solder also penetrated the holes made in the original bronze and in the corresponding repairs to obtain a better mechanical adherence.

It is important to point out that the discontinuous Roman solderings and the later cordlike solderings do not correspond to continuous, structural welding, as in the hard-soldering process. Ultrasonic tests have verified without a doubt that there is no structural continuity between soldering strips and joints in the metal sections (Canella et al. 1985), as denoted by the low thickness values (Fig. 3); these are joints of a mechanical kind instead.

Other assembly and repair techniques from Roman times and later have also been identified. The classification of types of dowels, plates, plugs, and cordlike strips is especially difficult because of the reuse of older elements in later repairs and the superimposition of subsequent restorations.

Particularly useful in this investigation was an instrument for the measurement of conductivity expressed in International Annealed Copper Standard (IACS) percentages (Medori 1983). The conductivity of the metallic walls was measured to a depth of a few millimeters by means of a magnetic field. If the material under examination is copper, the measured value will correspond to the maximum range (100%); for copper, tin, and lead alloys, the value will decrease from 100%, diminishing in proportion to the increase in the noncopper components.

Using this technique, about 18,000 measurements were carried out, thus allowing the clarification of doubtful cases and the partial aggregation of results of the quantitative analyses of alloys, prior to statistical analysis (Fig. 4). By the end of the experimental survey, it was possible to conclude that the original sections and the Roman repairs revealed IACS% conductivity values generally equal to or below 13%,

FIGURE 3. *Thickness measurements of soldering with cordlike strips.*

FIGURE 4. *IACS% conductivity measurements of original sections and repairs.*

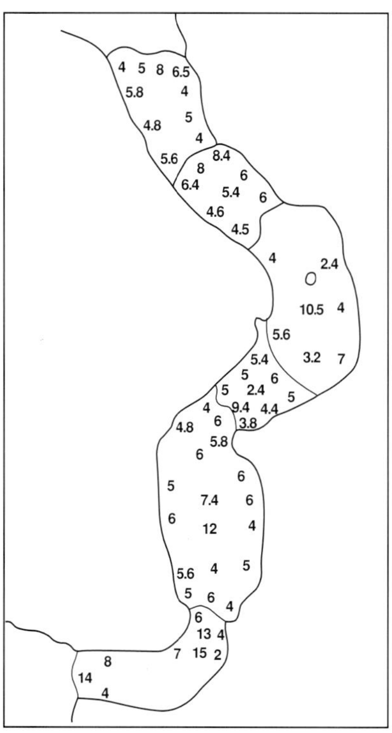

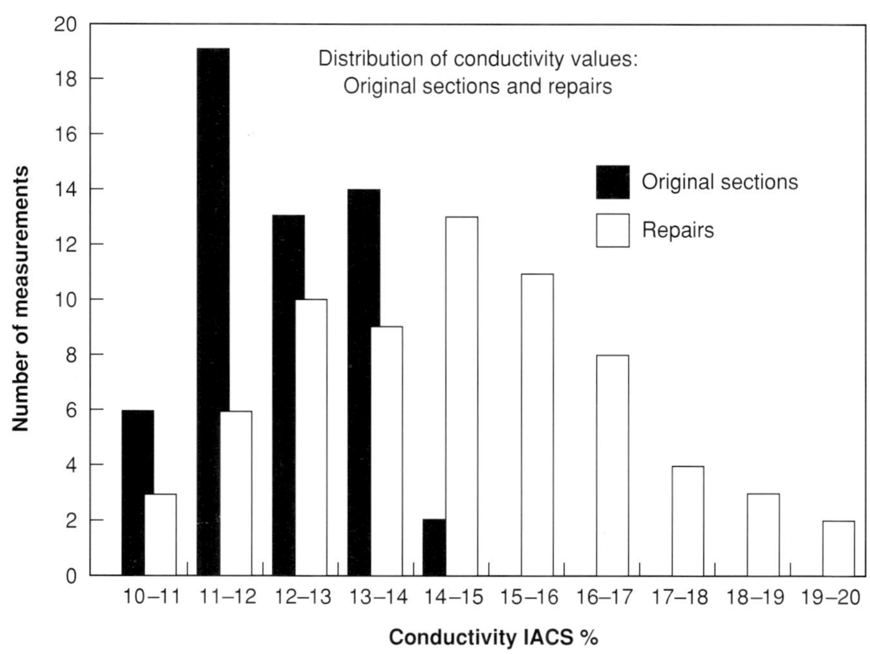

with some high points to about 15%, while the measurements of the later repair materials stayed mainly within a range of about 11–20%. The original sections of the horse and the horseman were chemically homogeneous, with rare exceptions.

Analysis of the Alloys

Quantitative analysis of the alloys was based primarily on two methods:

1. Dispersive X-ray fluorescence analysis for the principal elements (Cu, Sn, Pb)
2. Plasma spectrography for secondary trace elements (Ag, Zn, Fe, Ni, Co, As, Sb, Bi, Si)

The first technique, used largely in archaeometry, does not require particular elucidation. However, in this specific case, an original method for the preparation of the sample was developed. It consisted of dissolving about 50 to 100 mg of alloy (25–50 ml final solution), depositing 200 microliters onto a paper filter (ϕ 14 mm), and analyzing the spectrum of X-ray fluorescence obtained by the irradiation of the filter with a target of barium acetate excited by a primary X-ray source, Bα Kα = 32 KeV (Ferretti et al. 1989).

In plasma-emission spectrometry, the sample is introduced in aerosol form, into a flow of ionized argon at a range of 10,000–12,000 °C. Given the high temperature and the subsequent high level of excitation, sensitivities on the order of parts per billion or milligrams per liter are reached.

About one hundred specimens were studied in all. On initial examination the matrix of percentage values was difficult to interpret. Therefore, the values were reexamined and sequenced in light of two criteria: (1) the sources and analytical data available in the literature, and (2) the statistical elaboration of the data.

A very important passage on the description of bronze alloys used by the Romans, albeit somewhat ambiguous in part, is found in the *Natural History* of Pliny the Elder, book XXXIV, chapter 20 (1961:95–98). Pliny lists five types of bronze alloys: (1) *campana,* an alloy used for vases and utensils; (2) an alloy similar to the previous one, used for the same purposes; (3) an alloy for statues and bronze plaques; (4) *tenerrima,* an alloy for casting statues in molds; and (5) *ollaria,* an alloy for making vases.

Table 2 lists the components of these alloys according to Pliny's categories without interpretation. In the last few years, three interpretations have been given to the term *plumbum argentarium* cited by Pliny. According to Caley (1970), it is a 50/50 lead-tin alloy (Table 3). However, this interpretation seems unfounded, as Pliny refers to an alloy used for counterfeits, which "some call *argentarium*" (1961:95–98).

A second interpretation (Picon et al. 1967) identifies *plumbum argentarium* with tin (Table 4). This identification appears to be well founded because of the noticeable absence of tin in all of the alloys cited by Pliny, and because this interpretation may allow the different compositions to be typed and differentiated, as Picon et al. show rather clearly in two other publications (1966, 1969).

TABLE 2. *Alloys described by Pliny (1961).*

Alloy	*aes*	**a.c.*	***p.a.*	****p.n.*	*plumbum*
1. *Campana*, bronze alloy for vases and utensils	90.9		9.1		
2. Alloy similar to the previous one	92.6				7.4
3. Alloy for statues and bronze plaques	68.6	22.8	8.6		
4. *Tenerrima*, alloy for casting statues in molds	87.0		4.3	8.7	
5. *Ollaria*, alloy for vases	96.2–97.1		2.9–3.8		

*a.c. = *aes collectaneum*
**p.a. = *plumbum argentarium*
***p.n. = *plumbum nigrum*

TABLE 3. *Pliny's alloys according to Caley (1970).*

Alloy	*Copper*	*Tin*	*Lead*
1	90.9	4.5	4.5
2	92.6	7.4	
3	86.8	6.6	6.6
	81.2–81.3	8.7–9.7	9.1–10.0
4	81.4	6.8	11.8
	72.7	7.8	19.5
5	96.2–97.1	1.4–1.9	1.4–1.9

TABLE 4. *Pliny's alloys according to Picon et al. (1967).*

Alloy	*Copper*	*Tin*	*Lead*
1	90.9	9.1	
2	92.6	7.4	
3	87.0–89.0	11.0–13.0	
4	86.9	4.4	8.7
5	96.2–97.1	2.9–3.8	

The third hypothesis by the *Projektgruppe Plinius* (Plinius der Ältere 1984) identifies *plumbum argentarium* with lead (Table 5). This interpretation encounters two difficulties: First, tin does not appear as an alloy component, which would require an alloy containing tin to be identified with the term *aes* in every case. Second, in the formula for statuary bronzes (alloy no. 4) lead would have to be added and named twice—as *plumbum nigrum* and as *plumbum argentarium*, respectively—without substantial difference and therefore without apparent reason.

Nevertheless this very formula of no. 4 (13% Pb) should be very close to the lead-bronze formula commonly used by the Romans for sculptural works, according to a technical tradition that dates back to the fourth century B.C.E. It is probable that the use of lead bronze was slow to be accepted because the characteristics caused by

TABLE 5. *Pliny's alloys according to the* Projektgruppe Plinius *(Plinius der Ältere 1985).*

Alloy	*Copper-bronze*	*Lead*
1	90.9	9.1
2	92.6	7.4
3	68.6 + 22.8	8.6
4	87.0	13.0
5	96.2–97.1	2.9–3.8

the addition of lead to bronze alloys were not well known. In fact, large quantities of this metal led to the phenomenon of liquation and to the development of discolored patinas.

It is also likely that from the fourth century B.C.E. on, a technical tradition developed for the use of lead in controlled quantities in statuary, taking advantage of the metal already available on the market as a by-product of silver-working. This would explain an interesting observation concerning the statistical interpretation of the data. The results of the quantitative analysis were interpreted for various groups in order to obtain the average value, the standard deviation, the coefficients of correlation between the various elements, and the levels of statistical significance. Statistical elaboration of the data was carried out on characteristic groups of values corresponding to the types of alloys already identified by means of the preceding chemical analyses and nondestructive tests.

The logical process of the research may be summarized as follows: nondestructive testing plus visual examinations, initial identification of the alloys, sampling and chemical analysis, testing with measurements of conductivity IACS%, classification of analytical data in groups, and statistical analysis of the groups.

Several interesting conclusions can be drawn from the final results of statistical analysis, only partially shown in Table 6. First, the original sections show a negative correlation between copper and lead (−0.79), while there is no correlation between tin and either copper or lead. The standard deviation relative to the percentage concentrations of lead is relatively low. From this, one could deduce that the ancient founder was concerned about keeping the lead within a "safe" percentage by applying a formula of reference of the type:

100 − lead = *aes* + *aes collectaneum* (scrap copper and bronze) + tin

The tin does not correlate with copper and lead, probably because the percentage of tin in the *aes collectaneum* varied each time without a systematic point of reference.

Second, the addition of lead confers some specific characteristics on the alloy: the fusion point of the alloy diminishes and the cast becomes more fluid, while the surface of the bronze becomes more workable and polishable with scrapers, files, and pointed tools (although the workability by hammering declines).

Third, the absence of correlations between the other alloy elements shows that the various original sections, cast separately, were made with metal from different stocks, probably also using *aes collectaneum*.

TABLE 6. *Statistical analysis of the original (Roman) alloys of the* Marcus Aurelius.

Alloy	Average % Cu	Sn	Pb	Standard deviation % Cu	Sn	Pb	Minimum–Maximum % Cu	Sn	Pb
Roman sections	80.7	6.8	12.0	2.57	1.44	2.32	75.9 85.6	3.9 10.5	8.4 16.4
Roman soldering	74.0	6.6	19.4	1.75	2.16	1.74	71.6 77.0	3.8 10.2	16.9 23.1

Table 6 shows the statistical values for twenty-eight original Roman alloy specimens and for ten specimens of Roman soldering. For soldering, no correlation was found between copper, tin, and lead, suggesting a rather approximate mixture of principal components, the only restriction being that the cumulative percentage of tin plus lead must not drop below a certain level. In this case, the lead not only lowers the melting point and viscosity of the alloy but also acts as a true deoxidant for the soldering, forming with the tin dioxide (SnO_2) a compound (Pb_2SnO_4) that melts at 1060 °C (Steinberg 1973; Lechtman and Steinberg 1970).

The elaboration of the data for the repaired sections was still in progress in early 1992, with some difficulties of interpretation because of the great variety of alloys used for restoration (in collaboration with E. D'Arcangelo).

Thermal Behavior of the Monument

One particularly interesting area of study has been that of the environmental causes of deterioration. The exchange of thermal energy between the environment and the monument has been thoroughly investigated, revealing that the mechanical stresses suffered by the bronze in its position on Capitoline Hill have also been dependent on the daily cycles of expansion and contraction of the metal structure. A description of the thermal behavior of the material is useful for a better understanding of an important series of problems that are not only mechanical but also involve the electrochemical and chemical corrosion of the surface.

The Piazza del Campidoglio is located about twenty meters above traffic level and is enclosed on three sides by the Palazzi Capitolini. The monument of Marcus Aurelius is placed in the center and oriented toward the northwest by 60°; that is, toward the wide ramp designed by Michelangelo. The particular placement of the monument and the geometry of the plaza allow direct sunlight to strike the metallic surface unevenly, warming different sections of the bronze at different hours of the day. In order to analyze the thermal exchange between the monument, its stone base, and the surrounding air, continuous readings of the surface temperature in ten areas were taken during the summer, along with thermovision images of the monument. At the same time, a series of acoustic-emission measurements were taken to register incidents of deformation in the horse over a 24-hour period (Accardo et al. 1983).

This last technique, in particular, operates on the principle that structural deformations and the formation or increase of cracks release microquantities of elastic

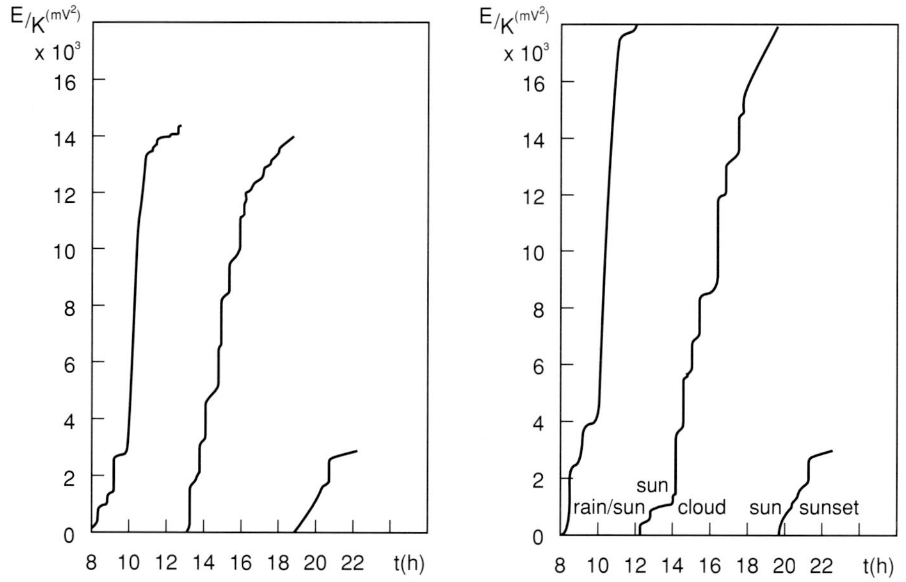

FIGURE 5a, b. *Registration of acoustic emission on (a) a clear day; and (b) a cloudy day.*

energy, causing propagation of mechanical pressure waves at a frequency greater than 10 MHz, which are picked up by a piezoelectric transducer and stored and analyzed by a sequential electronic apparatus.

Using these techniques, several important findings have emerged. First, the horse's left-front leg showed particular stress from direct solar radiation after ten o'clock in the morning. Of the two registrations in Figures 5a and 5b, the first shows the course of energy emitted on a clear day, while the second represents the phenomenon on a cloudy day with rain. It is evident that more energy is released under conditions of maximum irradiation as well as during rapid variations of surface temperature.

Second, because of its greater thermal inertia, the stone base maintains a surface temperature higher than that of the bronze alloy and keeps the lower part of the horse warmer during the night, while the hindquarters cool down through radiant emission toward the sky (Fig. 6).

In general the bronze surface responds quickly, because of its scant thermal inertia, to the temperature variations of the surrounding air. Exceptions may include the legs, which are filled with a lead-tin alloy (*metallone*) and the belly of the horse, because of its thermal exchange with the stone base. The thermovision images of the legs are certainly influenced by the greater thermal capacity of the volumes filled with lead-tin alloy, which show up as lighter (i.e., hotter), while the dark areas correspond to "empty" spaces (Accardo et al. 1983).

From the structural point of view, one may conclude that the low level of energy released by the structure corresponds to incidents of temporary (elastic) deformation, particularly involving that section of the left-front leg of the horse already subject to the mechanical stresses of the monument's weight.

Finally, in regard to the electrochemical aspects, climate certainly has a decisive influence on the kinetics of the bronze's corrosion. Given the rapid adjustment of the metal surface to the temperature of the surrounding air, the events of precipitation and capillary condensation are the primary elements that accelerate electrochemical corrosion.

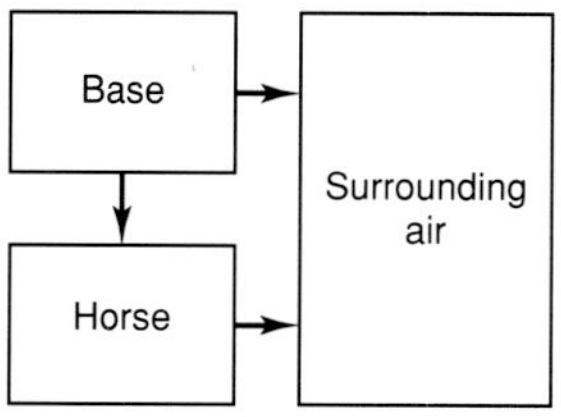

FIGURE 6. *Nocturnal heat exchange between the air, the monument, and the base.*

In recent years damage functions have been developed to calculate the electrochemical corrosion of a metal object over the course of a year, taking into account the amount of time the surface remains wet and the integrated fluxes of deposition of the more destructive airborne pollutants. For the Roman climate, the time of wetness (*tw*) of a metallic surface exposed outdoors is given as:

$$tw = tw1 + tw2$$

where *tw*1 equals the time of wetness of the surface caused by rainfall and *tw*2 equals the time of capillary condensation (Marabelli et al. 1988). Capillary condensation is linked to the shape and diameter of capillary pores in the patina and starts at a relative-humidity value well below 100% (corresponding to traditional surface condensation).

In order to measure the threshold value of relative humidity corresponding to the beginning of capillary condensation, the ICR Chemistry Laboratory developed a prototype consisting of an automated apparatus programmed by computer and capable of measuring surface conductivity and volume conductivity of patinas, which is variable in terms of relative humidity (Marabelli et al. 1988). This instrument comprises a measurement cell of conductivity, which clings to the metal surface and within which a particular hygrometric progression is produced in a predetermined way.

In the case of the Marcus Aurelius, by increasing the relative humidity (RH), it was possible to document, after some 160 experiments, that the surface conductivity increases rapidly above approximately 80% RH (Marabelli et al. 1988). Figure 7 shows a typical experimental curve corresponding to the surface electrical conductivity function, *f*(RH). The resulting theoretical function is:

$$y = 0.8 \cdot \exp\left(-\ [80-x]\ /\ 3\right)$$

where *y* equals the conductivity in microsiemens and *x* equals RH.

FIGURE 7. *Surface conductivity dependent on relative humidity.*

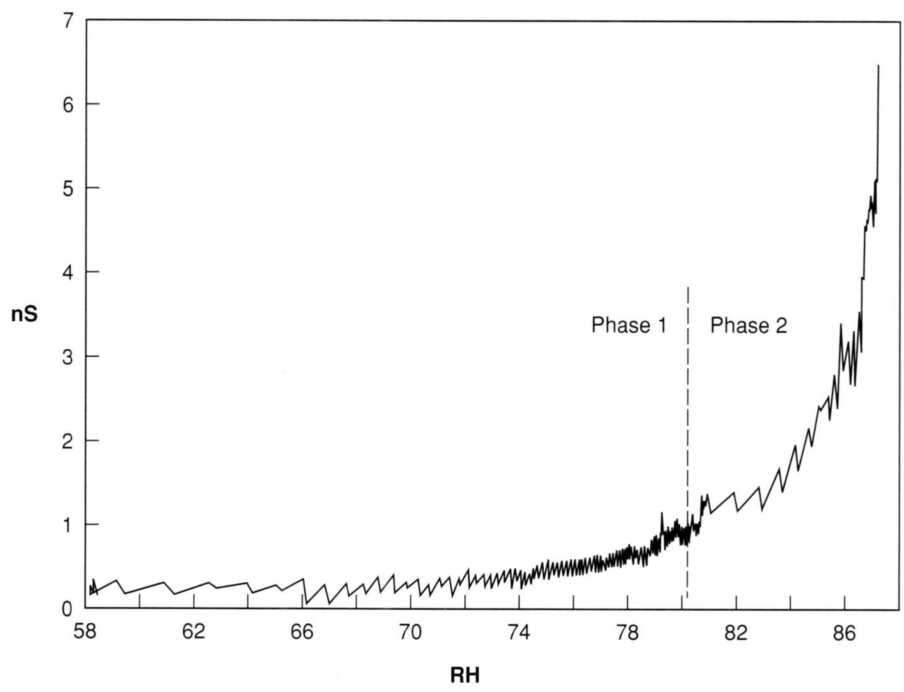

Knowing the daily distribution frequency of relative humidity over the course of a year, it was possible to determine that *tw*2 equals approximately 22.8 days, while *tw*1 can be roughly deduced from the monthly averages of pluviometric data for the historical center of Rome over a 10-year period. The total *tw* equals approximately 0.22 per year. This finding was used to calculate the corrosion velocity of the monument exposed outdoors by using a damage function developed by Benarie and Lipfert and slightly modified for this specific case (Marabelli 1992). The velocity of electrochemical corrosion, *Vc,* expressed in g/m^2 per year, equals:

$$Vc = tw \cdot 0.38 \cdot a\,(fSO_2 + 1.1 \cdot fCl^-)$$

where tw equals the total annual time of wetness; *a* equals the dilution factor of the pollutants, taking into account the elevated position of the plaza; and fSO_2 and fCl^- equal the integrated fluxes of deposition expressed in mg/m^2 per day.

As a result, it was possible to determine the velocity value of the electrochemical corrosion of the alloy at roughly 0.2 microns per year (Marabelli 1992). It would seem possible to extrapolate from these data encouraging indications for the conservation of the Marcus Aurelius outdoors. However, it must be remembered that the chemicophysical corrosion of the patina, caused by airborne acidic pollutants as well as rainfall, still causes a constant erosion of the surface with loss of gilding.

To better understand the conditions of the formation and transformation of corrosion products in relation to the climate and other environmental parameters in the broad sense, a series of samples of the patina differentiated by color, consistency, and orientation to sunlight and rainfall was taken and examined using X-ray diffraction.

Patinas and Types of Corrosion

The surface of the Marcus Aurelius reveals extensive sulfation, with the formation not only of brochantite but also antlerite and chalcanthite, a soluble copper sulfate. Since brochantite is stable between 3.5 and 6.5 pH, and antlerite is stable between 2.8 and 3.5 pH, the presence of the chalcanthite indicates that the pH level on the surface of the monument must have fallen below 2.8, probably as a result of microcondensation (Graedel 1987).

The partial dissolution of the patina evidently makes the already precarious mechanical adhesion of the gold even more unstable, to the point that even the application of a fixative may cause damage to the gilding. Furthermore, the gilding always appears so fragmentary and riddled with holes that water easily infiltrates the underlying patina (Fig. 8). The areas protected from the driving rain and from water runoff appear darker due to the accumulation of carbon substances and other components of the atmospheric particulate (gypsum, feldspars).

Conversely, the horizontal surfaces facing upward and those corresponding to the geodetic lines of rainwater appear lighter because of the absence of carbon particles. The alternation of darker (cathodic) stripes and lighter (anodic) stripes on the flanks of the horse form a typical zebra pattern (Fig. 9). Spots and whitish stains along with gray patinas covering the gold are rich in anglesite, present along with brochantite in almost all the samples. A few areas of the monument bear traces of a brownish surface coating, the composition of which has not yet been defined.

Finally, atacamite, a basic copper chloride, is present below the brownish-black patina deposits, indicating an electrochemical attack on the alloy in the presence of a chloride ion. This ion accelerates the corrosion and, in certain cases, promotes pitting. Its presence can be attributed either to the airborne chloride deposits (marine particulate, emissions from the combustion of plastics containing chlorine), or to the attack of the surface by chemical compounds containing chlorine.

Process and Condition of the Gilding

Not all of the tests have been completed for this important and complex monument. Study of the gilding process in particular is still in progress. The first phase of testing involves metallographic analysis of samples taken from the horse and from the mantle of the horseman to obtain information on the thickness of the gold leaf, the stages of application, and the extent of the corrosion process.

Figure 10 shows the metallographic section of one sample: two pieces of gold leaf rest on corrosion products that penetrate to a maximum depth of 0.3 mm; the pieces are completely detached from the metal and separated from each other by the same oxidation products. The thickness of the gold leaf varies from 3 to 9 microns. This measurement is consistent with the values cited by Oddy et al. (1979).

Three other characteristics of the gilding of the monument should be noted: (1) residual gilding is present almost exclusively on the Roman sections and repairs; (2) the surface of the horseman shows minute scoring in definite directions, suggesting that the alloy was textured in this way to anchor the gold leaf more effectively (see the term *concisuris* in Pliny 1961, book XXXIV, chapter 19); and (3) in two areas of the horse's hindquarters, which are covered by the horseman, a series of roughly square gold leaves with sides varying from 5 to 9 cm are visible. A similar square pattern is present on the *Horses of San Marco* (Galliazzo 1981) and on some bronze statues cited by Oddy et al. (1979). In Pliny's treatise two methods of gilding

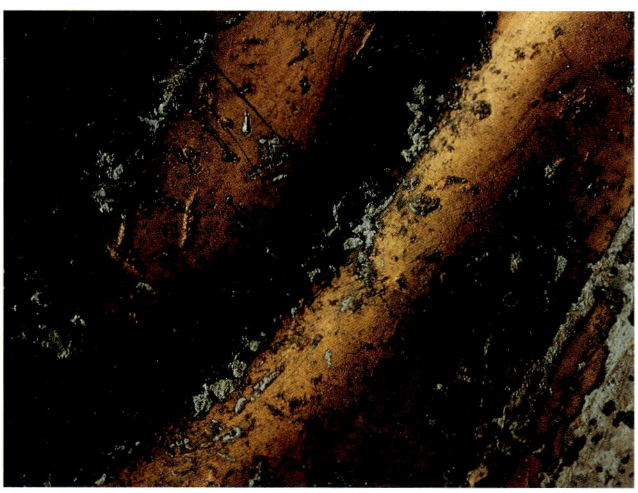

FIGURE 8. *Damaged gold surface.*

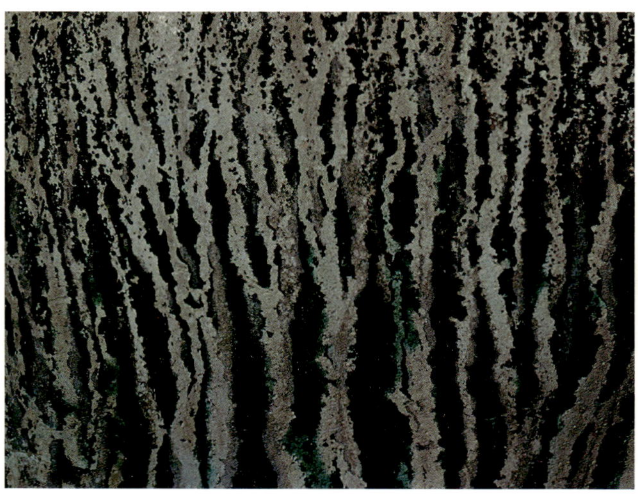

FIGURE 9. *Typical alternation of light and dark areas of surface corrosion.*

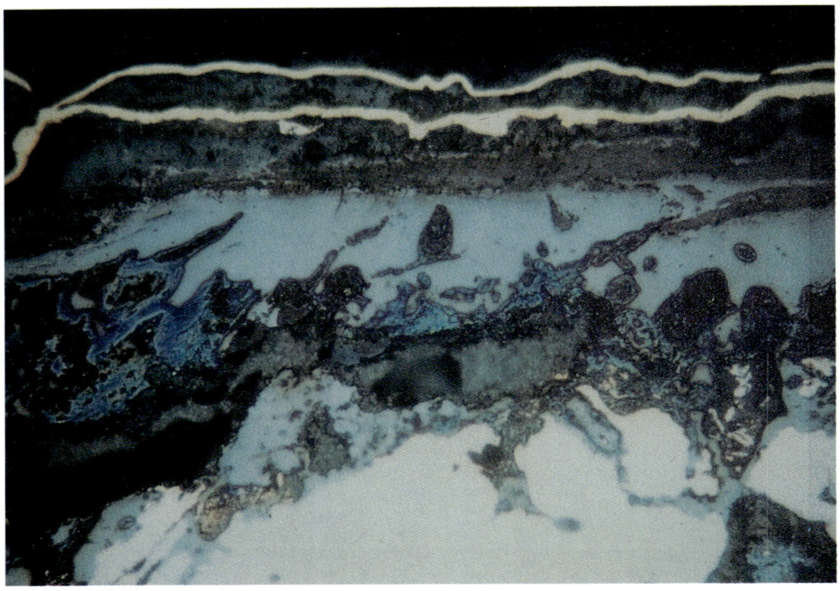

FIGURE 10. *Metallographic section of an alloy specimen from the horse.*

are cited directly (book XXXIII, chapter 20), and a third indirectly (book XXXIV, chapter 19), concerning a bronze statue of Alexander the Great, which the Emperor Nero later had gilded. Basically the methods involve gilding with cold mercury, gilding with proteic glue, and gilding with gold foil or gold leaf (*à l'hache*).

Oddy's hypothesis that fire (mercury) gilding began at the end of the second or beginning of the third century C.E. seems well founded (Oddy 1982). Both Oddy (1982) and Craddock et al. (1987–88) have published lists of bronzes that contain large quantities of tin and lead and were not gilded with mercury. On the other hand, lead bronzes (including the Marcus Aurelius) cannot be gilded with mercury, either by the cold process or the hot process. Therefore, discarding the hypothesis of gilding with proteic adhesive for the Capitoline monument, which was intended to be placed outdoors, only the *à l'hache* technique seems probable. However, the use of this process should be checked against both the analysis of alloy microsamples and the current foundry experiments.

Observations and Conclusions

During the restoration, several reagents and processes for cleaning the surface were perfected in collaboration with the restorer Paola Fiorentino. The practical experimentation was rather long and laborious, since the objective was essentially to remove the corrosion products on top of the gold without dissolving or detaching those underneath.

ICR and the Selenia Company, working in collaboration, carried out a test of eleven surface coatings for the conservation of bronzes outdoors (Marabelli and Napolitano 1991). At the end of the study, it was possible to establish that the best formula was provided by Incralac or Paraloid B72 as primer and Reswax WH (a mixture of a polyethylene wax and two microcrystalline waxes) as a protective finish.

Despite the studies completed thus far, a product capable of ensuring protection without extensive maintenance for a period of at least twenty to thirty years has not been developed. On the other hand, given the precarious adhesion of the gold, massive fixative treatments or cyclical surface cleaning are inadvisable for the Marcus Aurelius if it is relocated outdoors. At the present time, therefore, the best and most rational solution for the conservation of the two statues would be a climate-controlled museum environment that provides filtration of the atmospheric pollutants. This does not exclude the future possibility of returning the monument to its original position outdoors, if protection and maintenance operations could be assured with little or no damage to the patina and residual gold.

References

ACCARDO, G., D. AMODIO, P. CAPPA, A. BENNICI, G. SANTUCCI, AND M. TORRE

1989 Structural analysis of the equestrian monument to Marcus Aurelius in Rome. In *Structural Repair and Maintenance of Historical Buildings*, 581–92. C. A. Brebbia, ed. Southampton, U.K.: Computational Mechanics Institute.

ACCARDO, G., A. BENNICI, M. TORRE, D. AMODIO, P. CAPPA, AND G. SANTUCCI

1989 An experimental study of the strain fields on the statue of Marcus Aurelius. In *Proceedings of the 1989 SEM Spring Conference on Experimental Mechanics*, 534–37. Bethel, Conn.: Society for Experimental Mechanics.

ACCARDO, G., C. CANEVA, AND S. MASSA

1983 Stress monitoring by temperature mapping and acoustic emission analysis: A case study of Marcus Aurelius. *Studies in Conservation* 28:67–74.

ACCARDO, G., F. CAPOGROSSI, G. SANTUCCI, AND M. TORRE

1984 Determinazione del carico. In *Marco Aurelio: Mostra di Cantiere*, 60. Rome: Arti Grafiche Pedanesi.

ACCARDO, G., F. DE SANTIS, F. GORI, G. GUATTARI, AND J. M. WEBSTER

1985 The use of speckle interferometry in the study of large works of art. In *Proceedings of the 1st International Conference on Non-destructive Testing in Conservation of Works of Art* 4(1):1–12. Rome: Istituto Centrale per il Restauro (ICR) and Associazione Italiana Prove non Distruttive (AIPnD).

CALEY, E. R.

1970 Chemical composition of Greek and Roman statuary bronzes. In *Art and Technology: A Symposium on Classical Bronzes*, 37–49. Cambridge: MIT Press.

CANELLA, G., M. MARABELLI, A. MARANO, AND M. MICHELI

1985 Esame ultrasonoro della statua equestre del Marco Aurelio. In *Proceedings of the 1st International Conference on Non-destructive Testing in Conservation of Works of Art* 1(8):1–12. Rome: Istituto Centrale per il Restauro (ICR) and Associazione Italiana Prove non Distruttive (AIPnD).

CRADDOCK, P. T., B. PICHLER, AND J. RIEDERER

1987–88 Legierungszusammensetzung in naturwissenschaftliche Untersuchungen an der Bronzestatue Der Jüngling vom Magdalensberg. *Weiner Berichte über Naturwissenschaft in der Kunst* 4(5):262–95.

DE LACHENAL, L.

1989 Il monumento nel Medioevo fino al suo trasferimento in Campidoglio. In *Marco Aurelio: storia di un monumento e del suo restauro*, 129–55. Milan: RAS.

FERRETTI, M., R. CESAREO, M. MARABELLI, AND G. GUIDA

1989 The analysis of bronze alloys from the equestrian statue of Marco Aurelio by means of a thin sample XRF technique. *Nuclear Instruments and Methods in Physics Research* B 36:194–99.

FIORENTINO, P.

1989 La doratura: Note sulle tecniche di esecuzione e osservazioni sulla superficie del monumento. In *Marco Aurelio: Storia di un monumento e del suo restauro*, 263–77. Milan: RAS.

FITTSCHEN, K.

1989 Il ritratto del Marco Aurelio: considerazioni, critiche dopo il restauro. In *Marco Aurelio: Storia di un monumento e del suo restauro*, 75–78. Milan: RAS.

GALLIAZZO, V.

1981 *I cavalli di S. Marco*. Treviso: Canova.

GRAEDEL, T. E.

1987 Copper patinas formed in the atmosphere III. *Corrosion Science* 27(7)[special issue]:741–69.

LECHTMAN, H., AND A. STEINBERG

1970 Bronze joining: A study in ancient technology. In *Art and Technology: A Symposium on Classical Bronzes*, 5–35. Cambridge: MIT Press.

MARABELLI, M.

1979 *Scheda di analisi* 684. Rome: ICR.

1992 The environment and the future of outdoor bronze sculpture: Some criteria of evaluation. In *Proceedings of "Dialogue 89."* Baltimore: National Association of Corrosion Engineers (NACE).

MARABELLI, M., A. MARANO, S. MASSA, AND G. VINCENZI

1988 La condensazione capillare di vapore acqueo in patine di bronzi esposti all'aperto. In *Preprints of the 2d International Conference on Non-destructive testing, Microanalytical Methods and Environment Evaluation for Study and Conservation of Works of Art* 2(25):1–20. Rome: ICR and AIPnD.

MARABELLI, M., AND G. NAPOLITANO

1991 Nuovi sistemi protettivi applicabili su opere o manufatti in bronzo esposti all'aperto. *Materiali e Strutture* 1(2):51–58.

MEDORI, M.

1983 Utilizzazione del piano di impedenza nelle ispezioni Eddy Current. In *Preprints: Conferenza nazionale Prove non Distruttive*, 1–25. Brescia: AIPnD.

MICHELI, M.

1989 Le tecniche di esecuzione e gli interventi di riparazione. In *Marco Aurelio: Storia di un monumento e del suo restauro*, 253–62. Milan: RAS.

ODDY, W. A.

1982 Gold in antiquity: aspects of gilding and of assaying. *The Journal of the Royal Society of Arts* (October):1–14.

ODDY, W. A., L. BORRELLI VLAD, AND N. D. MEEKS

1979 The gilding of bronze statues in the Greek and Roman World. In *The Horses of San Marco, Venice,* 182–87. G. Perocco, ed. Milan and New York: Olivetti.

PICON, M., S. BOUCHER, AND J. CONDAMIN

1966 Recherches techniques sur des bronzes de Gaule Romaine I. *Gallia* XXIV (1):189–215.

PICON, M., J. CONDAMIN, AND S. BOUCHER

1967 Recherches techniques sur des bronzes de Gaule Romaine II. *Gallia* XXV (1):153–68.

1969 Recherches techniques sur des bronzes de Gaule Romaine III. *Gallia* XXVI (2):245–78.

PLINIUS DER ÄLTERE

1985 *Über Kupfer und Kupferlegierungen, herausg: Projektgruppe Plinius 1984*. Essen: Verlag Gluckauf.

PLINY THE ELDER

1961 *Natural History*. Reprint. London: Heinemann.

STEINBERG, A.

1973 Joining methods on large bronze statues: Some experiments in ancient technology. In *Application of Science in Examination of Works of Art: proceedings of the seminar: June 15–19, 1970,* 103–37. Boston: Museum of Fine Arts.

TORELLI, M.

1989 Statua Equestris Inaurata Caesaris: mos e ius nella statua di Marco Aurelio. In *Marco Aurelio: storia di un monumento e del suo restauro,* 83–102. Milan: RAS.

Biography

Maurizio Marabelli, chemist, is head of the Chemistry Laboratory at the Istituto Centrale del Restauro (ICR) in Rome, where he teaches chemistry at the ICR School of Restoration. He also teaches chemistry of restoration at the Faculty of Conservation of Cultural Property, University of Tuscia, Viterbo. Dr. Marabelli is the author of more than ninety papers in the fields of nondestructive technology, conservation of metals and mural paintings, and air-pollution control.

Restoration of the Monument of Marcus Aurelius: Facts and Comments

PAOLA FIORENTINO

Although the history and conservation of the gilt-bronze equestrian monument of Marcus Aurelius is already well known, it is appropriate to begin this discussion with a reminder that this monument is the only equestrian statue to have survived intact from ancient times (Fig. 1). The monument was already being discussed in the Middle Ages, when it stood in front of the cathedral of Rome as the image of Constantine, the first Christian emperor, symbolizing Rome's continuity of power and prestige from the pagan to the Christian world (*Marco Aurelio: Storia di un monumento e del suo restauro* 1989).

Early Restorations

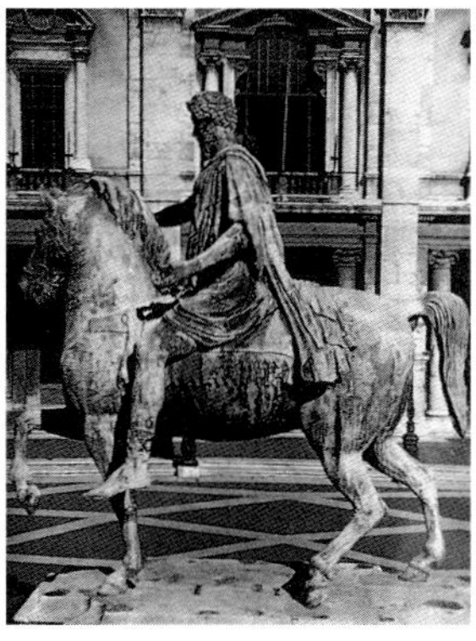

FIGURE 1. *The monument of Marcus Aurelius prior to 1981.*

Marcus Aurelius has always existed as a monument from the time it was first manufactured (around 176 C.E.). It was never buried or excavated; rather, it has gone through a series of relocations in the open environment. The pedestal on which it rests has often been altered; in fact, it has been completely replaced several times throughout history. The monument has gained and lost decorative and sculptural elements, such as the figure of a barbarian upon which the horse's raised hoof once rested. Some of these changes have been recorded, from multiple restorations in the twelfth century to the most recent restoration efforts in 1912, during which some 2,189 repairs were counted (Apolloni 1912). The monument was last moved during World War II.

Past restorations focused on the importance of the visual presence and appearance of the two bronzes (the horse and the rider). To maintain the association of the rider and the horse, repairs were limited to those areas where damage had been visually disruptive. Efforts also focused on those threats that caused immediate concern for the survival of the monument. Little attention was paid to the materials of the bronze, the previous repairs, and the interaction of the monument with the environment. Structural repairs were often roughly made or, at best, served only to reinforce older repairs that were in a state of collapse. New supports were added, however, such as the *metallone* (lead-tin alloy) casting in the horse's three load-bearing legs.

Other interventions more specific to the surface of the castings can still be recognized today. These include several regildings that took place up to the fifteenth century and the more recent applications of protective coatings with resinous films, which have certainly not helped the preservation of the bronze.

The entire monument is particularly predisposed to corrosion because of its extensively heterogeneous nature. This heterogeneity is due in large part to past structural and surface repairs, such as regilding, and the high lead content of the bronze alloy used for the original castings.

Current Condition of the Monument

Urban pollution has affected Roman monuments for more than a century and has further modified and accelerated the electrochemical corrosion process occurring on the Marcus Aurelius monument. The result has been a reduction in the thickness of the casting, with chemical attacks on the patina, causing a partial removal of the gilt layer. The monument also has many cracks and thin faults passing through the metal. These are particularly severe in the horse, which, as the bearing structure, undergoes load strain.

The extent of this damage, much more of which was revealed during the recent restoration, was partially hidden by a deposit of airborne particulate that, cemented with the alloy-alteration products, had grown 5–6 cm thick in the recesses less exposed to rain leaching (Fig. 2). Such concretions considerably altered the outline of the sculpture. In those areas with the most exposure to rain and the greatest loss of gilding, powdery patinas or the typical geodetic lines of the rain-washed patterns (anodic areas) have formed. The corrosion is clearly more extensive in these areas.

The Restoration Plan

Observed alterations and causes of degradation were investigated and experiments for deciding what restoration methods to use were undertaken. Seven chemical reagents for cleaning the surfaces were tested, of which trisodium EDTA, ammonium

FIGURE 2. *Trappings of the horse, detail showing particulate deposits.*

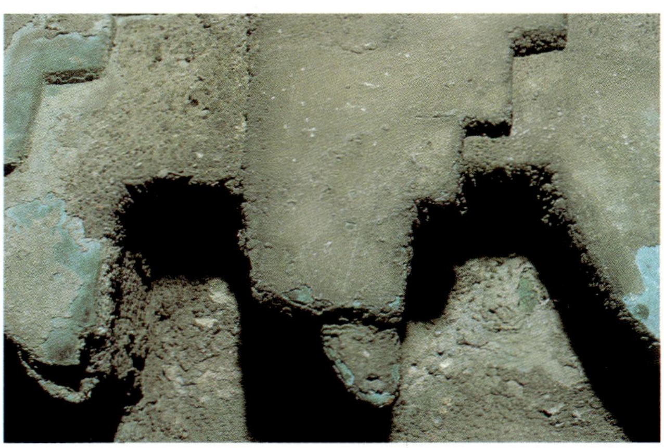

tartrate, and a cationic resin in acid form (RH) were found most suitable. The expediency of placing the monument in a controlled environment rather than depending on coatings or treatments—which might or might not inhibit corrosion and would surely require frequent maintenance—was also considered.

The restoration of a monument requires a detailed knowledge of its structure and the chemical and physical deterioration mechanisms it has undergone or is likely to undergo given its environment and the various stresses to which it is exposed. Restoration also requires a full identification and characterization of the materials originally used to manufacture the monument and any alteration compounds produced since its manufacture.

Considering this, the Marcus Aurelius can be seen as a unicum, or one-of-a-kind object. It may seem logical to compare it to the horses of St. Mark's Cathedral in Venice. Like the Marcus Aurelius, St. Mark's horses are gilt-bronze castings that have come down to us from antiquity and were continuously exhibited in the open (though, unlike the Marcus Aurelius, they were exposed to a marine as well as industrial atmosphere) until fifteen years ago. However, there are some important and striking differences between the two monuments. The St. Mark's horses are better preserved than the Marcus Aurelius and have undergone fewer repairs during their history. Of greater influence, however, was the fact that the horses were made of copper mixed with only about 2% secondary components. Ultimately then, the St. Mark's monument cannot serve as a specific reference model for the restoration of the Marcus Aurelius (Fiorentino and Marabelli 1977).

The team of experts that studied the Marcus Aurelius monument for two years was aware of the seriousness of the damage but based its research on the premise that the monument would remain in the Piazza del Campidoglio to which it is historically linked. Surveys were carried out to determine fusion, repair, and gilding techniques, following current practices. The studies pinpointed the causes and mechanisms of degradation. Climatic conditions around the monument and their effects were also studied. Calculations were made for the preparation of an internal consolidation structure which, as far as possible, would support the rider and relieve the load on the horse. The structure of Michelangelo's marble base and the dynamics of the corrosion process in relation to the microclimate conditions were examined.[1] Finally, research was done on reliable protective surface coatings for the preservation of gilt bronze. Specifically, new methods of evaluation were often applied to determine the suitability of coatings when applied to a bronze in a specific state of preservation and the ultimate effectiveness of these coatings in the open air (Marabelli and Napolitano 1991).

Restoration Methods

In 1981, following an initial series of examinations in situ, the Marcus Aurelius monument was transferred to a laboratory of the Istituto Centrale per il Restauro in Rome. There the first research workshop-laboratory was established that was solely dedicated to restoring the monument.[2] In these facilities, a series of evaluations was undertaken to clarify both the monument's structural integrity as well as the corrosion processes it had undergone. The study of the monument's corrosion history involved a full characterization of the corrosion products present on the surface of the sculpture (Marabelli herein). Tests to identify the alloy-alteration products were required, involving some sixty samples taken from the external and internal surfaces of the sculpture. These samples were chosen according to specific characteristics such as color—dark green, light green, gray, whitish, yellowish, light blue, black, earthy—as well as their physical characteristics, such as smooth and compact or powdery and voluminous.

Brochantite was by far the most common mineral identified for the light- and dark-green samples. Anglesite was predominant for the gray samples. In some samples atacamite predominated, while in others cassiterite was present. In the blue samples, taken mainly from the areas where rainwater gathered, chalcanthite was clearly present. Gypsum and feldspar composed most of the particulate deposits. The extremely widespread black alterations—probably formed of amorphous sulfides, carbon particles, and oxidized organic material—did not provide clear diffraction patterns, and their identification is inferred.

Finally, the presence of gypsum and copper oxalate was found in many samples of the yellowish corrosion products, while in the internal walls of the castings, at points where there was the greatest accumulation of particulate on the outer areas, cupric chloride in a typical pitting formation was found. These tests revealed the extensive surface sulfation caused by urban pollution, and the obvious accumulation of airborne particulate that retained humidity in some areas, encouraging cyclic corrosion involving cupric chloride.

The monument presented many different corrosion patterns, alternating even within quite small areas and requiring a special, if not unusual, set of treatment interventions for the monument's conservation. Using a method already tested on the St. Mark's horses (Fiorentino and Marabelli 1977:233–46), researchers identified and isolated twelve sample areas of 9 × 6 cm each (Fig. 3). These twelve areas were used to evaluate the efficacy and suitability of washing with demineralized water. The purpose of the washing was to extract the harmful soluble salts contained in the corrosion patinas, as well as to remove any residue from chemical cleaning agents. The use of demineralized water avoided any damage to the gilding and the more stable corrosion patinas.

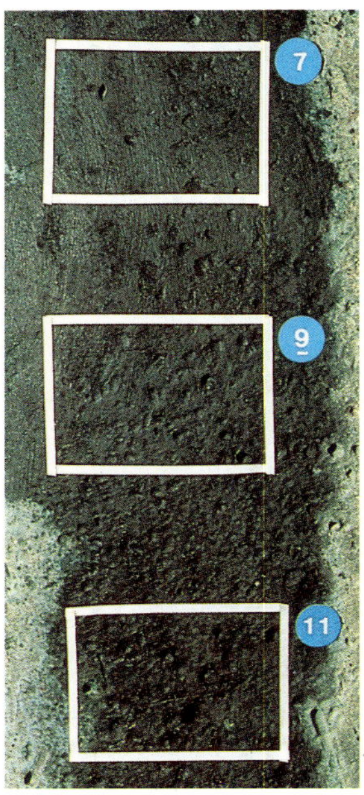

FIGURE 3. *Three sample areas chosen for the cleaning tests.*

The twelve areas chosen had the following characteristics:

- relatively compact and sufficiently visible gilding
- gilding clearly covered with black alterations
- alteration both exposed and not exposed to rain
- zones with geodetic lines
- alterations where rainwater converged
- alterations in the insides of castings

Washing was carried out with standard methods, using 100 ml fractions of demineralized water and applying brushes for five minutes. The extraction of total soluble salts was calculated for each fraction of water by conductivity measurements of the runoff. The washing was repeated until a reasonable water-conductivity value was achieved; in other words, not exceeding 20 mS cm^{-1}.[3]

The maximum number of washings for the external surface was 14 fractions (1,400 ml total) on a partially gilded area with powdery alterations, and the minimum was 5 fractions (500 ml total) on a gilded area with black alterations. Up to 18 (1,800 ml total) fractions of deionized water were needed for the internal surface.

Some assumptions can be made from these tests: Any minute detachment of patina particles that occurred due to the mechanical action of the brush could be considered acceptable, and no gold particles were noted in the solutions collected. In addition, the proportion of soluble salts removed from the external surface was lower than that found inside the monument, where the salts had accumulated—obviously because the interior was less exposed to rainwater—and also where, given the greater surface adherence, it was possible to carry out longer treatments under safer conditions. Finally, the black alterations were found to be the least soluble and less likely to be removed. The water collected was then used to identify the ions released, with particular reference to sulfate, chloride, copper, and lead ions (*Marco Aurelio, mostra di cantiere* 1984:83–84).

Subsequently, sixty smaller sample areas were chosen (24 × 36 mm each, the size of photographic film) in which the eight different corrosion patinas characterized by X-ray diffraction were represented as homogeneously as possible. The purpose was to compare seven reagents for their efficacy in removing the deposits and alterations that concealed the gilding. The reagents were chosen for their relative inability to react with the underlying bronze alloy and gold gilding layer. Each type of alteration was represented by several samples taken from the statues of both the horse and the rider, providing a series of similar samples for the experiment.

The reagents used for the cleaning tests were as follows:

1. deionized water
2. aqueous solution of 2% Tween-20
3. EDTA trisodium solution 12%
4. Rochelle salt in saturated solution
5. ammonium tartrate in saturated solution
6. mixed-bed ion-exchange resin (Rm)
7. cationic ion-exchange resin in acid form (RH)[4]

These treatments also followed standard procedures, which included applying the reagents in a gel form, using 3.0 g of carboxymethyl-cellulose as a suspending medium per 100 ml of solution.

For the resin tests, 7.0 g of dry resin in 17 ml of water were used. The gels were applied for fifteen minutes each and repeated three times on each area, so the action of the reagent could be checked each time the gel was removed. The applications were followed by washing with demineralized water and soft brushing as previously described. The results of the treatments and subsequent washing are summarized in Tables 1 and 2.

TABLE 1. *Reagents.*

Series	*Type of Alteration*	*Sample No.*	*Water*	*Tween-20*	*EDTA*	*Rochelle*	*Amm. tartrate*	*Rm*	*RH*
1	Dark green	10	IE	IE	SF	IS	IS	IS	IE
2	Uniform black	8	IE	IE	SF	IS	IS	IS	IE
3	Uniform black with underlying gold	8	IE	IE	IS	IE	IS	IE	ST
4	Nonuniform black with underlying gold	8	IE	IE	IS	IE	IS	IS	ST
5	Gray with underlying gold	8	IE	IE	SF	IE	SF	IS	IE
6	Whitish-gray with underlying gold	8	IE	IE	SF	IS	IS	IS	IE
7	Powdery light green	2	ST	—	—	—	—	—	—
8	Thick layer of gypsum deposits	8	IS	IS	IS	—	—	—	—

IE = ineffective

IS = insufficient

SF = sufficient

ST = satisfactory

TABLE 2. *Washing of areas treated with reagents, showing conductivity values (micro Siemens/cm).**

Series	*Type of alteration*	*No. wash*	*Water*	*No. wash*	*Tween-20*	*No. wash*	*EDTA*	*No. wash*	*Rochelle*	*No. wash*	*Amm. tartrate*	*No. wash*	*Rm*	*No. wash*	*RH*
1	Dark green	5	112–17.5	2	10–3.8	4	318–18	3	340–15.5	3	354–13.3	2	9–4	3	19–4
2	Uniform black	5	68–20	2	8.4–5.2	3	440–16	3	331–17	4	465–10.5	2	10–5	3	13–4
3	Uniform black with underlying gold	4	46–18	2	7.2–4.5	3	260–6	2	275–11.5	4	357–6	2	6–4	3	135–9
4	Nonuniform black with underlying gold	2	7.5–5.2	2	16–10	5	333–6	3	351–18.5	3	349–15	2	6–4	3	40–4
5	Gray with underlying gold	2	20–12	2	13.5–6	5	333–9.5	3	343–13.5	3	343–13.5	2	5.5–4.5	3	54–4.5
6	Whitish-gray with underlying gold	3	29–10	2	10–6	3	331–13.5	3	333–16.5	4	385–2.0	2	6.5–4	3	31–5
7	Powdery light green	4	42–15	2	—		—		—		—		—		—
8	Thick layer of gypsum deposits	9	118–23	4	200–20	13	480–19		—		—		—		—

*The results of the areas where the highest conductivity values have been obtained, followed by the lowest values, are shown for each type of alteration and for each reagent, preceded by the total number of washings.

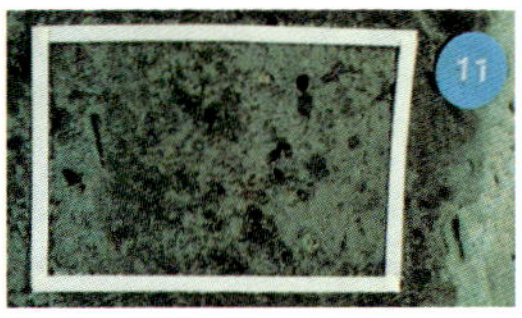

FIGURE 4. *An area of the monument after treatment with EDTA, showing the effectiveness of the reagent compared to the untreated region outside it.*

FIGURE 5. *Detail of the rider, left side, showing folds of the tunic before cleaning, below left.*

FIGURE 6. *Same area as in Figure 5, after cleaning, below right.*

Series No. 7 was treated only with water since the result was satisfactory. Series No. 8 was treated only with the first three reagents, since they were more specific for the deposits present there, which were essentially composed of gypsum and oxidation products of the alloy.

On visual inspection for series Nos. 3 and 7, the effectiveness of the reagents appeared quite satisfactory (Figs. 4–6). Therefore, some larger areas (about 30 × 30 cm) were chosen to check the various treatments on a working level; that is, areas considered representative for treatment of the monument. The test included all the above-mentioned alteration products and was used to assess both the possibility of repeating the various treatments and prolonging the washing, as well as the efficacy of subsequent drying by ventilation. A cleaning methodology was worked out on the basis of the different requirements of the surfaces of the two bronzes. Using the reagents found to be suitable (water, EDTA, ammonium tartrate, RH) it was possible to treat the whole surface except for the areas with thick and tenacious accumulations of particulate. Mechanical means—such as chisels, dentists' drills, or Cavitron—had to be adopted for these areas to reduce the layers. The various reagents were applied after the surfaces were freed of encrustation.

The treatment procedures, conducted with extreme caution, enabled all the existing gilding to be saved, and also revealed subtle and previously hidden aspects of the sculptural form, which in many cases had been concealed by thick encrustation. Inside the castings, various details of the fusion or assembly techniques were revealed. This provided new information regarding the fabrication techniques and repair methods used both in ancient times and at the times of the various restorations and repairs. Obviously, the restoration of such a degraded and mistreated monument involved other, less exacting operations, such as a more thorough electrochemical cleaning of the internal areas with pitting,[5] or retouching the patina of the Renaissance repairs which, being of a different and better-preserved alloy, were darker and did not match that of the restored monument.

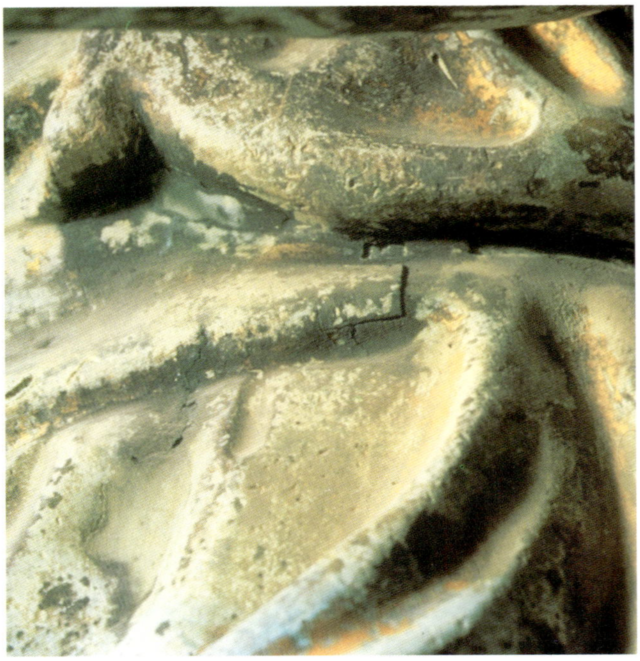

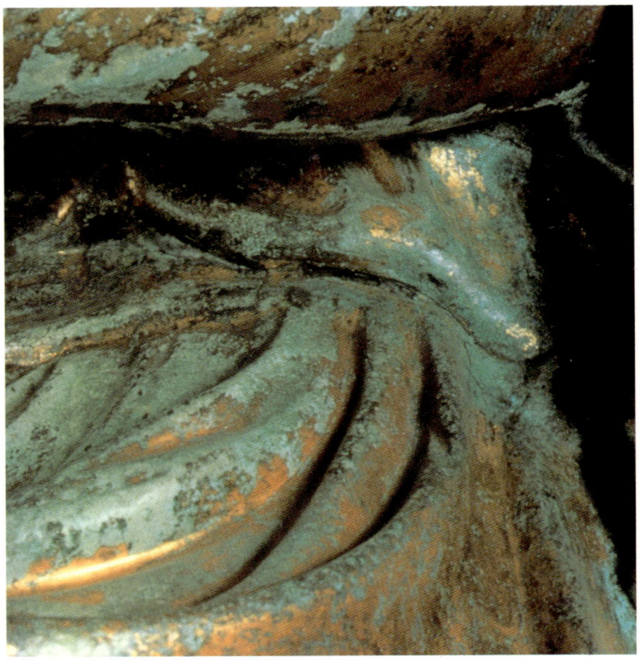

Observations and Conclusions

The cleaning treatments used in this restoration of the statue of Marcus Aurelius have made the monument more aesthetically pleasing and, at the same time, have revealed some unresolved conservation problems. The surface of the monument remains porous and cracked, and the gold is not stable. The micro- and large fissures, previously concealed by encrustation, now allow rainwater to enter and spread to a greater extent and absorb water (rain and condensation). Closing them with repairs would once again require a brutal grafting on already fragile and nonhomogeneous castings. As an alternative, synthetic materials might be applied. Such materials would have to be proven suitable for the project, stable with regard to the main chemicophysical points of view, and resistant outdoors. These substances, if used as sealants, would result in a virtual plastification of the monument, which is contrary to any conservation principle. For these reasons, fractures, holes, and gaps have not been repaired.

For the most part, the surfaces have been freed of polluting salts by cleaning and washing, and are thus in a more balanced and stable state. But despite the treatments, cupric chloride is still present inside the crystalline structure of the alloy, and a corrosion-inhibition treatment would probably be more harmful than not because of the volumetric and chemicophysical modifications to the patinas, with negative consequences on the gilding. In any case, if the bronze were to be exposed in the open again, such a stabilization treatment could only be effective for a brief period.

The possibility still exists of finding a coating that, by remaining unaltered for a reasonable time, would postpone maintenance for as long as possible, even if this alone would not be enough to defend the monument from rain infiltration and the consequences of mechanical and thermal stress. But such maintenance of the Marcus Aurelius would also mean the replacement of the coating, and removing the coating would damage the corrosion patina permeated by the resin. In addition, for correct maintenance, it would be necessary to separate the two bronzes, but the maintenance would then be extremely difficult.

In addition to these concerns, one must keep in mind that it was precisely the damage caused by the old coatings that prompted the team restoring the Marcus Aurelius to reflect on whether it was advisable to continue to use these substances. Traces of two different materials remain: the older coating, perhaps dating back to the early years of this century, was recognized in samples of hardened and oxidized material under the microscope (Fig. 7). Because the material was fractured and partially detached, it had formed blackish stains (cathodic areas), which were higher than the surrounding anodic areas, marked by powdery alterations. It was only possible to remove the remains of this by-product with careful, lengthy, and mechanical action, since solvents had no effect on it.[6]

FIGURE 7. *Microscopic view of the first (older) oxidized coating, which is partially detached.*

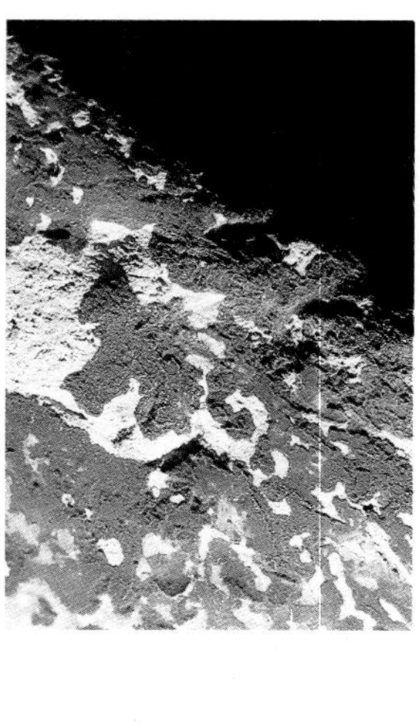

It was clearly evident that the more recent coating, perhaps applied in the last twenty to thirty years, had shrunk and was tearing off the corrosion patina (Fig. 8). Mechanical means were also used to remove this patina, since solvents only restored a little elasticity. A new synthetic resin, selected from those currently in use and recently studied, applied on a cracked surface exposed to climatic variations would

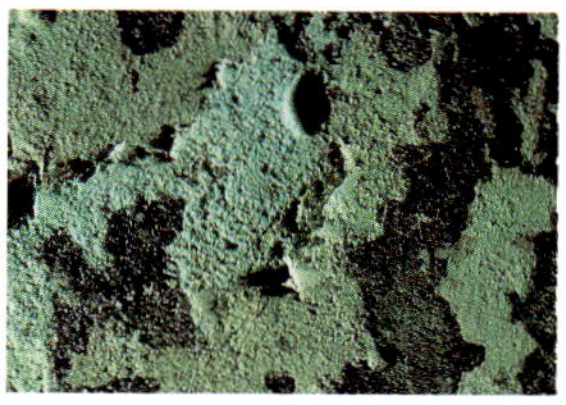

FIGURE 8. *Area of the monument showing both older and more recent coatings with sections of the latter peeling away from the older corrosion patina.*

soon behave like the previous resins, and could also lead to worse damage for the remaining gilding, which is now entirely exposed. At the most, a gentle consolidation of the corrosion patinas was necessary to prevent their continuous crumbling. A film of Paraloid B72 (concentration of 3% in trichloroethane) was applied as a fixative and not as a surface coating for the bronze alloy.

As the surveys and restoration gradually progressed, the decision was finally reached not to repair the lesions of the castings as well as not to protect the surface. This decision may at first seem defeatist. But there are various fundamental aims in the conservation of a work of art, such as respect for the historical value and the elimination or partial inhibition of the causes of degradation. A careful evaluation of the risk factors is always necessary.

For the Marcus Aurelius monument, the causes of decay have only been partially removed (Fig. 9). Continuing to work against the preservation of the monument is the environment of the Piazza del Campidoglio, which has not been improved and could rapidly reactivate the alteration processes if the equestrian statue were to be returned to the same location. The conservation of this monument mainly entails preventing, insofar as possible, any further work on or handling of the castings. Thus, the solution of conserving it in an air-conditioned environment, protected from dust, rainwater leaching, mechanical and thermal stress, as well as the avoidance of any introduction of a new support system between the rider and the horse, should not be considered a hasty measure but the most important conservation action carried out on the monument.

Even ignoring the mechanical causes of the deterioration, the extent of water-vapor absorption inside the surface and the speed with which the patina would continue to be corroded and leached if the monument were to be placed outside once again (Marabelli et al. 1988), tally with what can be directly observed on its surface.

FIGURE 9. *The monument of Marcus Aurelius after restoration.*

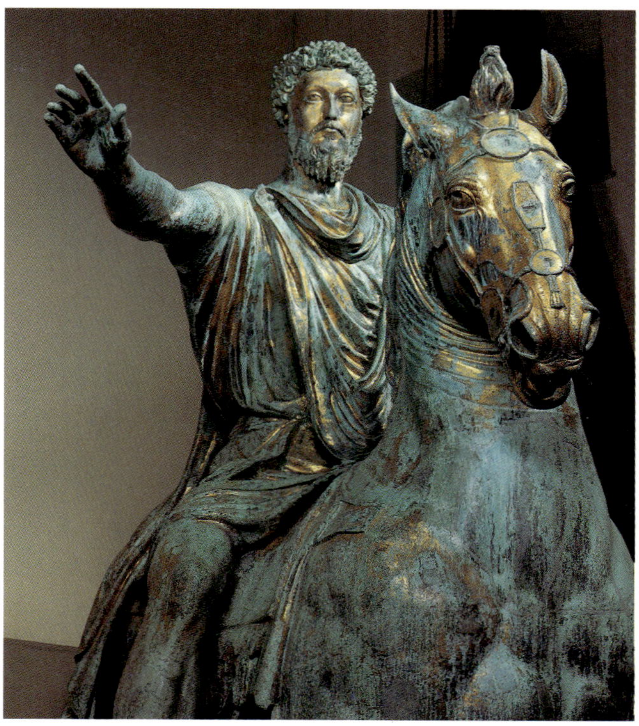

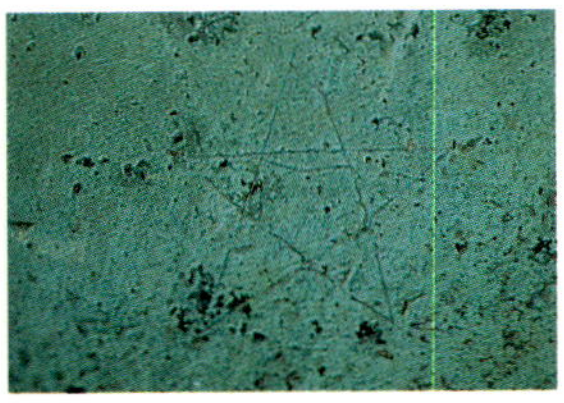

FIGURE 10. *Detail of the horse's raised hoof showing medieval graffiti in the shape of a star.*

In 1912 Apolloni, who was restoring the monument at that time, carefully recorded the presence of ancient graffiti on its surface in the form of letters, crosses, symbols, and various figures left by pilgrims who visited Rome in the Middle Ages (Apolloni 1912). Of all those graffiti, only one remains: a barely perceptible star on the horse's raised hoof (Fig. 10). This causes one to contemplate the remarkable survival of this monument thus far and the loss for future generations if it were to be erected again in the open air and this link to the past were thereby destroyed.

NOTES

1. For the study of the structure of the monument, see Accardo, Amodio et al. 1989; Accardo, Bennici, et al. 1989; Accardo, Caneva et al. 1983; Accardo et al. 1985; Accardo and Santucci 1988. For corrosion, see Marabelli et al. 1988.
2. The restoration was begun in 1987 and took 18 months to complete. Four restorers and the students of ICR's school of restoration participated under the author's technical management.
3. Demineralized water was used with conductivity values of 1.5 $\mu S\ cm^{-1}$. The conductivity values fell within the 175–4.5 $\mu S\ cm^{-1}$ range for external surfaces and 700–4.0 $\mu S\ cm^{-1}$ for internal surfaces.
4. The following reagents were used:
   - Tween 20-Merck (poliossietilensorbitanmonolaurato)
   - A 0.5 M (pH 6.5) solution of trisodium EDTA, obtained from 37.2 g bisodium EDTA + 43.4 g tetrasodium EDTA, in 1,000 ml of water
   - Cationic Bio-Rad Ag50W–X8 resin in acid form, 100–200 mesh, pH 5; mixed-bed resin made up of the above resin + Bio-Rad Ag1–X8 resin in OH^- form, 100–200 mesh, proportion 1:1.6 washed up to pH = 5.5

   For a similar use of the Rm resin see Fiorentino et al. 1982.
5. A localized treatment was carried out with repeated applications of 1 g of agar-agar and 6 g of glycerine in 80 ml of water + aluminium foil at 60–80 °C. Retouching was done with watercolors.
6. Alcohol, acetone, toluene, benzene, and xylene were used for this purpose.

REFERENCES

ACCARDO, G., D. AMODIO, P. CAPPA, A. BENNICI, G. SANTUCCI, AND M. TORRE

1989 Structural analysis of the equestrian monument to Marcus Aurelius in Rome. In *Structural Repair and Maintenance of Historical Buildings,* 581–91. C. A. Brebbia, ed. Southampton, U.K.: Computational Mechanics Institute.

ACCARDO, G., A. BENNICI, M. TORRE, D. AMODIO, P. CAPPA, AND G. SANTUCCI

1989 An experimental study of the strain fields on the statue of Marcus Aurelius. In *Proceedings of the 1989 SEM Spring Conference on Experimental Mechanics,* 534–37. Bethel, Conn.: Society for Experimental Mechanics.

ACCARDO, G., C. CANEVA, AND S. MASSA

1983 Stress monitoring by temperature mapping and acoustic emission analysis: A case study of Marcus Aurelius. *Studies in Conservation* 28:67–74.

ACCARDO, G., P. DE SANTIS, F. GORI, G. GUATTARI, AND J. M. WEBSTER

1985 The use of speckle interferometry in the study of large works of art. In *Proceedings of the 1st International Conference on Non-destructive Testing in Conservation of Works of Art* 4(1):1–12. Rome: Istituto Centrale per il Restauro (ICR) and Associazione Italiana Prove non Distruttive (AIPnD).

ACCARDO, G., AND G. SANTUCCI

1988 Metodo di calcolo agli elementi finiti e misure estensimetriche per l'analisi strutturale dei manufatti storico-artistici. In *2d International Conference on Non-destructive Testing: Microanalytical Methods and Environment Evaluation for Study and Conservation of Works of Art* 1(3):1–18. Rome: ICR and AIPnD.

APOLLONI, A.

1912 *Vicende e restauri della statua equestre del Marco Aurelio: Atti e Memorie*. Rome: Accademia di San Luca.

FIORENTINO, P., AND M. MARABELLI

1977 *I cavalli di San Marco*. Venice: Procuratoria di San Marco.

FIORENTINO, P., M. MARABELLI, M. MATTEINI, AND A. MOLES

1982 The condition of the Door of Paradise by L. Ghiberti: Tests and proposals for cleaning. *Studies in Conservation* 27:145–53.

MARABELLI, M., A. MARANO, S. MASSA, AND G. VINCENZI

1988 La condensazione capillare di vapore acqueo in patine di bronzi esposti all'aperto. In *Preprints of the 2d International Conference on Non-destructive Testing: Microanalytical Methods and Environment Evaluation for Study and Conservation of Works of Art* 2(25):1–20. Rome: ICR and AIPnD.

MARABELLI, M., AND G. NAPOLITANO

1991 Nuovi sistemi applicabili su opere o manufatti in bronzo esposti all'aperto. *Materiali e strutture* 1(2):51–58.

MARCO AURELIO, MOSTRA DI CANTIERE.

1984 Rome: Arti Grafiche Pedanesi.

MARCO AURELIO: STORIA DI UN MONUMENTO E DEL SUO RESTAURO.

1989 Cinisello Balsamo (Milan): RAS.

Biography

Paola Fiorentino is chief restorer at the Istituto Centrale del Restauro (ICR) in Rome. She received her degrees at the Arts Academy of Rome and at the ICR School of Restoration, where she has taught metal-restoration techniques since 1966. She has worked for the ICR since 1961, specializing in metal preservation and the restoration of important monuments.

Bronze Objects from Lake Sites: From Patina to "Biography"

FRANÇOIS SCHWEIZER

Over the last decades, archaeologists have made extensive use of scientific methods to investigate, analyze, and interpret excavated artifacts. Apart from dating techniques, botanical, zoological, and sedimentological studies have contributed to a better understanding of the cultural and ecological development of ancient populations. As far as metal artifacts are concerned, research has been centered mainly on the examination of metal alloys and the history of technologies.

Considering all these investigations, it is surprising that little attention has been paid thus far to the composition and structure of corrosion layers on metals as an opportunity for archaeometric research. The aim of this contribution is to show that there is a close link between the composition of patinas and the environments in which they are formed. If one understands the relationship and interaction between soil types and the formation and stability fields of corrosion products on metals, one may be able to tell under what sort of environmental conditions these materials have grown. This information should provide investigators with the possibility of writing the *biography* of ancient metal artifacts.

This paper was written for the use of archaeologists; however, the results of the author's investigations on the composition and structure of corrosion layers of ancient bronzes from lake settlements are also included.

Few studies have been published on the effect of the soil type on the composition of patinas (Geilmann 1956; Tylecote 1979; Robbiola et al. 1988; Robbiola 1990).

Archaeological Context

The lake of Neuchâtel in the western part of Switzerland contains many important lake settlements that have been excavated over recent decades by the archaeological unit of the Canton of Neuchâtel under the directorship of Michel Egloff. The site of

Hauterive-Champréveyres was long occupied during the late Bronze Age, as dated by dendrochronology from 1050 to 870 B.C.E. (Benkert and Egger 1986). During the recent excavation, more than 5,900 bronze objects (needles, pins, bracelets, etc.) were found in the different occupation layers (Rychner 1991). The archaeologist in charge of the metal artifacts, Annemarie Rychner-Faraggi, was struck by their different appearances. She distinguishes two main groups of patina on bronze objects:

1. *Lake* patina—a smooth, dense, brown-yellow patina (approximately 70%)
2. *Land* patina—a thick, green-blue patina containing quartz grains

A few objects contained both patina types.

The great number of bronzes with land patina is very unusual for a lake settlement. Therefore, the question was raised as to whether this settlement was originally on dry land or on damp or wet ground. In approximately 750 B.C.E., the water level of the lake of Neuchâtel rose. From that time until their recent excavation, all the objects had remained underwater.

In collaboration with Rychner-Faraggi, five questions were formulated:

1. Are the different patinas due to different bronze-alloy compositions?
2. What is the composition and stratigraphy of each—the green-blue land and the brown-yellow lake—patina?
3. Under what sorts of environmental conditions (on dry land, in wet soil, in the water) were the patinas formed?
4. Are they primary corrosion products or were they formed later by chemical reactions with the soil?
5. Is it possible to retrace the *history* or the *corrosion biography* of an individual bronze object after its use?

Origin and Type of Bronze Material Analyzed

To determine the origin and type of bronze material on the objects, five small bronze objects were initially studied: four pins and a fishing hook. Later, four more bronzes were added (Table 1).[1]

The first series of objects was analyzed using the five questions outlined above as a central focus. The second series contained objects that were used for metallographic examinations and for the investigation of corrosion mechanisms.

The site of Hauterive-Champréveyres contains five different archaeological layers (Rychner 1991):

Layer 1: Yellow, sandy layer of recent origin, probably formed by washing out the lower (older) layers, and containing artifacts from these layers. Its pH is 7.55.

Layer 2: Lake sediment of sand and clay.

Layer 3: Layer containing different strata of organic material due to human activities. Its pH varies between 7.1 and 7.7.

TABLE 1. *Data from analysis of several bronze objects.*

Lab MAH Genève	*Inv. No. Neuchâtel*	*Object*	*Patina type*	*Archaeological layer*
1st series				
85-27	17'773	pin	lake	1
85-28	3'389	pin	lake and land	3
85-29	3'967	fishing hook	lake	1
85-194	3'071	pin	lake and land	1
86-77	18'603	pin	lake and land	1
2nd series				
87-194	18'152	pin-needle	lake	3
87-195	3'031	pin-needle	land	1
87-196	6'567	metal piece	lake	3
87-197	6'246	metal piece	lake and land	1

Layer 4: Sandy stratum.

Layer 5: Layer rich in organic remains. This stratum is related to human activities during the Bronze Age and is on top of a neolithic lake sediment.

Table 1 indicates that the bronze objects examined are from layers 1 and 3.

Experimental Method

Different techniques were used to characterize the bronzes and their corrosion products, as follows:

To determine the chemical composition of the bronze alloys, first X-ray fluorescence analysis was used on the uncleaned surface to establish the alloy type.[2] Inductively coupled plasma (ICP)-atomic-emission-spectrometry[3] was employed for major and trace elements. A tungsten drill was used to sample between 30 and 50 mg of bronze from the uncorroded metal core.

To analyze the metallographic structure of the bronze alloys, sections across the samples were removed with a jeweler's saw, embedded in a polyester resin, ground on carborundum paper up to grade 1000, and polished with diamond pastes of 6μ, 3μ, 1μ, and 0.25μ. After observation, the samples were etched with alcoholic $FeCl_3$ solution.

For composition analysis of the corrosion product by X-ray diffraction, some grains were removed with a steel blade, mounted on a glass needle, and exposed in a Gandolfi camera (114.5 mmø) for 12–16 hours to Fe kα radiation, 30 kV, 20 mA, with no filter. Some samples were also examined[4] with the Debye-Scherrer camera using Fe radiation for 8 hours. Quantitative analysis on polished cross sections of the corrosion layers of the lake patina were undertaken with an electron microprobe.[5] The distribution of different elements in the corrosion layers (element mapping) was examined with the electron microanalyzer.[6]

Analytical Results

The X-ray fluorescence analysis on the surface revealed that all objects are copper-tin bronzes containing a number of minor elements such as arsenic, nickel, iron, and antimony. The results of the ICP spectrometry are listed in Table 2.

The bronzes are classical, copper-tin alloys with minor constituents that were certainly not added intentionally. There is no systematic difference between bronzes with a lake patina (87-194 and 87-196) and those with a land patina (87-195 and 87-197).

The four objects analyzed showed a similar microstructure: a network of fairly regular twinned grains. Close to the surface, some grains contain slip lines. There was probably a series of working and annealing regimes after the casting process. In a final phase, they were again slightly cold-worked.

The corrosion products of the land patina are, essentially, basic copper carbonates and basic copper sulfates, as indicated below:

malachite $CuCO_3Cu(OH)_2$	ASTM 10-399
antlerite $CuSO_4(OH)_4$	ASTM 7-407
posnjakite $Cu_4SO_4(OH)_6 \cdot H_2O$	ASTM 20-364

Whereas malachite and antlerite are quite common corrosion products, with the latter especially prevalent in polluted urban areas, to the author's knowledge this is the first time that the presence of posnjakite—$Cu_4SO_4(OH)_6 \cdot H_2O$—on archaeological bronzes has been reported. Posnjakite is a light-blue mineral closely related to antlerite and brochantite. It was described first by Komkov and Nefedov (1967). Geologically, it is associated with auricalcite and other secondary minerals near oxidized chalcopyrite. The X-ray diffraction pattern is presented in Table 3.

The identification of the corrosion products of the lake patina proved to be more difficult than expected. In the author's preliminary publication (Schweizer 1988), the presence of an unusual mineral, sinnerite ($Cu_6As_4S_9$) which has an X-ray diffraction pattern close to chalcopyrite ($CuFeS_2$) was reported. Stephan Graeser of the Natural History Museum in Basel, who analyzed one of the samples, presumed the presence of colusite [CU_3 (As, Su, V, Fe) S_4] ASTM 9–10.

The difficulty of interpreting X-ray diffraction patterns of complex copper sulfides is well illustrated in Table 4, in which the specimen is listed together with reference minerals and American Society for Testing and Materials (ASTM) patterns. It was only by quantitative analysis of the chemical composition of the corrosion layer (as will be discussed herein) that the presence of chalcopyrite could be ascertained. The difficulties of interpreting X-ray diffraction patterns of archaeological corrosion products are fully discussed by Fabrizi and Scott (1987).

TABLE 2. *Results of the ICP spectrometry.*

Lab MAH No.	*Cu*	*Sn*	*Pb*	*As*	*Sb*	*Ag*	*Ni*	*Co*	*Zn*	*Fe*
87-194	89.22	9.57	0.34	0.19	0.26	0.15	0.05	0.06	0.05	0.09
87-195	91.29	5.65	0.51	0.55	1.00	0.22	0.69	0.06	0.01	0.02
87-196	87.52	8.02	1.46	0.60	0.81	0.21	1.04	0.25	0.03	0.05
87-197	89.85	8.02	0.34	0.34	0.60	0.18	0.55	0.10	0.01	0.02

TABLE 3. *X-ray diffraction lines of posnjakite $Cu_4SO_4(OH)_6 \cdot H_2O$ found on a bronze needle (Lab MAH 87-27, inv. 17773) from the site of Champréveyres. Gandolfi camera 114.5 mmø, 30 kV, 20 mA, 11 hours, Fe unfiltered radiation.*

Sample 87-27/001	*Film No. 303*	*Reference*	*ASTM 20-364*
d (A)	I	d (A)	I
7.67	40		
6.95	100	6.94	100
5.25	30	5.25	8
		5.15	4
4.84	10	4.85	6
4.65	5	4.77	4
3.80	15	3.74	2
3.46	50	3.47	30
		4 weak lines	
2.70	50	2.70	25
2.61	30	2.614	16
		2.576	2
2.41	50	2.422	25
2.33	25	2.334	12
2.25	5	2.260	8
2.01	40	2.018	12
1.95	25	1.952	6
1.86	15	1.870	4
		1.734	2
1.66	15	1.662	4
1.61	15	1.616	2
1.58	15	1.585	4
1.54	45	1.541	10

On one sample (85-197), chalcocite (Cu_2S) and djurleite ($CU_{1.93}S$) were also found.

A small section of needle 87-194 was examined by different techniques to get a better understanding of the formation mechanism of the chalcopyrite lake patina. The copper-tin alloy was attacked locally, resulting in a fingerlike structure. The thickness of the corrosion layer was found by metallographic examination to vary between 100 and 150 µm. The layer is separated into three zones. The first zone, close to the metal, shows evidence of pseudomorphic replacement of metal grains by corrosion products (Fig. 1a). The second zone, clearly visible in dark-field illumination (Fig. 1b) is very regular and free of any inclusions or holes. The third layer is quite porous. The number and size of the pores increase toward the surface.

After etching with alcoholic $FeCl_3$ solution, one can clearly see the crystalline appearance of the structure of the corrosion layer on top of the corroded α-phase grains (Fig. 1c). The corrosion proceeds into the metal through the grains like a root.

To gain a better understanding of the formation of the corrosion layer, the element and its distribution were analyzed.[7] Analysis showed the area represented in Figures 2a–d to be the same as that in Figure 1a. The results may be summarized as

TABLE 4. *X-ray diffraction lines of corrosion products from bronzes from the site lake of Hauterive-Champréveyres, Switzerland, and of the minerals Sinnerite ($Cu_6As_4S_9$) and Chalcopyrite ($CuFeS_2$).*

Sample/ reference mineral	*Corrosion on bronze LabMAH 85-28*		*Corrosion on bronze LabMAH 85-28*		*Sinnerite $Cu_6As_4S_9$ Lengeubach, Binn*		*Sinnerite $Cu_6As_4S_9$ ASTM 25-264*		*Chalcopyrite $CuFeS_2$ Westphalia*		*Chalcopyrite $CuFeS_2$ ASTM 35-732*	
Notes (below):	*a*		*b*		*c*		*d*		*e*		*f*	
	d(Å)	I	d(Å)	I	d(Å)	I	d(Å)	I	d(Å)	I	d(Å)	I
			5.0525	10b								
											4.715	1
					4.2581	<10						
			4.1349	10b								
	3.34	60	3.3578		3.3627	40	3.34	20	3.33	30		
	3.02	100	3.0586	100	3.0288	100	3.02	100	3.03	100	3.038	100
					2.6897	<10						
	2.62	20	2.6598	10	2.6305	70	2.611	40	2.64	10	2.644	5
											2.606	2
					2.5333	40	2.538	40				
					2.4812	40						
											2.308	<1
					2.1787	40						
					2.2460	40	2.237	10b				
					2.1787	40	2.177	5b				
							2.149	5b				
					2.1266	40						
							2.109	5				
					2.0586	40	2.054	10	2.04	20		
							1.915	5b				
	1.85	80	1.8705	90	1.8598	90	1.852	80	1.86	90	1.8697	22
											1.8570	37
	1.75	30	1.7651	50	1.7466	20			1.75	10		
							1.667	5				
							1.640	5				
							1.614	5				
	1.59	60	1.6030	80	1.5849	80	1.581	70	1.59	40	1.5927	27
											1.5753	14
							1.556	20				
					1.5222	30						
							1.514	5			1.5192	1
	1.32	20	1.3265	10	1.3211	30	1.329	5			1.3219	3
					1.3132	30	1.312	20				
							1.297	5			1.3027	<1
							1.288	5				
	1.21	30	1.2067	10	1.2067	40	1.205	30	1.212	10	1.2125	3
									1.204	10	1.2052	5
											1.1998	3
					1.1769	30	1.172	5				
	1.08	40	1.077	10	1.0743	40	1.072	20	1.079	10	1.0770	5
									1.070	10		
							1.060	5				
											1.0452	<1
									1.019	20	1.0173	4
					1.0118	40	1.010	10b			1.0128	5

[a] Corrosion product from bronze needle (Lab MAH, Inv. 3389). Gandolfi camera 114.5 mm, Fe radiation unfiltered, 30 kV, 14 mA, 15 hours. Film Lab MAH No. 302.

[b] Same sample as noted above X-rayed by S. Graeser, Natural History Museum, Basel. Fe radiation, 8 hours. Film Lab MAH, No. G 451p.

[c] Sinnerite sample from the Lengenbach mine near Binn in the Valais, Switzerland. X-rayed by S. Graeser, National History Museum, Basel. Fe radiation, 8 hours. Film Lab MAH No. G 452p.

[d] Ref. Makovisky and Skinner, *Am. Mineral.* 57:824–34 (1972), synthetic crystal.

[e] Sample from Victoria Mine, Westphalia, obtained from J. Deferne, Natural History Museum, Geneva. XRD: Gandolfi camera 114.5 mmø, Fe radiation unfiltered, 30 kV, 14 mA, 17 hours. Film Lab MAH No. 427.

[f] Ref. Nat. Bur. Stand. (U.S.) Monogr. 21 (1984). Sample from Merkur Mines, Germany.

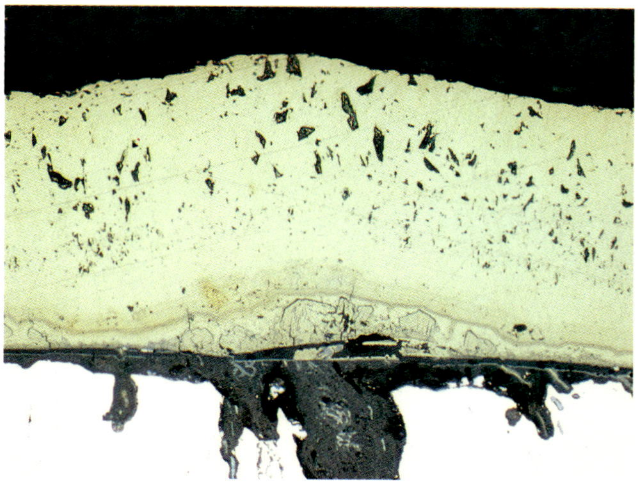

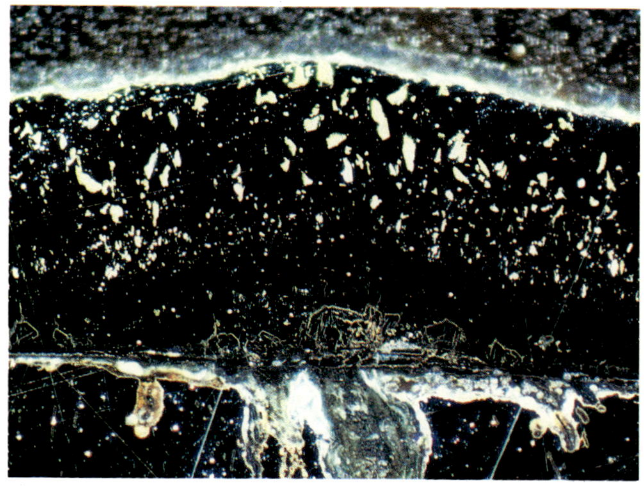

FIGURE 1a–c. *Cross section of a chalcopyrite lake patina on a bronze needle, showing (a) above left, unetched, bright field with layered structure on top of the metal; (b) above right, unetched, dark field with different zones in the corrosion layer; and (c) right, region etched with alcoholic* $FeCl_3$*. Note grainlike area of the corrosion layer on top of the corrosion pit and its rootlike bottom in the latter image. Lab MAH 87-194.*

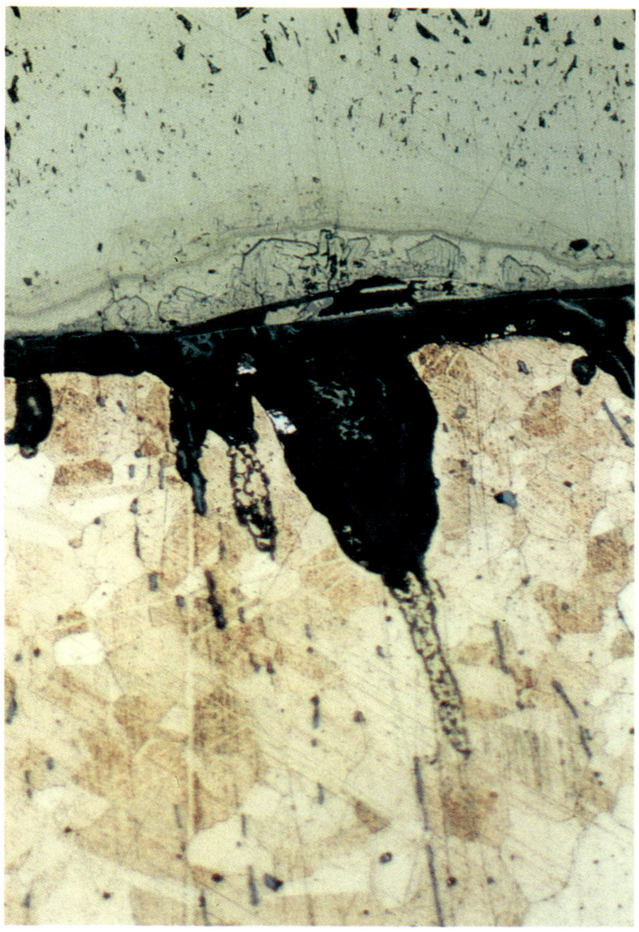

follows. The bronze alloy corrodes by selectively eliminating copper, which is redeposited as chalcopyrite on the surface (Fig. 2b). There is an enrichment in the tin content due to the preferential corrosion of copper in the alloy (Fig. 2c). The corrosion layer does not contain any tin. Sulfur and iron are distributed in exactly the same manner (Fig. 2d). They have diffused slightly into the corroded areas of the bronze and are evenly distributed in the corrosion layer. An element scan shows that the corrosion layer contains only copper, iron, and sulfur. No other elements can be detected.

To analyze the chemical composition of the lake patina, a small sample of bronze needle 87-194 was studied with the electron microprobe.[8] The results are listed in

FIGURE 2a–d. *X-ray mapping of the same cross section as in Figure 1 showing (a) secondary electron image, (b) Cu X-ray image, (c) Sn X-ray image, and (d) Fe X-ray image. The S X-ray image is identical to the Fe X-ray image. The bar on the left-hand side is equal to 100 µm.*

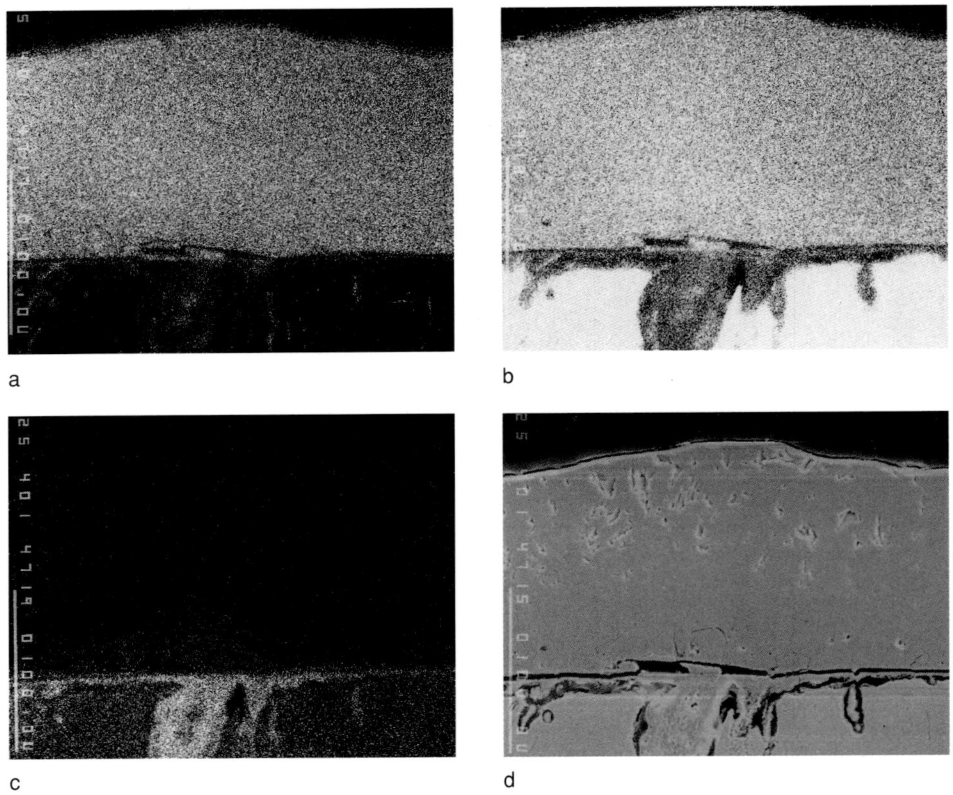

Table 5. The first column gives the composition of the bronze. Some differences in the analyses by atomic-emission spectrometry are probably due to inhomogeneities in the alloy (Table 2). At the interface bronze-corrosion layer, the tin content increases and traces of sulfur are present. The iron values are still very low. Optically, two layers may be distinguished in the corrosion crust. Their compositions are listed in columns 3 and 4 of Table 2. One may observe that there is no significant difference between them. Both layers have a composition consistent with that of chalcopyrite. The corrosion layer contains no tin and only very little arsenic and zinc.

Most publications on microbial corrosion of metals under anaerobic conditions concern the corrosion of ferrous metals (Panter 1980). A short review of literature

TABLE 5. *Electron-microprobe analyses of the corrosion layer of a bronze needle from the lake site of Hauterive-Champréveyres[a] (Lab MAH 87-194).*

	Bronze	*Bronze int. crust*	*Int. crust[b]*	*Ext. crust[c]*	*Chalcopyrite* $CuFeS_2$
Fe %	0.17	0.77	30.69	30.25	30.43
S %	nd[d]	0.03	33.14	33.76	34.94
Pb %	nd	nd	nd	nd	____
As %	1.70	1.59	0.66	0.14	____
Zn %	0.13	0.13	0.02	0.01	____
Cu %	92.40	88.34	34.52	35.44	34.62
Sn %	5.58	9.10	nd	nd	
	99.98	99.96	99.03	99.60	99.99

[a] Analyses by R. Oberhänsli, Mineralogical-petrographical Institute of the University of Berne, Switzerland.
[b] Mean values of 3 measurements.
[c] Mean values of 5 measurements.
[d] nd=not detected.

on microbial corrosion of nonferrous metals has been published by McDougall (1966). For archaeological copper-tin alloys, Tylecote published a short note in his excellent paper on the effect of soil conditions on the long-term corrosion of buried bronze alloys (1979:352).

More recently, Duncan and Ganiaris (1987) reported an interesting investigation on bronze and lead alloys found on London waterfront sites. For the formation of the gold- and black-colored sulfides, they postulate two possibilities: (1) precipitation reactions with copper and ions in the soil, and (2) direct action of hydrogen sulfide on the metal-oxide surface.

Based on the foregoing, and in the absence of electrochemical investigations, the formation of the chalcopyrite corrosion layers may be described as follows:

1. Sulfate-reducing bacteria produce hydrogen sulfide in the organically rich soil close to the metal artifacts. It is unlikely that they grow on the bronze alloy, as they normally cannot resist more than 2,000 ppm of copper (McDougall 1966:11).
2. Copper is dissolved preferentially in the α-copper-tin grains, leaving a tin-enriched phase.
3. Copper ions from the metal, iron ions from the surrounding soil, and sulfur ions combine and precipitate as chalcopyrite.
4. The chalcopyrite forms a fairly uniform protective layer on the copper alloy.
5. The growth rate of the chalcopyrite layer must slow down after some time due to increased diffusion times of the copper ions through the corrosion layer.

It is also important to consider the relationship of the different patinas and their composition and appearance on the bronze objects. As Table 6 shows, one may find different types of patinas on the same object. Their different composition is not related to the chemical composition of the alloys nor to the methods of manufacture.

It is obvious that the different compositions of the corrosion products are related to the environmental conditions in which they were formed. The presence of land-type and lake-type patinas on the same object (Figs. 3, 4) still leaves the question of which one was formed first. Is it even possible that both were formed simultaneously?

Even after all the corrosion products have been identified, it is still not possible to answer this question nor those posed here initially. In order to write the *biography* of

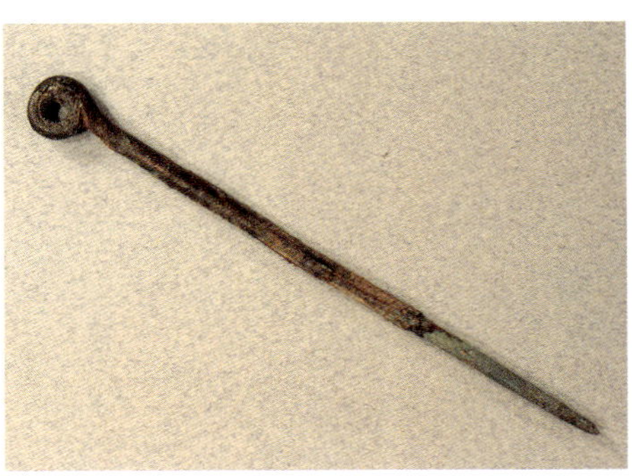

FIGURE 3. *Bronze pin with lake and land patina. Lab MAH 86-77.*

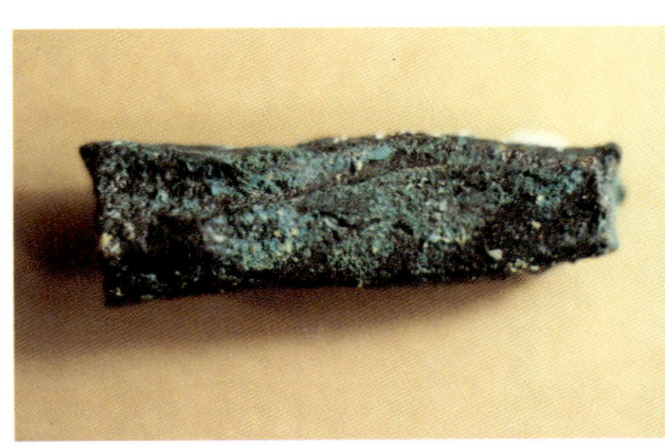

FIGURE 4. *Small bronze shaft with lake and land patina. Lab MAH 86-77.*

TABLE 6. *Appearance and composition of patinas on the bronze objects.*

Lab MAH No.	*Object*	*Patina*	*Composition*
85-194	needle	brown, shiny green, granular white, granular	chalcopyrite $CuFeS_2$ malachite $CuCO_3 \cdot Cu(OH)_2$ calcite $CaCO_3$
85-27	needle	blue, granular green, granular	posnjakite $Cu_4SO_4(OH)_6 \cdot H_2O$ malachite $CuCO_3 \cdot Cu(OH)_2$
85-77	needle	brown, shiny blue-green, granular	chalcopyrite $CuFeS_2$ antlerite $Cu_3(SO_4)(OH)_4$
85-29	fishing hook	brown, shiny	chalcopyrite $CuFeS_2$
85-28	needle	brown, shiny	chalcopyrite $CuFeS_2$
87-194	pin-needle	brown, shiny	chalcopyrite $CuFeS_2$
87-195	pin-needle	green-blue granular	malachite $CuCO_3 \cdot Cu(OH)_2$
87-196	metal piece	brown, shiny	chalcopyrite $CuFeS_2$
87-197	metal piece	green-blue granular and brown, shiny	posnjakite $Cu_4SO_4(OH)_6 \cdot H_2O$ chalcocite Cu_2S djurleite $Cu_{1.93}S$

the bronze objects, the environmental factors affecting corrosion-mineral formation and their fields of stability must first be considered.

Environmental Considerations

When a new copper or bronze object is deposited in the soil, its corrosion behavior will depend on different factors:

- composition of the alloy
- acidity of the soil (pH)
- oxidation and reduction potential of the environment—dry (sandy) and oxygen-rich soils or wet, anaerobic soils rich in organic materials
- cations and anions present in the soil

To take a simple case first, the corrosion products that can be formed in the presence of copper, water, and carbon dioxide are tenorite, CuO; cuprite, Cu_2O; malachite, $CuCO_3 \cdot Cu(OH)_2$; and azurite, $2CuCO_3 \cdot Cu(OH)_2$. Which products will actually form depends on the pH of the environment and its oxidation-reduction potential (Eh). Every aqueous system can therefore be characterized by four areas (Pourbaix 1977): oxidizing and acidic, oxidizing and alkaline, reducing and acidic, and reducing and alkaline (Fig. 5).

The fields of stability of the minerals constituting the corrosion products are published (Pourbaix 1977) and are illustrated in a simplified diagram in Figure 6. One can see that alkaline and reducing conditions favor the formation of tenorite (CuO), whereas azurite is stable in an oxidizing and slightly acidic environment. The region between the two oblique lines represents the range of stability of water. Fields of stability for some copper sulfides are shown in Figure 7.

FIGURE 5. *Potential versus pH diagram showing acid, alkaline, reducing, and oxidizing areas in aqueous solutions.*

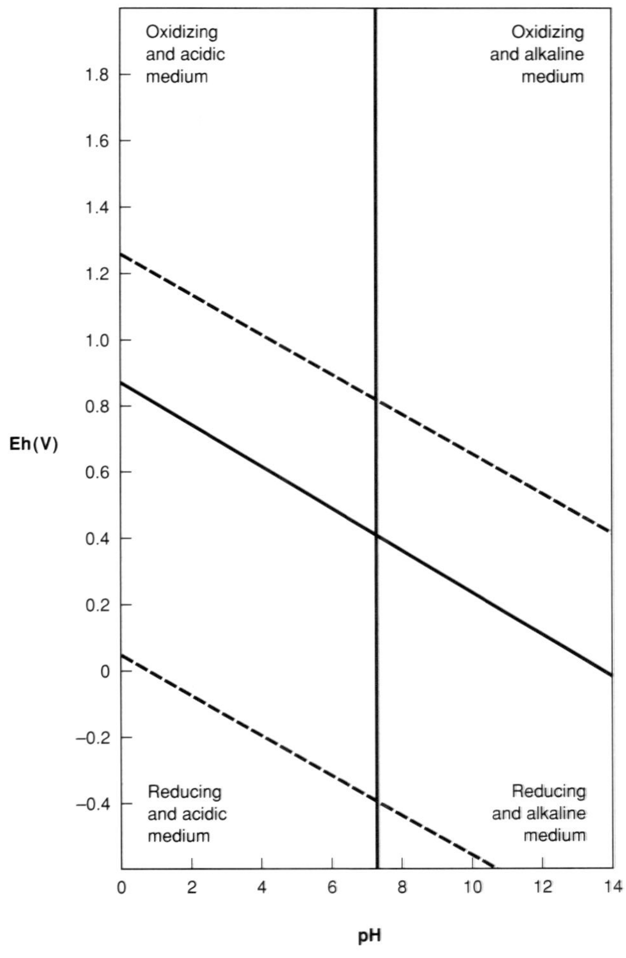

FIGURE 6. *Simplified pH-Eh diagrams for the ternary system Cu-CO_2-H_2O (after Pourbaix 1977).*

FIGURE 7. *Simplified pH-Eh diagrams for the ternary system Cu-S-H_2O (after Pourbaix 1977).*

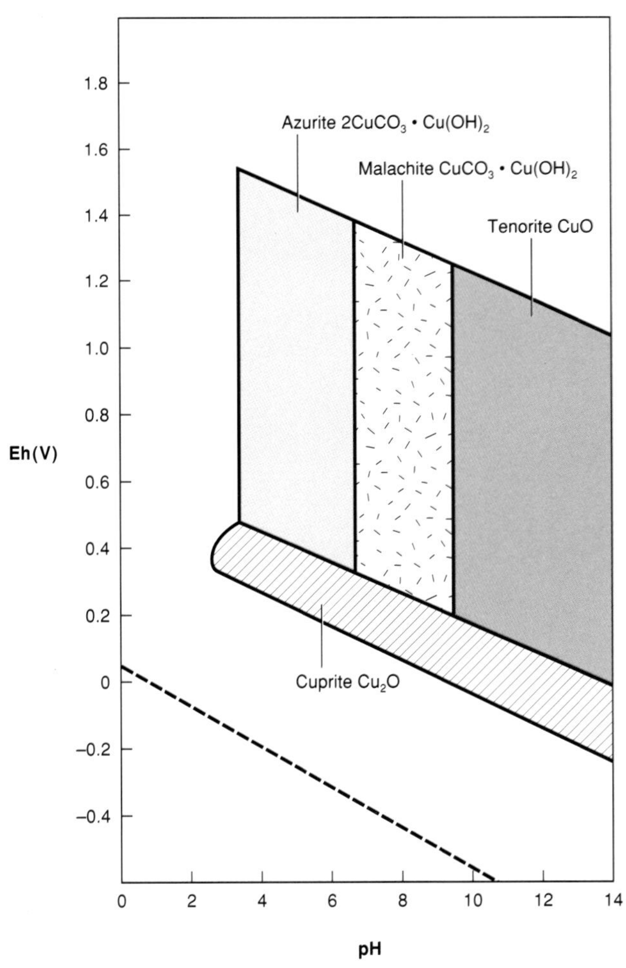

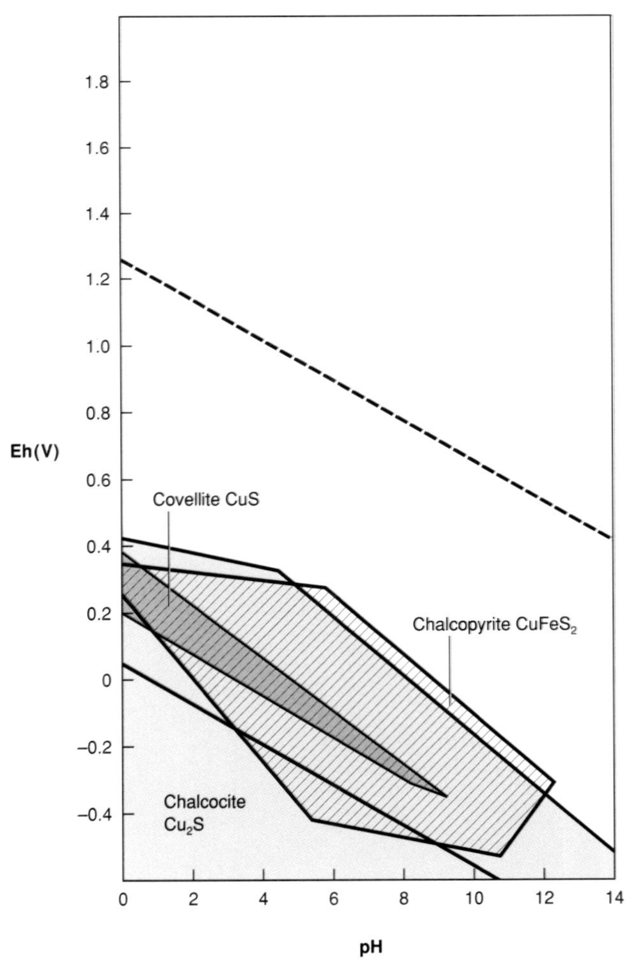

As the patinas of the bronze objects were formed in the soil, a relation between the fields of stabilities (pH-Eh diagrams) of the corrosion products and different types of soil must now be established. This should indicate the sorts of environmental conditions in which the bronzes acquired their patina.

Soil Types

The formation of soils over long periods of time is a very complex process and is influenced by many factors: composition of the original rocks, degree of erosion, sedimentation, clay content, organic substances, bacteriological activity, type and concentration of soluble salts, and water content. All these factors lead to different soil types.

The earth sciences have tried to characterize soils by measurable parameters. Apart from the chemical, mineralogical, and biological composition, the acidity (pH) and the redox potential (Eh) have proven to be of great interest. A large number of pH-Eh diagrams of different soils have been published (Garrels and Christ 1965; Baas Becking et al. 1960). Figure 8 shows the position in the pH-Eh diagrams of some natural environments.

Basically there are three main regions:

- well-aerated soils in contact with the atmosphere
- transitional environments
- soils isolated from the atmosphere

Clearly, it is possible to characterize the formation areas and fields of stability of copper corrosion products as well as soil types by using pH-Eh diagrams. It is therefore logical to combine these two diagrams.

The information in Figures 6 and 7 is combined in Figure 9. This new diagram may be compared with Figure 8 showing the pH-Eh diagram of soil types. One can see clearly that the fields of stability for copper sulfides correspond to pH-Eh regions of soils, which are isolated from the atmosphere. Malachite and azurite, however, are formed in well-aerated soils.

With these considerations in mind, the formation of the patinas on the different bronze objects may now be considered.

Formation "Biographies"

The objects are subdivided into three groups according to their patina composition in Table 6:

- objects with lake patinas only
- objects with lake and land patinas
- objects with land patinas only

Four objects (85-29, 85-28, 87-194, and 87-196) have a patina of chalcopyrite, or lake patina, only (Figs. 10a, b). This copper sulfide must be the primary corrosion product and the objects must have been deposited soon after their use in an anaerobic, damp, and humus-rich soil. The chalcopyrite could not have been formed later

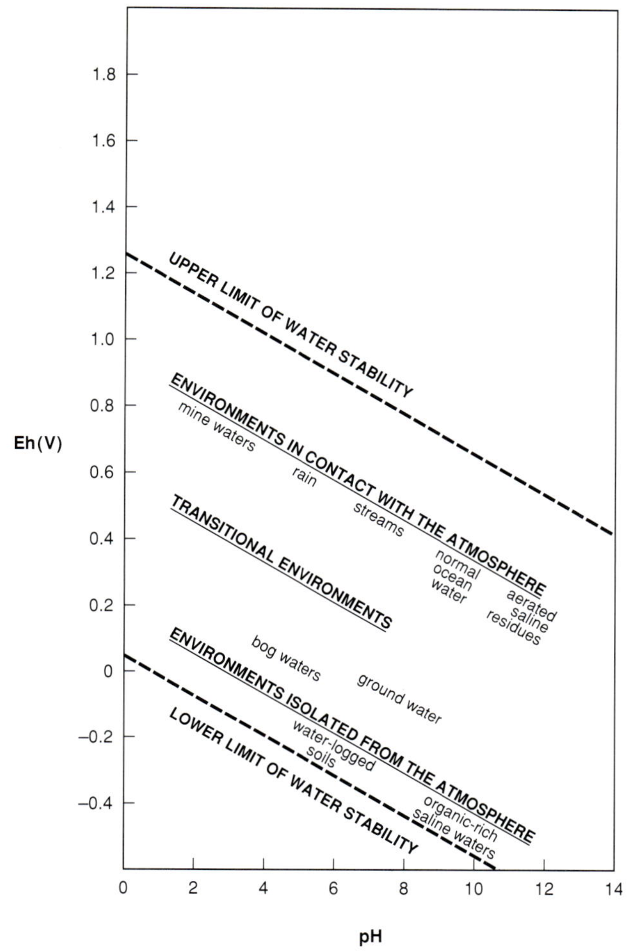

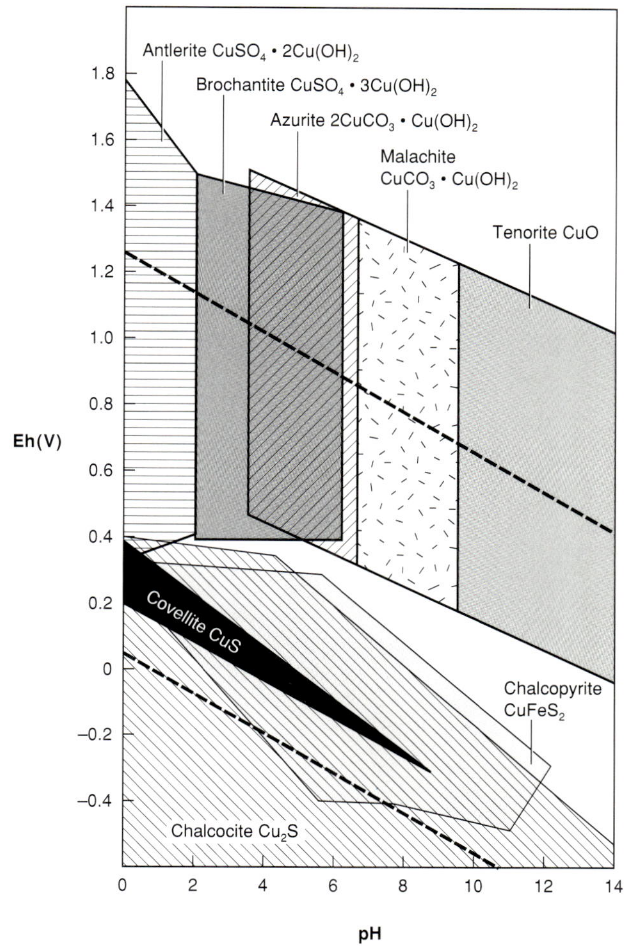

FIGURE 8. *Approximate position of some natural environments as characterized by pH and Eh, above left (after Garrels and Christ 1965).*

FIGURE 9. *Fields of stability of the corrosion products identified on the bronze objects, above right (after Pourbaix 1977 and Garrels and Christ 1965).*

from basic copper carbonates or sulfates. The latter corrosion layers have a granular microstructure, whereas the chalcopyrite layers are smooth and adhere closely to the surface of the metal.

Three objects show both lake and land patinas. They contain corrosion products in which formation conditions and fields of stability correspond to different environmental soil types. On needle 85-194, chalcopyrite and malachite were found; on needle 85-77 (Fig. 3), chalcopyrite and antlerite; and on the metal piece 87-197 (Fig. 4), chalcocite, djurleite, and posnjakite (see Table 6).

Three hypotheses may therefore be formulated:

1. All corrosion products were formed simultaneously. This possibility can be excluded as the fields of stability for the copper sulfides and the basic copper carbonates and sulfates are too different (Fig. 9).

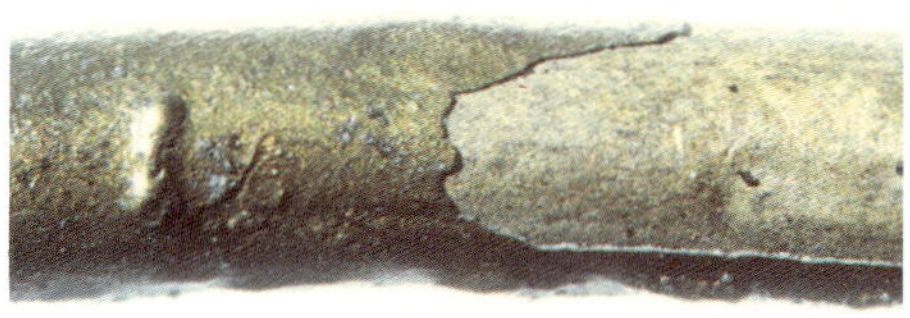

a

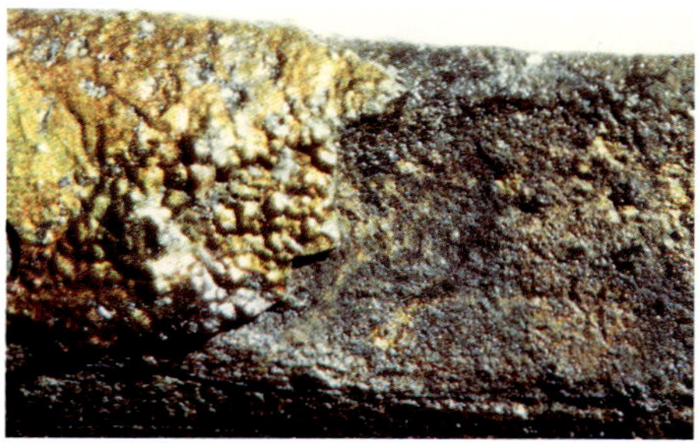

b

FIGURE 10a, b. *Chalcopyrite corrosion layer on a pin-needle showing (a) the apparent good condition of the underlying bronze metal, and (b) detail of chalcopyrite corrosion layer typical for a lake patina. Lab MAH 87-196.*

2. Malachite, antlerite, or posnjakite was formed and subsequently reduced to copper sulfides. This possibility seems unlikely. First, the transformation of basic copper carbonates and sulfates into copper sulfides requires the presence of sulfate-reducing bacteria and a humus-rich anaerobic environment. This condition is certainly not fulfilled in layer 1, which is sandy. Second, one would expect to see granular sulfide layers in the metal cross sections and not compact ones, which adhere closely to the metal.
3. Copper sulfides were formed under anaerobic conditions and subsequently oxidized to basic copper carbonates and sulfates. This type of transformation is well known, and many minerals are formed in the earth from copper sulfides in the so-called oxidation zone. The mechanism of this transformation has been studied by Sato (1960) and Garrels (1954). The formation of malachite and azurite from copper-sulfide minerals was investigated in the 1930s (Schwartz 1934). Posnjakite does occur near oxidized chalcopyrite deposits (Komkov and Nefedov 1967). The formation of these minerals is activated in well-aerated soils and by partial dehydration.

One may conclude that the formation of the first corrosion products (copper sulfides) took place under anaerobic conditions. The formation of malachite, antlerite, and posnjakite is secondary.

Two objects with land patinas, a needle (85-27) and a pin-needle (87-197), contain no copper sulfides. Needle 85-27 has a corrosion layer of malachite and posnjakite, pin-needle 87-197 only of malachite.

As explained previously, posnjakite is a mineral formed by oxidation of copper sulfides. It is, therefore, probably a secondary corrosion product. Malachite can also be formed directly or by oxidation of copper sulfides. When looking at the cross section, structures observed under the malachite crystals appear to resemble copper-sulfide layers. It is therefore very likely that these two objects were first exposed to anaerobic conditions (wet and humus-rich soil) before they were oxidized.

Summary

The five archaeological questions posed earlier in this paper may now be addressed:

Question 1: Are the different patinas due to different bronze-alloy compositions?

Answer: The compositions of the bronze alloys are very similar and have no influence on the formation of a land or lake patina.

Question 2: What is the composition and stratigraphy of each—the green-blue land and the brown-yellow lake—patina?

Answer: The green-blue land patina contains basic copper carbonates (malachite and azurite) and basic copper sulfates (posnjakite and antlerite). It has a granular structure with quartz-grain inclusions. The brown-yellow, smooth and shiny lake patina is composed of copper sulfides. The presence of chalcopyrite is predominant with some djurleite. The patina adheres closely to the metal surface. It is quite uniform and measures 100–150 μ.

Questions 3, 4: Under what sorts of environmental conditions were the patinas formed? Are they primary corrosion products or were they formed later by chemical reactions with the soil?

Answer: The lake patina was formed under anaerobic conditions in a soil rich with organic matter and in the presence of sulfate-reducing bacteria. It is a primary corrosion product. The land patina was formed in aerated soil in contact with the air. It is composed of secondary corrosion products, which were formed by oxidation from the primary sulfite layers.

Question 5: Is it possible to retrace the *history* or the *corrosion biography* of an individual bronze object after its use?

Answer: After their use by lake-dwelling settlers, the bronze objects studied must first have fallen into damp, waterlogged soil or sunk to the bottom of the lake close to shore. They must have remained there for quite a long time in close contact with organic remains and sulfate-reducing bacteria. Under these conditions, the chalcopyrite corrosion layers were formed. After the water level of the lake receded, parts of the objects must have been exposed to oxidizing conditions. The copper sulfides were transformed into basic copper carbonates and basic copper sulfates.

The inverse process seems extremely unlikely. The transformation of granular and porous copper carbonates and sulfates to sulfides would not have resulted in a smooth and dense layer, which adheres closely to the metal. A simultaneous formation of both patina types seems also unlikely. The formation conditions and fields of stability are too different.

After the lake water level rose, around 750 B.C.E., the objects with the land patina must have remained underwater in the washed-out upper layers.

Acknowledgments

The research presented here was stimulated by the questions of A.-M. Rychner-Faraggi. The author thanks her for profitable discussions. Most of the analytical work was carried out by Martine Degli Agosti, who also did the artwork. The author is grateful to R. Oberhänsli of the University of Berne for analysis with the microanalyzer. Professor St. Graeser of the Natural History Museum in Basel kindly helped identify the corrosion products. P. O. Boll performed the analyses with the electron microprobe at the Swiss Federal Laboratories for Materials Testing and Research at Dübendorf. Finally, the author thanks Danielle Matter for typing the manuscript.

Notes

1. The samples in the left-hand column of Table 1 are from the Laboratoire de recherche, Musée d'art et d'histoire (Lab MAH), Genève, and all the photographs in this chapter are reproduced by courtesy of the same.

2. Energy-dispersive X-ray fluorescence spectrometer used was a Kevex tube 40 kV, 0.3 mA, 0.8 mm Ø collimator, Seforad Si (Li) detector with a Tracor Northern 5400 multichannel analyzer.
3. This analysis was carried out by the Swiss Federal Laboratories for Materials Testing and Research, Dübendorf.
4. This was done by S. Graeser of the Natural History Museum, Basel.
5. This was done by R. Oberhänsli of the Mineralogical-petrographical Institute, University of Berne.
6. See note 3 above.
7. This was done by P. O. Boll of the Swiss Federal Laboratories for Materials Testing and Research, Dübendorf.
8. This was carried out at the Mineralogical-petrographical Institute of the University of Berne by R. Oberhänsli.

References

BAAS BECKING, L. G. M., I. R. KAPLAN, AND D. MOORE

1960 Limits of the natural environment in terms of pH and oxidation-reduction potentials. *Journal of Geology* 68(3):243–84.

BENKERT, A., AND H. EGGER

1986 Dendrochronologie d'un site du bronze final Hauterive-Champréveyres (Suisse). *Bulletin de la Société Préhistorique Française* 83(11–12):486–502.

CHRISTEN, B. H.

1981 The mineralogy of bronze corrosion. BSC diss., Institute of Archaeology, London University.

DUNCAN, S. J., AND H. GANIARIS

1987 Some sulphide corrosion products on copper alloys and lead alloys from London waterfront sites. In *Recent Advances in the Conservation and Analysis of Artifacts,* 109–18. J. Black, comp. London: Summer School Press.

FABRIZI, M., AND D. A. SCOTT

1987 Unusual copper corrosion products and problems of identity. In *Recent Advances in the Conservation and Analysis of Artifacts,* 131–33. J. Black, comp. London: Summer School Press.

GARRELS, R. M.

1954 Mineral species as functions of pH and oxidation reduction potentials, with special reference to the zone of oxidation and secondary enrichment of sulphide or deposits. *Geochemica and Cosmochimica Acta* 5:153–68.

GARRELS, R. M., AND C. L. CHRIST

1965 *Solutions, Minerals and Equilibria.* A Harper International Student Reprint. New York: Harper and Row.

GEILMANN, W.

1956 Verwitterung von Bronzen im Sandböden Angew. *Chemie* 68:201–12.

KOMKOV, A. I., AND E. I. NEFEDOV

1967 Posnjakite, a new mineral. *Zap. Vses. Mineralog. Obschch.* 96:58–62 (in Russian). Summary in *American Mineralogist* 52:1582–83, 1972.

MCDOUGALL, J.

1966 Microbial corrosion of metals. *Anti-corrosion Methods & Materials* 1(8):9–13.

PANTER, I. F.

1980 The role of sulphate-reducing bacteria in the corrosion of iron from anaerobic environments. BSC diss., Institute of Archaeology, London University.

POURBAIX, M.

1977 Electrochemical corrosion and reduction. In *Corrosion and Metal Artifacts: A Dialogue Between Conservators and Archaeologists and Corrosion Scientists.* B. F. Brown, H. C. Burnett, W. T. Chase, M. Goodway, J. Kruger, M. Pourbaix, eds. NBS Special Publication 479:1–16. Washington, D.C.: U.S. Department of Commerce, National Bureau of Standards.

ROBBIOLA, L.

1990 Caractérisation de l'altération de bronzes archéologiques enfouies à partir d'un corpus d'objets de l'âge de bronze: Mécanismes de corrosion. Ph.D. diss., University of Paris 6.

ROBBIOLA, L., I. QUEIXALOS, L. P. HURTEL, M. PERNOT, AND C. VOLFOVSKY

1988 Etude de la corrosion des bronzes du Fort-Harrouard: Altération externe et mécanisme d'altération stratifiée. *Studies in Conservation* 33:205–15.

RYCHNER, A.-M.

1991 Hauterive-Champréveyres: Métal et parure au bronze final. Typescript.

SATO, M.

1960 Oxidation of sulphide ore bodies II: Oxidation mechanisms of sulphide minerals at 25 °C. *Economic Geology* 55:1203–31.

SCHWARTZ, G. M.

1934 Paragenesis of oxidized ores of copper. *Economic Geology* 29:55–75.

SCHWEIZER, F.

1988 Die Patina von Bronzen aus Seeufersiedlungen: Eine Biographie? *Arbeitsblätter für Restauratoren,* 221–34.

TYLECOTE, R. F.

1979 The effect of soil conditions on the long-term corrosion of buried tin-bronzes and copper. *Journal of Archaeological Science* 6:345–68.

Biography

François Schweizer received his doctorate in chemical engineering in 1965 in Zürich. Since 1974, he has been head of the laboratory and keeper of conservation at the Musée d'art et d'histoire in Geneva. From 1983 to 1989, he directed the National Research Programme 16, Methods for the Preservation of Cultural Properties, of the Swiss National Science Foundation. His main subject of interest is the ancient technology and conservation of metals, ceramics, and stained glass. Schweizer is a fellow and a member of the Council of the International Institute of Conservation and a member of the Swiss Commission of Ancient Monuments, the ICOM Committee of Conservation, and the Technical Committee of the Corpus Vitrearum.

The Royal Art of Benin: Surfaces, Past and Present

JANET L. SCHRENK

An understanding of the surfaces of ethnographic metal objects pre- and post-collection is important for conservation and curatorial considerations. Historical references to the appearance of copper-alloy objects of Benin Kingdom (Nigeria) prior to 1897 are scarce. However, there is some evidence for clean surfaces on the objects and the application of blood to objects used on altars. Since the removal of these objects from Africa in 1897, they have been subjected to a variety of coatings and other surface treatments. These treatments have left the surfaces discolored, corroded, and saturated with oils and waxes, and thus not generally representative of the original West African aesthetic.

The Benin collection in the National Museum of African Art at the Smithsonian Institution provides an interesting array of surfaces that vary according to the object's collection history. Analysis of samples of coatings and corrosion from these objects reveals a variety of corrosion products, modern pigments, oils, and waxes. A frequently observed problem is the presence of turquoise, fatty-acid metallic salts due to the breakdown of the oils and subsequent attack of the metal surface.

Background and History

Copper-based alloys were cast throughout West Africa. The most impressive of these were cast in Nigeria, particularly in the kingdom of Benin. The metal casters in Benin City produced an amazing array of copper-alloy plaques, commemorative heads, hip masks, vessels, and other objects between the fourteenth and nineteenth centuries for the exclusive use of the *oba* (king) of Benin.

In 1897 the British launched a so-called punitive expedition against the kingdom of Benin. Oba Ovaramwen was exiled, and the British removed more than 4,000 copper-alloy objects from the royal palace. Some of these objects went to expedition participants, and others were sold at government auction to dealers, private collectors, and museums. Due to the circumstances of their collection[1] and their exquisite craftsmanship, these Benin objects were highly prized as curios. Today they are dispersed throughout the world's museums and private collections.[2] The National Museum of African Art has twenty-two of these objects, upon which this study is based.

The impetus for this research was the observation of a direct correlation between the surface appearance of these twenty-two objects and their past collection history.[3] For example, objects previously in the private collection of the Pitt-Rivers family[4] have a dark, saturated surface that is different from that of other copper-alloy objects owned by the National Museum of African Art, such as the waxy, dark-green corrosion characterizing objects from the J. P. Howe collection. This research was initiated to learn more about the surface condition of the National Museum of African Art objects—including identification of coatings, corrosion, and "investment"—and the resulting interactions of these materials with the metal surface. This information will help facilitate conservation treatments and identification of past surface treatments.

Alloy Composition

The Benin copper-alloy objects are predominantly brass, although a few bronze objects exist. Objects in the National Museum of African Art collection were examined by X-ray fluorescence spectroscopy and found to have compositions that range from 64 to 94% copper, 0 to 7% tin, 1 to 17% lead, and 0.6 to 26% zinc.[5] Small quantities of arsenic, silver, antimony, and nickel were also detected. Iron contents of 0.4–5% are probably due, at least in part, to surface dirt.

Objects are occasionally ornamented with copper sheet, iron and copper nails, and cast bronze. An example is the pendant hip mask (NMAFA 85-19-5) in the National Museum of African Art collection (Fig. 1). Although the surface is heavily obscured by treatments that occurred prior to entering the museum's collection, X-ray fluorescence reveals that the decorative strip down the nose is in excess of 98% copper. The pupils of the eyes were fashioned from iron, presumably iron nails. Finally, every other mudfish around the neck is cast in bronze (Table 1). Because each of these alloys would have a higher melting point than the brass base, which only contains 71% copper,[6] these adornments could be laid into the wax form and cast into place through the lost-wax casting process. This is clearly evident in close examination of the "rivets" that hold the bronze mudfish in place.

TABLE 1. *Elemental analysis of pendent hip mask (NMAFA-85-19-5).*

Item	*%Cu*	*%Zn*	*%Pb*	*%Sn*	*%Fe*	*Other*
Brass base	71	26	1	trace	1	—
Mudfish	93	1	2	2.5	1	trace As, Ag
Copper strip (nose)	98.5	0.4	0.6	trace	0.9	trace As, Ag
Eyes (pupils)	—	—	—	—	~100	trace Ca, As, Pb

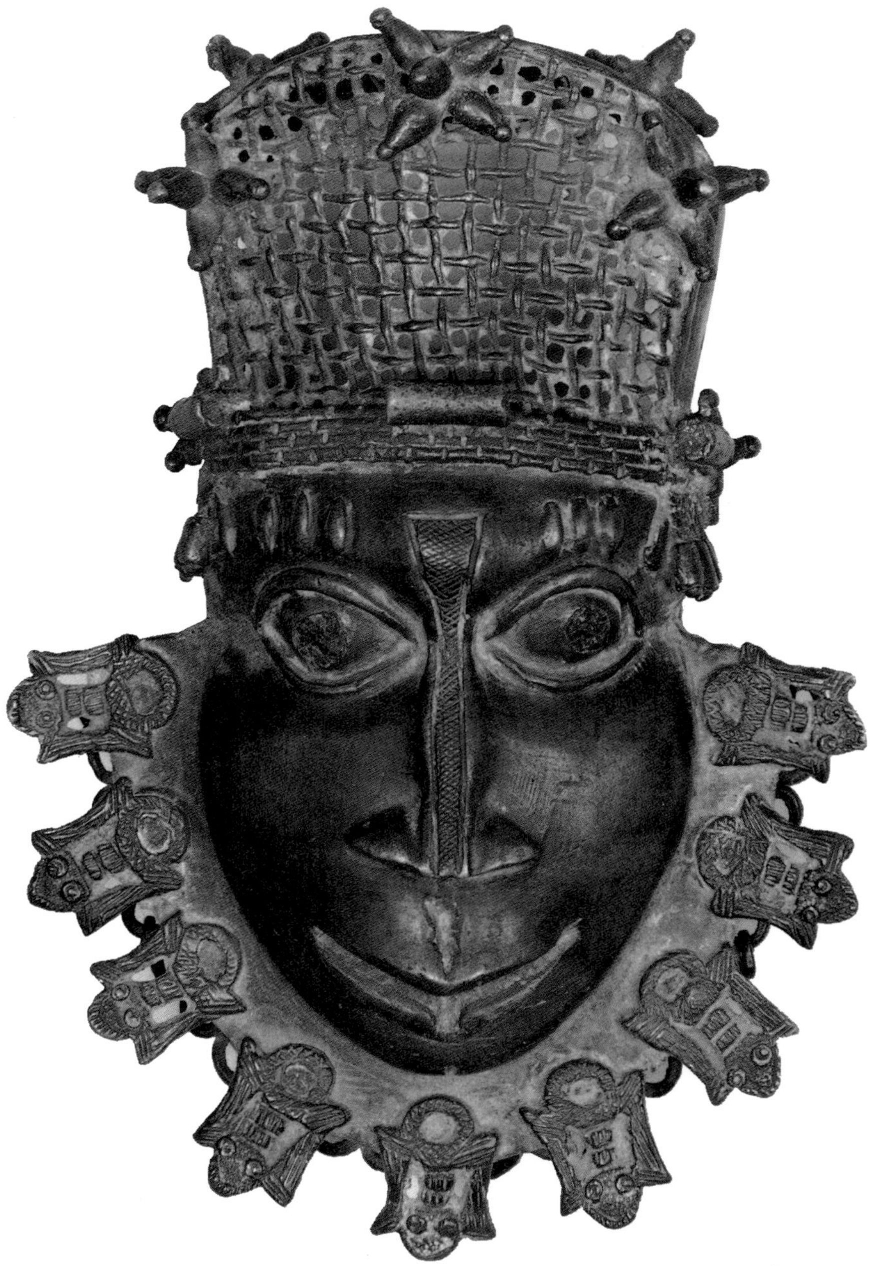

FIGURE 1. *Pendant Mask of the Edo Peoples, Benin Kingdom, Nigeria, cast-copper alloy. H:21.6 cm. NMAFA, 85-19-5.*

Surface Condition Before 1897

The variation in materials suggests that importance was placed on surface color and appearance in these objects prior to their entering Western collections. Throughout much of West Africa, the color red was highly prized and carried much significance. Today in Benin, copper's red color and shiny surface are seen as beautiful as well as frightening (Ben-Amos 1980:15). Throughout Benin history, the red stone and coral beads have been highly prized and an important part of the *oba's* regalia (Freyer 1987:57; Read and Dalton 1899:22). In 1862 Burton observed that the doors of the Ogboni secret society house were decorated in yellow, red, and black (Read and Dalton 1899:9). If completely cleaned, the pendant mask would have all three of these colors. The copper nose inlay and the bronze mudfish would appear red compared to the yellow brass, and the iron pupils would appear dark, almost black. It seems reasonable to believe that these differences were an important part of the Benin aesthetic.

Very little has been written concerning the original appearance of these objects. The earliest mention of the plaques occurs in 1668 by Dapper, a Dutch author who compiled accounts by visitors to Benin City (Roth [1903] 1968:160). He writes that within the palace there were

> beautiful and long square galleries about as long as the Exchange at Amsterdam, but one larger than another resting on wooden pillars, from top to bottom covered with cast copper on which are engraved pictures of their war exploits and battles, and are kept very clean.

The phrase "kept very clean" conjures an image of a shiny, polished surface, without the red-clay "investment" material that is so strongly associated with Benin art. Later accounts merely confirm the continued presence of plaques hanging on pillars in the palace.[7]

Commemorative heads and other objects were placed on sheltered altars outdoors in the palace courtyard.[8] Some accounts report the application of blood from sacrifices to the altars, including the ivory tusks and the commemorative heads. While virtually all the accounts prior to the punitive expedition mention animal and/or human sacrifice, observations of blood on these objects seem to date from the period of the expedition.[9] It is possible that some exaggeration may have occurred to fuel the sensationalism surrounding the works of art and the punitive expedition.

Even if the objects were coated with sacrificial blood, according to Benin studies expert R. E. Bradbury, "Traditionally the cleaning and refurbishing of shrines was an institutionalized art of ritual performance" (Dark 1973:31). Also according to Bradbury, "Much of the loot removed from Benin in 1897 was in a good state of preservation, showing that care had been taken of it. In fact, there were people in the palace whose job it was to look after its treasures" (Dark 1973:29). Routine cleaning might be expected to remove not only blood but also dirt and any residual investment material.

Many of the Benin copper-alloy objects have a red-clay soil in the crevices, unless scrubbed completely clean. It is unclear whether this is an intentional addition by members of the Benin Kingdom to enhance the surface details, or material left over from the casting process, or merely accumulated dirt. This red clay serves to highlight the intricate details of the design and *could* represent a change in aesthetics from the time of Dapper's description of "very clean" surfaces.

The red soil-like appearance extends beyond the crevices of objects in some collections. H. O. Forbes, director of the Liverpool Museums, wrote in 1897 concerning two new acquisitions (1897:57):

> Both of the tusk holders [commemorative heads] like some of the other pieces in the collection, are of so rich a terra-cotta color, that, they might easily pass, on superficial inspection for clay. Whether this color results from a fine coat-

ing of laterite, from the clay molds in which they were cast—which would of course be removed from the chiseled portions—or is a patina artificially produced or naturally arising from long exposure to the air, is not yet determined. If these (and other) figures be of antiquity, which there is some evidence to show that they are, it appears rather surprising to find, after so long an exposure to the air and weather, any clay adhering to them; and practically no oxidation of the metal.

Surfaces that fit this description and on first glance appear almost paintlike may be found in museum collections such as those in Vienna's Museum für Volkerkunde.[10]

Other objects are heavily corroded. In 1897 Commander Bacon wrote: "The [King's] storehouses contained chiefly rubbish. . . . But buried in the dirt of ages, in one house, were several hundred unique bronze plaques, suggestive of almost Egyptian design, but really superb castings." The objects were probably not buried but had accumulated dirt while in the building. This may have been the house and burial chamber of an *oba,* and the objects closely associated with his reign (Read and Dalton 1899:9). This could account for the observation of heavily corroded surfaces on some objects, generally in large museum collections established immediately after the punitive expedition and thus less likely to have been heavily cleaned.[11]

It is likely that at least some of the objects had a cuprite layer on their outer surface at the time they were removed from Nigeria. Cuprite is the corrosion product that one would expect to form first on a copper surface. Ancient objects typically have a cuprite layer adjacent to the remaining core metal. Cuprite has been identified on a number of these objects. In addition, the objects that have copper fatty-acid salts on their surface (discussed later) contain flakes of cuprite within those extruding crystals. Small areas of other copper corrosion products, such as malachite and copper chlorides, have also been identified on a few of the objects from the National Museum of African Art collection.

Surface Modifications Since 1897

Dealers clearly believe the red material "belongs" in the crevices of these objects. For example, many areas on the pendant hip mask are obscured by a combination of cadmium sulfide, hematite, and calcium sulfate pigments.[12] Since cadmium pigments were not available until after the punitive expedition—they were first produced in Germany in 1925 (Wehle 1975:89)—this cannot be an indigenous addition. A textile impression on the pigments at the proper right edge of the right eye may have been intended to mimic the roughness of the clay material observed on other objects.

Many inappropriate surfaces, such as those found on the pendant mask, exist on the Benin objects. These often include the presence of modern pigments. For example, one of the plaques (NMAFA 85-19-19) in the collection has major portions of its surface obscured by the pigment Prussian blue.[13]

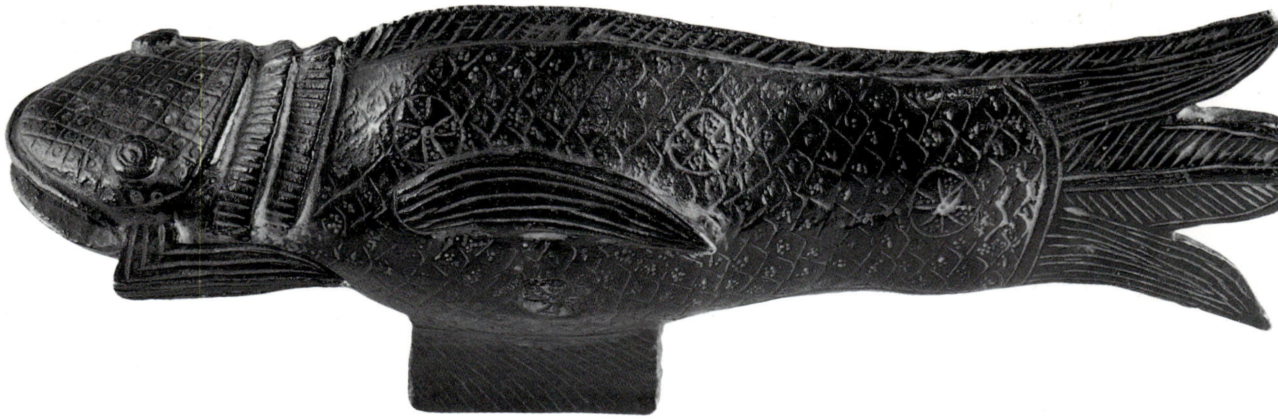

FIGURE 2. *Figure of a fish, Edo Peoples, Benin Kingdom, Nigeria, mid-sixteenth century, cast-copper alloy, copper inlay. H:16.5 cm. NMAFA, 85-19-8.*

Other objects have chemically or electrochemically stripped surfaces, such as the collection's unique figure of a mudfish (NMAFA 85-19-8) (Fig. 2). Not only is the surface badly etched, but a flat-black waxy coating completely obscures the surface (Fig. 3). The surface details include inlaid sheet-copper decorations. An arsenic-rich green pigment, emerald green, has been applied to all of the crevices.[14] One has to wonder whose aesthetic this was designed to please.

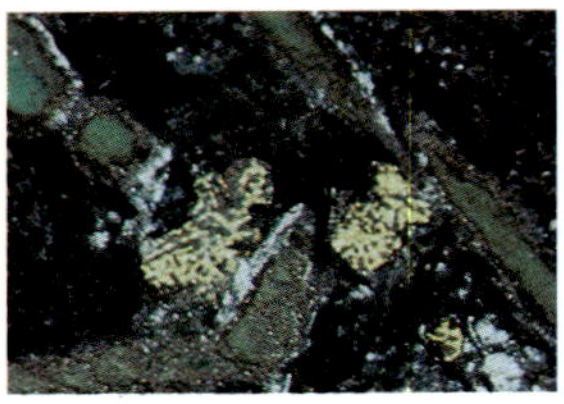

FIGURE 3. *Surface detail from area immediately below proper right fin of object shown in Figure 2. Etched metal surface, waxy black coating, and powdery green pigment in the crevices are visible.*

Application of "Protective" Coatings

The above examples are deliberate post-punitive-expedition modifications of the appearance of the Benin objects. The application of oils to these objects may have been intended to enhance their appearance or may have been intended to protect their surfaces. In either case the application of oils to metal objects is consistent with European housekeeping practices in the earlier part of this century. Unfortunately, the application of oils has resulted in saturated, discolored, and corroded surfaces. In general, as these objects appear today they cannot represent the original West African aesthetic.

For example, reports indicate that the Benin objects in the private collection of the Pitt-Rivers family were coated with neat's-foot oil, an oil typically used on leather, possibly because of plans to exhibit these pieces outdoors (Fagg, pers. comm. 1989).While the examination of the composition of fatty acids in the coating is inconclusive at this time, the objects previously in the Pitt-Rivers collection have surfaces that are dark and saturated. The tarry black coating is frequently pooled in the crevices, and in some cases dark, disfiguring drips of the oily coating material can be seen (Fig. 4). A goal of this ongoing research is the identification of the oils used on the various collections of Benin objects based on the ratio of fatty acids present.

Despite the oily appearance, it is often presumed that these surfaces are stable. As a consequence, the choice may be made not to treat the object. Unfortunately, these coatings often mask the severe corrosion occurring underneath. It is disconcerting to realize, as in the example of the Pitt-Rivers objects, that the dark, tarry

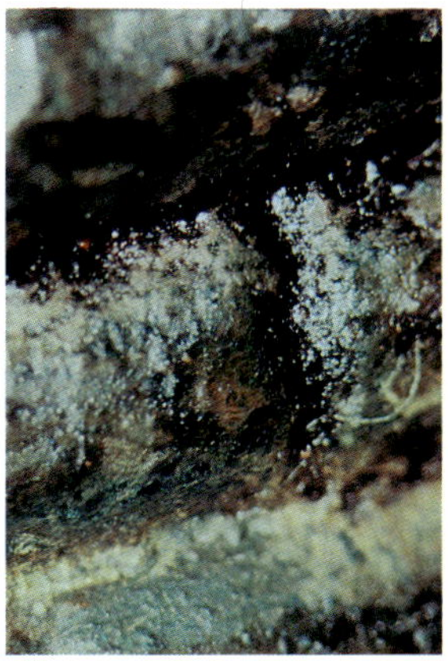

FIGURE 4. *Detail from the neck rings on a plaque of men with leopard-skin bags, showing the extremely saturated surface with thick accumulation of aged oil in the crevices. In the lower crevice where the pooled oil was removed, there is waxy turquoise corrosion. Edo Peoples, Benin Kingdom, Nigeria, mid-sixteenth to seventeenth century, cast-copper alloy. H:47 cm. NMAFA 85-19-13.*

appearance is due to a coating that is now amber-colored over extensive turquoise corrosion. As discussed in the following section, this turquoise corrosion is a result of the breakdown of the oils and subsequent reaction with the metal surfaces.

These coatings contain part of the history of the object. But the presence of these coatings and other surface modifications influences one's perspective of the Benin aesthetic. These materials also may obscure inlays, details of surface finishing, and other features. The removal of coatings without recording what is present may result in the loss of important information necessary to the understanding of that object, its appearance, and possibly its authentication. The choice of whether to remove a coating will also affect the long-term stability of the object.

Fatty-Acid Metallic Salts

The most widespread problem identified on the Benin objects in the National Museum of African Art is the presence of the fatty-acid salts of copper and, occasionally, zinc and lead. These salts have been found on at least thirteen of the twenty-two objects, representing a minimum of four different European collections. This corrosion originates with the breakdown of the oils, followed by attack of the metal surface, as illustrated in Table 2. All oils and fats have three ester linkages, which may be hydrolyzed (addition of water across the bond) to give free fatty acids (Reaction 1). This process occurs slowly but may be either acid or base catalyzed. The result is a free fatty-acid molecule available to attack the metal surface. If the metal is oxidized, then fatty-acid salt will be formed (Reaction 2). The reaction of free fatty acids with copper is an extremely efficient process.[15] This efficiency strongly suggests that the slow step is hydrolysis of ester linkages in the oil to produce the free fatty acids.

On undisturbed surfaces, this corrosion may extrude from the metal surface as if forced through a press (Fig. 5), or it may be a fuzzy blue-gray material, almost

TABLE 2. *Analysis of the corrosion of Benin objects, caused by the hydrolysis of fats and oils (Reaction 1) which leads to fatty-acid attack on copper (Reaction 2).*

Reaction 1: The hydrolysis of fats and oils

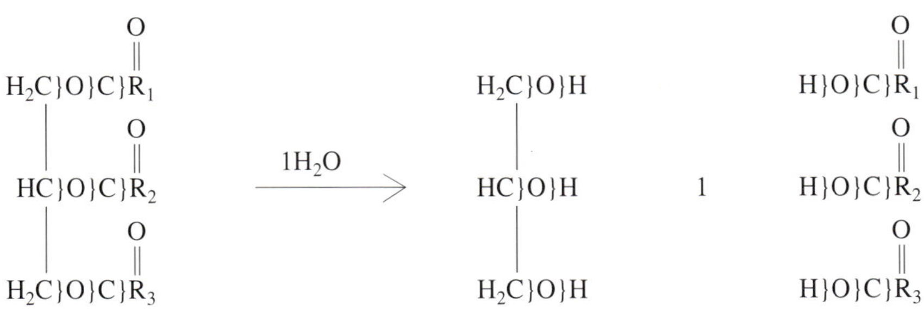

R_1, R_2, R_3 are long hydrocarbon chains, typically containing 12, 14, 16, 18, or 20 carbons.

Reaction 2: Fatty-acid attack on copper

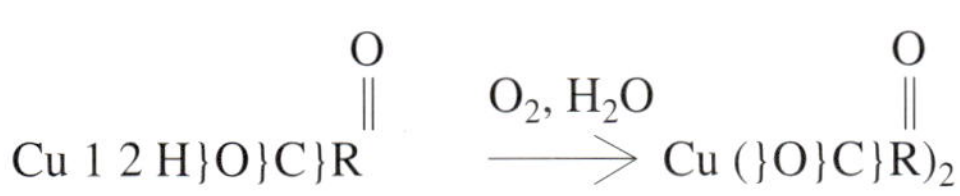

FIGURE 5. *Detail, near right, from Benin relief sculpture, area between the tassels over the central figure's wrapper, showing the extruding turquoise crystals. Edo Peoples, Benin Kingdom, Nigeria, probably seventeenth century, cast-copper alloy. H:46 cm. NMAFA 82-5-3.*

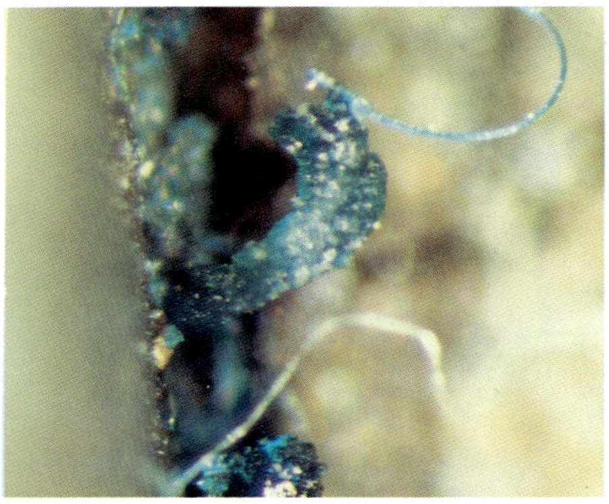

FIGURE 6. *Detail, far right above, from Edo musketeer figure, area between the figure's legs, showing the "fuzzy" blue-gray corrosion. Edo Peoples, Benin Kingdom, Nigeria, nineteenth century, cast-copper alloy, iron rods. H:51.4 cm. NMAFA 85-19-15.*

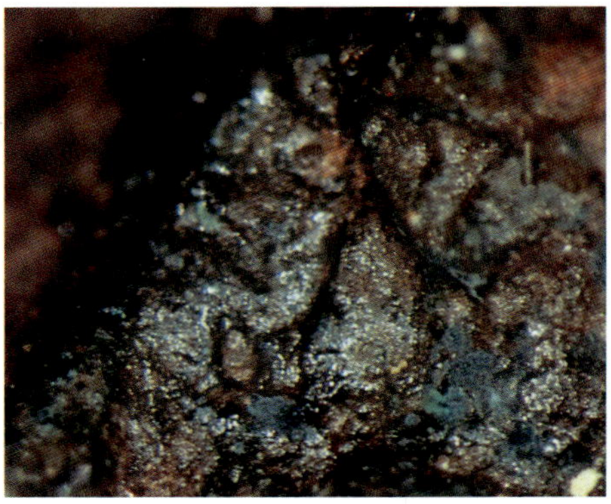

FIGURE 7. *Detail, above right, from the reverse of plaque of multiple figures, showing waxy blue-green corrosion. Edo Peoples, Benin Kingdom, Nigeria, mid-sixteenth to seventeenth century, cast-copper alloy. H:48.9 cm. NMAFA 85-19-18.*

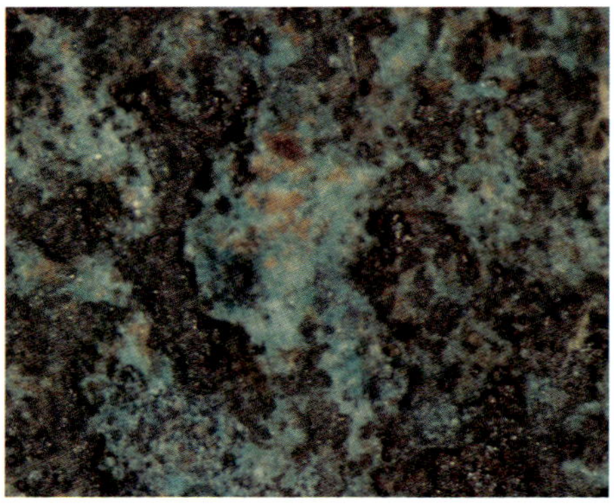

moldlike in appearance (Fig. 6). When disturbed or on other surfaces it appears to be very soft and waxy (Fig. 7). Scanning electron micrographs illustrating these morphologies have been previously published (Schrenk 1991:807–8).

Scanning electron micrographs illustrate combinations of these morphologies. Figure 8 shows the smooth, waxy form in combination with the almost hairlike surface of the fuzzy blue-gray. Figure 9 illustrates the extrusion of a turquoise crystal out of a more waxy base. Dirt particles, at the right-hand edge of the photograph, have been pushed away from the metal surface. What is controlling the morphology is unclear, though it may be the particular combinations of fatty acids or metal ions present.

A combination of analytical techniques has been used in the identification of the fatty-acid salts of copper and other metals. These include Fourier transform infrared

FIGURE 8. *Scanning electron micrograph of a corrosion sample of a figure of an* oba, *taken from under the figure's proper left arm, showing a combination of the waxy and fuzzy forms of the fatty-acid metallic salts. Edo Peoples, Benin Kingdom, Nigeria, nineteenth century, cast-copper alloy. H:41 cm. NMAFA 85-19-12.*

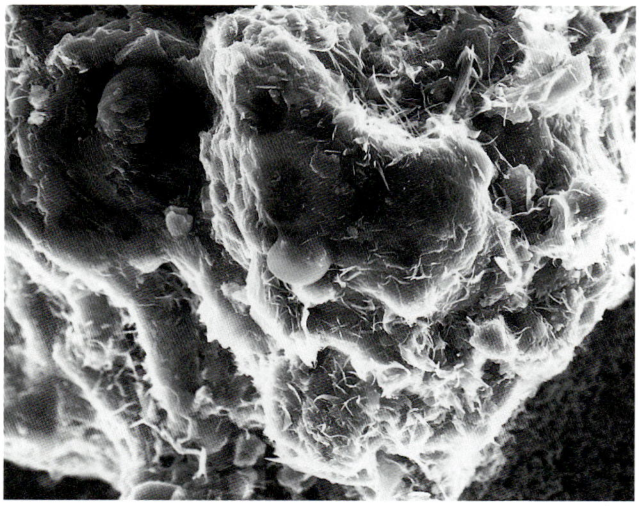

FIGURE 9. *Scanning electron micrograph of a corrosion sample taken from plaque of man with* eben, *proper right side of the figure below the waistband, showing both waxy and extruding forms of the fatty acid metallic salts. The right-hand side of the micrograph shows dirt particles which would have been displaced by the corrosion from the object's surface. Edo Peoples, Benin Kingdom, Nigeria, mid-sixteenth to seventeenth century, cast-copper alloy. H:44.1 cm. NMAFA 85-19-20.*

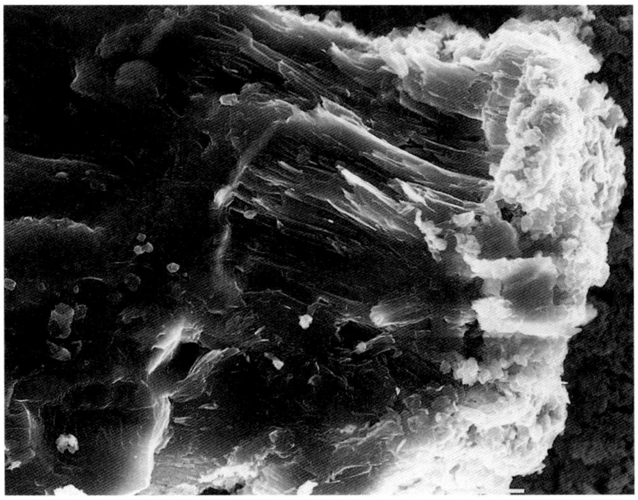

(FTIR) spectroscopy, X-ray diffraction (XRD), gas-liquid chromatography (GLC), and scanning electron microscopy with energy-dispersive analysis (SEM-EDS). Details and representative spectra have been published previously (Schrenk 1991).

FTIR spectroscopy is a good diagnostic tool because of the carbonyl stretching band that occurs in the region between 1500 and 1800 cm^{-1}. In a fat, oil, or wax this band occurs around 1730–1740 cm^{-1}. The free fatty acids may be found at about 1700 cm^{-1}. In sharp contrast, the band for the copper fatty acid salt is shifted to 1588 cm^{-1}, while in the zinc salt it appears at 1538 cm^{-1}, and in the lead salt at 1514 cm^{-1}.[16]

The beautiful morphologies probably occur due to the long periods of time in which the crystals have formed. In contrast, samples obtained from a chemical supply house are very waxy and produce much poorer XRD patterns. The typical XRD pattern for these corrosion products consists of one or more d spacings between 10 and 15 Å. A quick scan of the powder diffraction files shows relatively few copper complexes with such large d spacings.

Confirmation of the metals present was obtained by SEM-EDS. Typical spectra for the copper fatty-acid salts have two characteristic features: (1) sharp peaks for copper (and any other metals present), and (2) a large, broad band at lower energies indicative of organic scattering. In some cases aluminum and silicon appear due to the presence of clay materials on the object's surface.

GLC confirms the presence of fatty acids in both the coatings and the corrosion products. Typically the most prominent peaks correspond to palmitic, oleic, and stearic fatty acids. It is hoped that continued research in this area will lead to identification of the particular oils and waxes used on the objects.

CONCLUSION

The twenty-two objects in the National Museum of African Art show a wide variety of surface conditions predominantly due to their past collection history. Many of these surfaces clearly misrepresent the West African aesthetic, especially those containing modern pigments and oils and those that have been electrochemically or chemically stripped. Perhaps most alarming is the identification of metallic fatty-acid salts on at least half of the collection. The identification of cuprite on several objects, as well as in the extruding turquoise corrosion of some metallic fatty-acid salts, suggests that at least some of the objects had cuprite on their surface when they left Nigeria. Understanding the appearance of the objects prior to the punitive expedition of 1897 is critical to the proper understanding and interpretation of the Benin aesthetic.

Ongoing work will continue to address the issues related to the surface appearance and condition of the royal art of Benin Kingdom, both past and present. Particular issues being addressed are the identity of the coating materials, examination of the metallic fatty-acid corrosion products, the origin of the red-clay material characteristic of these objects, and historical documents that provide clues to the appearance of the objects prior to leaving Africa.

Acknowledgments

The author thanks the many individuals at the National Museum of African Art and the Conservation Analytical Laboratory of the Smithsonian Institution, especially Stephen Mellor, chief conservator at the National Museum of African Art; and David von Endt and David Erhardt, organic chemists at the Conservation Analytical Laboratory. The author also thanks the Smithsonian Institution for the Postdoctoral Research Fellowship in Conservation Science, and the National Museum of African Art for continued funding.

Notes

1. The London newspapers sensationalized the events leading up to and including the punitive expedition which resulted from the British government's desire to end the practice of human sacrifice. Few could believe "savage" people could produce such beautifully crafted works of art.
2. An attempt to catalogue these objects and their location is available (Dark 1982).
3. This observation was first made by Bryna Freyer, curator, and Stephen Mellor, chief conservator, at the National Museum of African Art.
4. The private collection of General Pitt-Rivers should not be confused with the collection of the Pitt-Rivers Museum in Oxford, England, which he established prior to collecting Benin art.
5. Unpublished data obtained by X-ray fluorescence examination of 20–40 areas (approximately one-quarter inch in diameter) over the surface of the object. Quantification was completed using software developed by W. Thomas Chase and the Freer and Sackler museums.
6. The melting points of pure copper and pure iron are 1035 °C and 1538 °C, respectively. Based on binary-phase diagrams, a copper-tin alloy (bronze) containing 93% copper would have a melting point of about 1050 °C, while a copper-zinc alloy (brass) containing 71% copper would have a melting point of about 950 °C. While the alloys are more complex than this indicates, it is reasonable to expect a lower melting point for the brass (Askelund 1985:250, 552).

7. For example, based on the 1630 accounts of Peter de Marcez, Ogilby writes: "But all [s]upported on Pillars of Wood, cover'd from the top to the bottom with melted Copper, whereon are Ingraven their Warlike Deeds and Battels, and are kept with exceeding Curi[os]ity" (Ogilby 1670:470). In 1820 Lt. King "noticed that the flat ceilings were traversed by beams covered with metal and ornamented with various figures" (Read and Dalton 1899:8).
8. For example, an 1891 photograph of a palace altar is published in Roth ([1903]1968:79). A 1970 photograph by Eliot Elisofon of a palace altar is published in Freyer (1987:19).
9. According to Roth, "Neither Dapper nor the officials say that the blood of the human victims was sprinkled on the ivories and metal work on the altars, but Landolphe noticed it, and Dr. Allman, who was on the punitive expedition, informs me he found blood and human entrails on the altars, and my brother likewise medical officer to the expedition, tells me the same. The native officials, no doubt, have some diffidence in giving the whole truth on this matter" (Roth [1903]1968:72). According to Sir R. Moor, a punitive expedition participant, "Their blood was smeared over the altar and allowed to run down the steps in front . . . brass and ivory [figures] were not painted, but covered with blood, human or otherwise" (Forbes 1897:56).
10. Excellent color photographs of objects in Vienna's Museum für Volkerkunde have been published (Duchateau 1990).
11. The author has observed such surfaces on objects at the Museum of Mankind (British Museum) in London and the Field Museum in Chicago.
12. Identification of cadmium and relatively large quantities of iron by X-ray fluorescence, along with a list of red pigments (Wehle 1975), facilitated the interpretation of the XRD pattern of the mixture of pigments.
13. Prussian blue was identified based on FTIR spectroscopy, in particular the distinctive C≡N stretching frequency at about 2090 cm^{-1}. This was confirmed by XRD and SEM-EDS.
14. X-ray fluorescence revealed the presence of arsenic, and a list of green pigments (Wehle 1975) facilitated the interpretation of the XRD pattern.
15. This efficiency was demonstrated by the application of a mixture of stearic, palmitic, and oleic acids on a sheet of polished copper. The reaction was followed by FTIR. Within four days there was a 50% conversion of the free fatty acids to the copper fatty-acid salts (Schrenk 1991:811).
16. Values are based on commercial samples of stearic acid, copper stearate, zinc stearate, and lead stearate.

References

ASKELUND, D. R.

1985 *The Science and Engineering of Materials*, alternate ed. Boston: PWS Engineering.

BACON, R. HUGH

1897 *Benin: The City of Blood*. London: Edward Arnold.

BEN-AMOS, P.

1980 *The Art of Benin*. New York: Thames and Hudson.

DARK, P. J. C.

1973 *An Introduction to Benin Art and Technology*. Oxford: Clarendon Press.

1982 *An Illustrated Catalogue of Benin Art*. Boston: G. K. Hall.

DUCHATEAU, A.

1990 *Benin Tresor Royal, Collection du Museum für Volkerkunde, Vienna*. Paris: Editions Dapper.

FORBES, H. O.

1897 On a collection of cast-metal work, of high artistic value, from Benin, lately acquired by the Mayer Museum. *Bulletin of the Liverpool Museums* 1:49–70.

FREYER, B.

1987 *Royal Benin Art in the Collection of the National Museum of African Art*. Washington, D.C.: Smithsonian Institution Press.

OGILBY, J.

1670 *Collection of African Travels*. Based on accounts written about 1630 by Peter de Marcez. London.

READ, C. H., AND O. M. DALTON

1899 *Antiquities from the City of Benin and Other Parts of West Africa in the British Museum*. London: British Museum.

ROTH, H. L.

[1903] 1968 *Great Benin: Its Customs, Art and Horrors*. Reprint. New York: Barnes and Noble.

SCHRENK, J. L.

1991 Corrosion and past "protective" treatments of the Benin "bronzes" in the National Museum of African Art. In *Material Issues in Art and Archaeology II; Materials Research Society Symposium Proceedings*, vol. 185:805–12. P. Vandiver, J. Druzik, and G. Wheeler, eds. Pittsburgh: Materials Research Society.

WEHLE, K.

1975 *The Materials and Techniques of Painting*. New York: Van Nostrand.

Biography

Janet L. Schrenk received her doctorate in inorganic chemistry from the University of Minnesota. In 1985 she joined the faculty of the University of Delaware/Winterthur Museum Art Conservation Program. In 1988 she joined the Research and Testing Laboratory of the National Archives and Records Administration. The following year she received the Smithsonian Institution's Conservation Analytical Laboratory Post-doctoral Research Fellowship in Conservation Science to pursue research on the Benin collection of the National Museum of African Art. Since 1990 she has served on the faculty of The University of the South in Sewanee, Tennessee.

Considerations in the Cleaning of Ancient Chinese Bronze Vessels

JANE BASSETT AND W. T. CHASE

Cast as symbols of power and wealth, Chinese bronze vessels in antiquity were collected, given in tribute, used in ceremony, and buried—both in sacrifice and with the dead. Few, if any, of the vessels known today survived the ensuing centuries without being buried. Within the last thousand years, vessels unearthed throughout China have been collected and distributed and are greatly prized and valued. Throughout these centuries of use, burial, rediscovery, restoration, and redistribution, great changes have occurred in their bronze surfaces.

The cleaning of any archaeological bronze surface is an irreversible process, demanding careful consideration before treatment is undertaken. In the case of Chinese bronze vessels, this decision is complicated by the potential wealth of information provided by an artifact's surface. The aesthetic and ethical implications of the cleaning of ancient Chinese bronze vessels should, therefore, be considered in light of the complexity of the surfaces which are presented.

As with the cleaning of all bronzes—especially when the surface no longer exists in its original form—the point at which cleaning begins and ends is arbitrary and is often left to the discretion of the individual conservator. Although some conservators may intuitively sense many implications of the cleaning of ancient Chinese vessels, the factors that should guide their judgment are not clearly articulated in the literature and are often conflicting.

Evaluating Stability

The primary objective in any cleaning treatment should be long-term stabilization. The first step must be to determine through examination whether or not the bronze is strong enough to allow cleaning, as highly mineralized bronzes may not be able to withstand even the gentlest of cleaning actions. Careful visual examination, including X-radiography and ultraviolet-light examination, will indicate cracks, repairs, and degree of mineralization, which will help in making this decision. A miniaturized metal detector, made by attaching a small (2–3 mm diameter) coil to an inexpensive metal detector can enable the conservator to detect totally mineralized bronzes.

A variety of safe cleaning methods is available today. All rely on careful removal, under low magnification, of soils and corrosion products using mechanical rather than chemical means. Due to the porosity of the corrosion products as well as that of the corroded bronze itself, absorption of cleaning agents such as dilute acids and complexing agents cannot be controlled. Therefore, only the gentlest of mechanical methods should be considered appropriate for the cleaning of bronze surfaces.

Numerous examples exist of Chinese bronze vessels damaged in the past due to acid cleaning. Perhaps the most extreme damage was caused by stripping procedures carried out to remove chlorides. In the 1940s, the Shang Dynasty *ding* (Fig. 1) was electrolytically cleaned in order to remove chloride salts. With the exception of a few small patches of cuprite, all surface accretions and corrosion products were removed from the exposed bronze surface, leaving a deeply fissured and pitted vessel that is a dull mustard-yellow color. Today such stripping methods are considered inappropriate, yet pieces such as this *ding* remain as examples of the importance of balancing

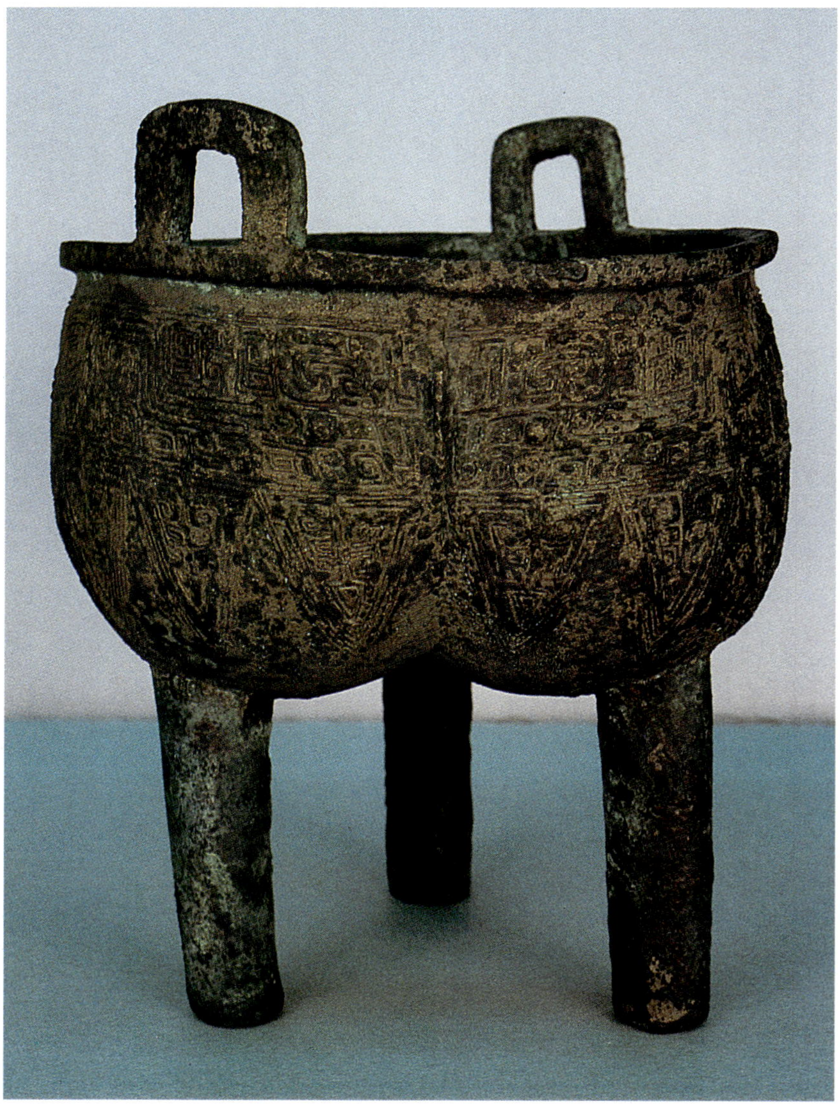

FIGURE 1. *Shang Dynasty* ding, *bronze, shown after electrolytic stripping. H:18.6 cm, D:17.9 cm. Honolulu Academy of the Arts, Honolulu; gift of Mrs. T. A. Cooke, 1939; no. 4776.*

scientific advances with other considerations. If we look at metallographic sections such as those of the Michigan mirror (see Chase herein) or those of the Kelley bronze (Gettens 1969:128–29), it becomes clear that electrolytic cleaning down to sound metal will, in the case of a heavily corroded bronze, leave a very rough and fissured appearance. Any evidence of the original surface (such as a cuprite marker layer) will be removed. Deep penetration of corrosion, as at pits or fissures, will also be removed, leaving these features clearly visible. A more careful consideration of the actual condition of this deeply corroded vessel might have prevented this damage.

Numerous examples of more subtle damage due to chemical cleaning may be found. The original surface on Chinese bronzes is often replaced by a smooth and even tin oxide corrosion layer. Acid cleaning agents can attack the corrosion below the tin oxide, causing pitting and damage to the tin oxide layer. Patchy, broken tin oxide layers may be seen on a *ding* in the Arthur M. Sackler Gallery (S1987.50), especially on the bottom of the body where acid has etched below the surface (Delbanco 1983:73; Bagley 1987). A well-known set of four *fang-ding* vessels also shows the same appearance. Etching is most notable on one leg of the *fang-ding* (FGA50.7) that belongs to the Freer Gallery of Art (Pope et al. 1965:191). One of the counterpart vessels from the same set in the National Palace Museum, Taiwan, has been treated more vigorously and shows patches of redeposited copper in many places.

Even cleaning treatments that are more conservative may effect the stability of bronze surfaces. The mechanical removal of slowly formed, compact corrosion layers, for example, may allow active corrosion to occur in areas that previously had achieved some degree of stability. Although this phenomenon may be a lesser problem with Chinese bronzes than with those from Southeast Asia and other areas, the problem should not be overlooked. One advantage of removing soil from an archaeological bronze is that a potential source of moisture and salts is eliminated.

Determining Objectives

After considering the possible effects of cleaning on the stability of the bronze surface, the next step is to determine what information can be gained by removing obscuring soils and corrosion products. One of the primary advantages of cleaning is that it may allow access to surface details, enabling the study of the vessel as a technological record. A growing interest is developing among art historians and collectors in the production processes of Chinese bronze vessels. The numbers of vessels being excavated in China at present underline the importance of bronze production in Shang and Zhou societies. Communication, trade, and labor systems can be better understood through research into the processes of mining and smelting ores and the casting of bronze vessels. The Chinese system of casting using clay molds is extremely complex and has many variables. Although radiography will reveal many of the casting details hidden by surface accretions, cleaning may allow an even more detailed study of a piece. Casting flaws, ancient repairs, inlay, decorations, finishing marks, and inscriptions all bear subtle differences and clues to help conservators understand the individual piece and therefore the technology as a whole.

Yet the conservator must be aware that indiscriminate cleaning may obliterate the remnants of surface technologies. Traces of original, intentional surface patinas may remain only in the corrosion layers, and their removal may eliminate the only chance for analysis. Other surface-coloration techniques must be anticipated and preserved, including black-silica and carbon inlays, which are found most often on Shang Dynasty and early Western Zhou vessels, as illustrated in Figures 2 and 3 (Gettens 1969:197–204). The same precautions apply to more unusual surface-coloration techniques such as the painted decoration found on Han period burial vessels (Nelson 1980:127–54). Other surface-patination techniques will undoubtedly come to light in time, such as the unidentified red coloration on the Eastern Zhou silver-inlaid animal figure exhibited in the United States in 1988–89 in conjunction with the Son of Heaven exhibition (Thorp 1988:134).

Another important concern must be the potential loss of archaeological evidence through cleaning. The majority of bronzes in Western collections were unearthed without the benefit of scientific excavation. Any information that can be retrieved may help to determine an object's original burial location or circumstances and its ceremonial uses. Soils and corrosion products can provide information about burial location (Holmes and Harbottle 1991:165–84; Smith 1978:123–33) and environment. Surface accretions, such as carbonaceous material on the bottom of cooking vessels, may reveal information about usage, as illustrated on the *xian* steamer vessel illustrated in Figure 4. These accretions may also contain remnants of materials placed adjacent to the vessel in burial, such as wood (Keepax 1989:15–20), textiles (Jakes and Sibley 1984:402–22), and cinnabar, which was sometimes scattered in Shang Dynasty tombs (Thorp 1980:51–64). Accretions on the interior of ceremonial cooking vessels may also contain remnants of offerings such as food and wine.

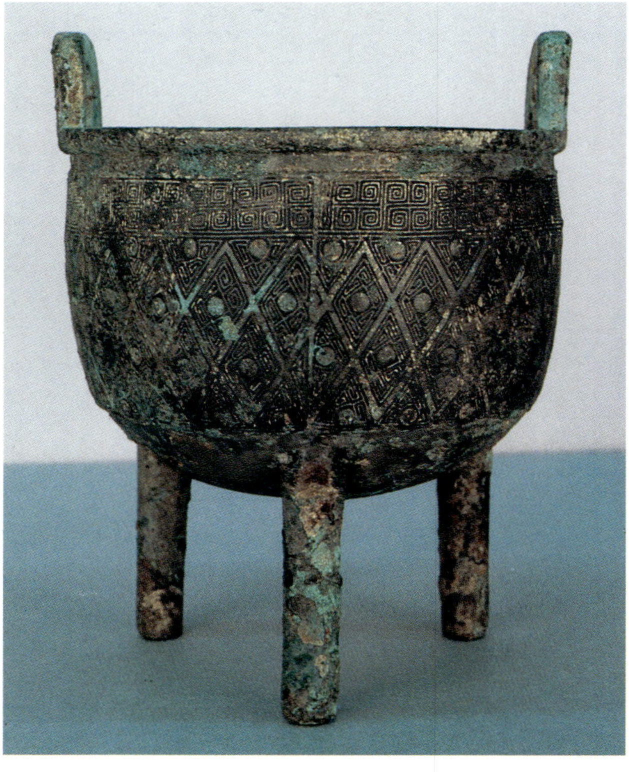

FIGURE 2. *Shang Dynasty* ding, *bronze, middle Anyang period. H:18.4 cm, D:16.5 cm. Honolulu Academy of the Arts, Honolulu; acquired by exchange, 1973; no. 4150.1.*

FIGURE 3. *Detail, Shang Dynasty* ding *(Fig. 2). Note black inlay.*

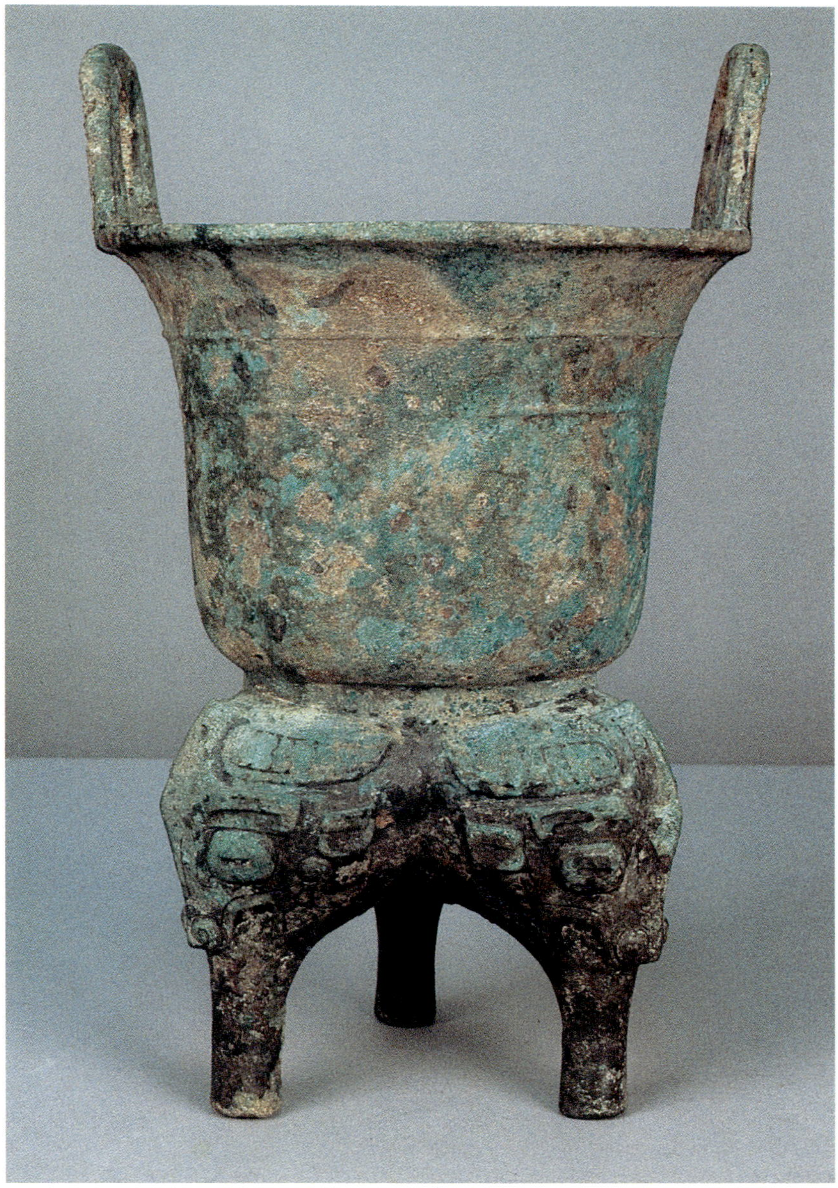

FIGURE 4. *Shang Dynasty* xian *steamer vessel, bronze. Note black soot on bottom. Honolulu Academy of the Arts, Honolulu, purchase no. 5850.1.*

Since materials removed can never be put back, present and future analytical capabilities need to be anticipated before cleaning is carried out. Colors of corrosion products should be recorded both photographically and with color analysis such as Munsell scales. Photomacrographs can be used to record corrosion patterns. Written descriptions of the texture and hardness of the corrosion products should also be recorded. In addition to photographic and written documentation, samples removed during cleaning should be saved and carefully labeled, with location references noted on corresponding photographs. Samples should include soils and corrosion products removed from the interior as well as the exterior of the vessel. Whenever possible, a section should be left uncleaned in an inconspicuous area (Jedrzejewska 1976:101–14). Although mineralized remains of organic materials on the surface can be photographed, and impressions can be made and samples saved, the limits that their removal will place on future research must be considered.

The conservator must also weigh the effects of treatment on the vessel as a cultural artifact. The vessel was made to reflect the power and wealth of the owner and the owner's ancestors, and the presence of soils and corrosion products may be considered to increase the strength and value of the artifact, as these attributes attest to the object's great age. From this viewpoint, the surface of the excavated bronze should not be violated, as the removal of corrosion products would deny the sanctity of its antiquity.

Aesthetic Considerations

The effect of cleaning on the aesthetics of the vessel as a work of art is of considerable importance in designing treatment. However, aesthetic expectations for treatment may be based more on an interest in certain anticipated appearances than on a desire for increased technological or archaeological understanding of the piece. In considering the aesthetic repercussions of treatment, two factors should be recognized: (1) The majority of vessels now in Western collections have been treated previously, usually involving some degree of cleaning, which may limit current treatment options, and (2) aesthetic standards of appearance are subjective by their very nature and will change with time; therefore, one must be conservative in the application of these standards in treatment. The demands of making vessels aesthetically pleasing to the viewer must be tempered by the conservation principles of reversibility and minimum intervention.

The following examples demonstrate how varying aesthetic expectations may influence treatment. An example of the cleaning of bronzes adopted on a large scale can be found in the Chinese Imperial Palace collections. Imperial Palace bronzes, such as the Western Zhou *ding* illustrated in Figure 5, were treated by removing soil, polishing down corrosion products, and applying a shiny wax or varnish to the

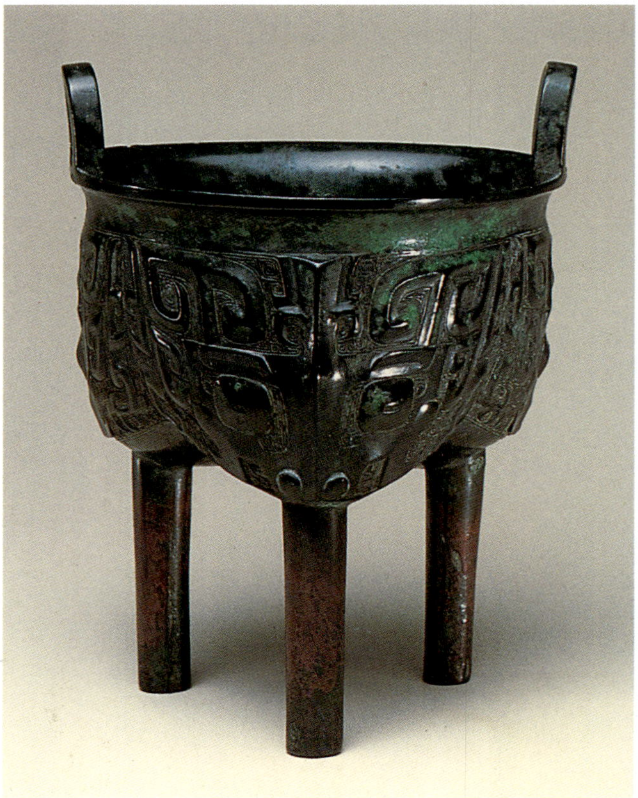

FIGURE 5. *Western Zhou Fugeng Li* ding, *bronze. 1100–1000* B.C.E. *H:21.2 cm, D:16.9 cm. Arthur M. Sackler Gallery, Washington, D.C., no. S1987.303.*

surface, probably for stabilization purposes (Lawton 1987–88:51–79). It may be assumed that this treatment, particularly in environments of extreme and uncontrolled humidity, did provide a degree of protection for the bronze surfaces. Whatever the initial purpose for the treatment, the sheer number of similarly treated objects in the imperial collections indicates that an aesthetic of darkened, polished, and shiny surfaces ensued, setting an aesthetic standard for a huge number of bronze vessels. These vessels can presently be found in large numbers in the National Palace Museum in Taiwan and in smaller numbers in museums in Beijing and throughout North America and Europe. This heavily cleaned, coated, and polished appearance became so accepted, in fact, that later archaistic reproductions were sought to duplicate this aesthetic.[1]

In Western collections today, viewers are more accustomed to the uneven, matte, and colorful appearance of natural burial accretions. A well-known example of this contrast in aesthetics can be illustrated with an early Zhou *hu* now in the collections of the Freer Gallery of Art (Gettens 1969:227). The black-and-white photograph of the *hu* (Fig. 6) was taken before the object entered the Freer collections. The photo shows a glossy, even-valued surface. In contrast, the present surface as illustrated in Figure 7 is a matte light-green color with areas of dark, highly contrasting corrosion products where the lid joins the vessel. It is thought that when the *hu* entered this country the vessel retained a shiny, dark patina indicative of Imperial Palace bronzes. In a New York dealer's workshop, this dark patina was altered to give an appearance more similar to the colorful corroded surfaces to which Western collectors had become accustomed. In keeping with the client's expectations, the dealer was able to create the anticipated surface appearance through chemical treatment.

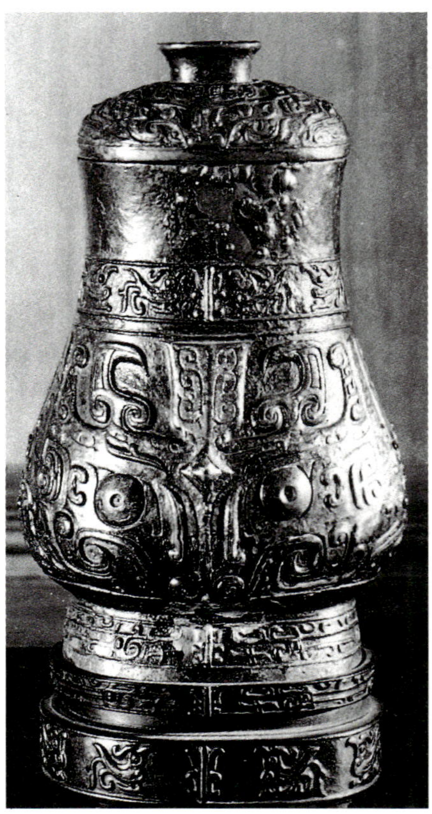

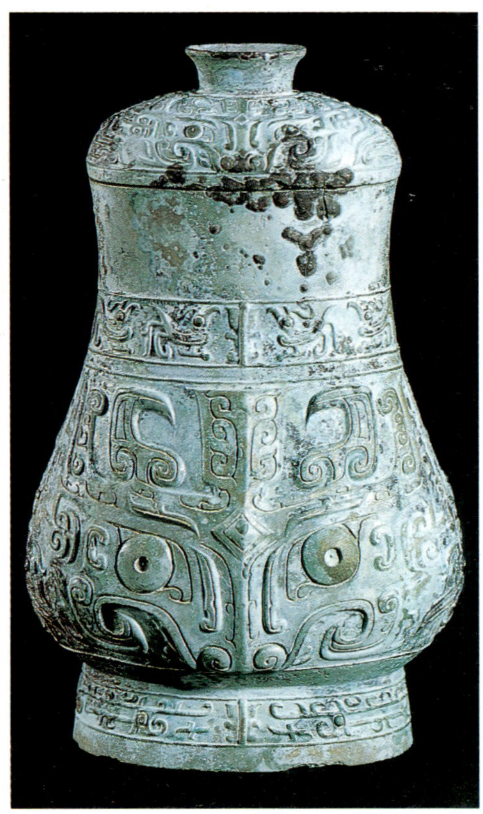

FIGURE 6. *Early Western Zhou Dynasty* hu, *bronze, near right, eleventh to early tenth century* B.C.E. *Photographed before acquisition by the Freer Gallery of Art (Gettens 1969:226). H (with lid):24.5 cm, W:15.5 cm. Freer Gallery of Art, Washington, D.C., no. 59.14.*

FIGURE 7. *Same Western Zhou Dynasty* hu *as in Figure 6, photographed after acquisition by the Freer Gallery of Art. Note darker area of corrosion products where lid joins vessel.*

A less drastic approach to achieving an acceptable aesthetic appearance can be seen with the Zhou Dynasty *gui* (Fig. 8). When this vessel entered the collection of the Freer Gallery of Art, it retained its shiny, dark surface which can be seen on the left in this photograph of the vessel taken during treatment. The right side of the vessel illustrates the results of swab-cleaning with organic solvents. When completed, the treatment revealed the state of the original materials unobscured by the darkened, glossy coating. Although any potential for analysis of archaeological remains was removed when the surface was polished, samples of the removed waxlike coating were saved for later analysis. One day, evidence hidden in such surface coatings may enable conservators to trace vessels to certain collectors or locations.

Present aesthetic standards can be illustrated with Eastern Zhou inlaid pieces. Gold, silver, copper, and gemstones such as malachite and turquoise were incorporated into bronze surfaces beginning in the sixth century B.C.E. In Western collections today, these stone surfaces are often meticulously cleaned to reveal the full contour of the inlay. Similarly, gold and silver inlay are brought to a high luster by the removal of surface accretions and tarnish. Although cleaning silver inlay permits an appreciation of the intricacy of its patterns, an arbitrary value relationship is established which may have little to do with the artifact's original appearance. Although one may rightfully assume that the silver was polished to a high luster when the object was first made, the original coloration of the surrounding bronze cannot be known for certain. In the case of these pieces, intentional cleaning of one metal surface to a higher degree than the other was possible simply because of differences in the corrosion processes.

FIGURE 8. *Zhou Dynasty* gui, *bronze, shown during treatment with cleaned section on the right. Freer Gallery of Art, Washington, D.C., no. 68.29.*

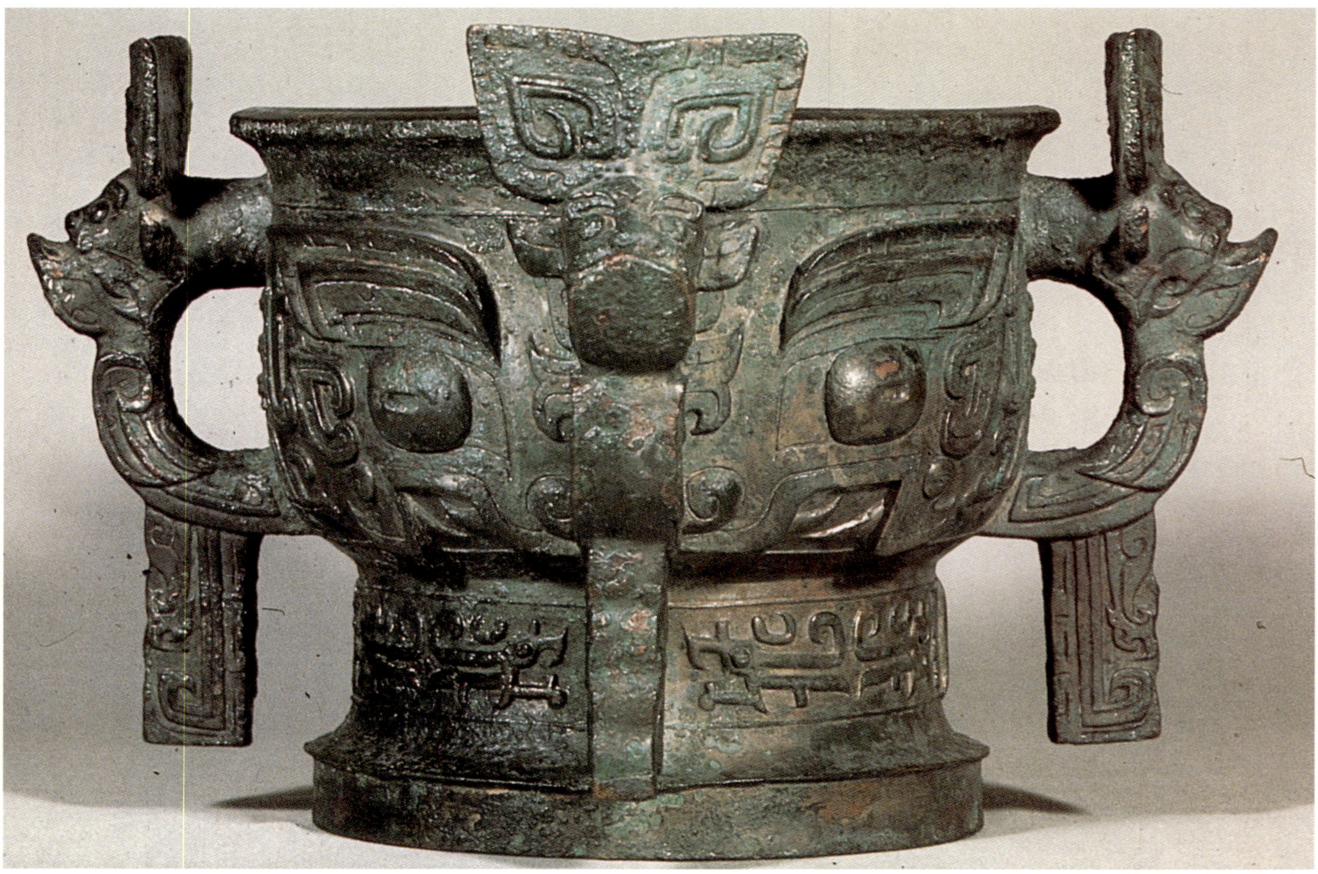

In contrast to gold and silver inlay, copper is generally not taken to a high polish. For example, early cleaning tests on the copper inlay in the lid of an Eastern Zhou *fang-hu* at the Freer Gallery (FGA61.32) were considered unsuccessful; the polished copper was thought too bright in color and therefore inappropriate. Two points are of interest here: First, the public has come to accept polished silver surfaces on ancient pieces in Western museums, yet a polished copper inlay was, in this instance, unacceptable; one day a highly polished archaeological silver surface may be considered extreme. Second, in light of recent work on the patination of such inlaid pieces, it is fortunate that cleaning of the copper was not undertaken (see Chase herein).

In determining the appropriateness of cleaning, the final consideration must be whether an aesthetically acceptable cleaning is possible. In reaching this decision the conservator must work in close cooperation with the curator, as the definition of an acceptable aesthetic standard may vary considerably. The conservator must be able to evaluate each surface before cleaning is undertaken and be able to communicate what that surface will look like after treatment. Experience with similar surfaces will be a considerable advantage.

Certain corrosion products may be indicative of a well-preserved surface. The cuprite marker layer may retain original surface detail. Additionally, corrosion layers such as botryoidal malachite may indicate the presence of a well-preserved surface below. Less desirable aesthetic effects of cleaning may include surface blotchiness where different corrosion layers have been cut through, or the complete loss of relief details due to extensive, blistering corrosion. These conditions should be diagnosed accurately before cleaning is attempted; in some cases it is simply better to leave the bronze alone. During the process of mechanical cleaning, the initial area may clean very nicely, but then deep pitting of an area that does not clean well mechanically may be encountered. In this case, the surface will probably look better with the original patina left in place; aesthetic sensitivity is expected of the conservator who is wielding the chisel.

Contextual Issues

Once the pros and cons of cleaning are understood in general terms, each artifact must be examined to determine the appropriate course of treatment. To do this, the function of the individual artifact in today's surroundings needs to be defined. Although research may help clarify the ancient function and meaning of Chinese bronze vessels, these objects are now unearthed, out of context, removed in most instances from their place of origin. The continuum of their ceremonial use has been broken by generations of change within China. Whether in private collections, anthropology museums, or art galleries, conservators and curators must decide what role each individual piece fulfills in present-day context, asking such questions as, How does this vessel work within the goals of the collection in which it is housed? and How can it add to our knowledge of Chinese bronze vessels in general? Each vessel may be seen as an archaeological or technological record; it may be viewed as a cultural artifact, valued for the build-up of wear and accretions of age, or simply appreciated as a unique work of art.

The process of determining an artifact's function should involve input from as many experts as possible, with the goal of increased knowledge for all. During such discussions the conservator's responsibility is to explain the various possible degrees of cleaning, the appearances that might result from each degree of cleaning, and the risks involved. The art historian or archaeologist has a similar responsibility to educate the conservator, sharing such knowledge as the rarity of the vessel type or decoration and its possible ceremonial or burial circumstances.

Conclusions

The keys to responsible treatment of ancient bronzes are as follows:

1. Communication with other specialists
2. Thorough documentation
3. Thorough examination to determine the extent of corrosion and previous treatments the piece may have undergone
4. Evaluation of the function of the vessel in today's context, including an understanding of the archaeological and technological information that may be lost in cleaning, as well as an appreciation of the aesthetic repercussions of treatment
5. Respect for the vessel and the people who made it

With these essential points in mind, the conservator can proceed with enlightened, successful, and conservative cleaning of ancient vessels.

Acknowledgments

Initial research was carried out during an Andrew W. Mellon fellowship awarded to Jane Bassett at the Los Angeles County Museum of Art in 1990–91. The authors would like to thank Pieter Meyers for his invaluable assistance in the preparation of this paper.

Note

1. Examples include a *hu* (FGA09.254) and a *yu* (FGA09.260) in the Freer Gallery of Art study collections. For a survey of historic references to the forgery of Chinese bronze vessels, see Barnard 1968.

References

BAGLEY, R. W.

1987 Shang Ritual Bronzes in the Arthur M. Sackler Collections. *Ancient Chinese Bronzes in the Arthur M. Sackler Collections*, vol. 1. Washington, D.C. and Cambridge: Harvard University Press.

BARNARD, N.

1968 The incidence of forgery amongst archaic Chinese bronzes: Some preliminary notes. In *Monumenta Serica*, 91–168. H. Busch, ed. Los Angeles: Monumenta Serica Institute, University of California.

DELBANCO, D. H.

1983 *Art from Ritual: Ancient Chinese Bronzes from the Arthur M. Sackler Collections*, catalogue no. 24. Cambridge: Arthur M. Sackler Museum, Harvard University.

GETTENS, R. J.

1969 *The Freer Chinese Bronzes, Volume II: Technical Studies*. Oriental Studies, no. 7, Smithsonian Publication 4706. Washington, D.C.: Smithsonian Institution, Freer Gallery of Art.

HOLMES, L. L., AND G. HARBOTTLE

1991 Provenance study of cores from Chinese bronze vessels. *Archeomaterials* 5:165–84.

JAKES, K. A., AND L. R. SIBLEY

1984 Examination of the phenomenon of textile fabric pseudomorphism. In *Archaeological Chemistry*, J. B. Lambert ed., *Advances in Chemistry* 205:402–22. Washington, D.C.: American Chemical Society.

JEDRZEJEWSKA, H.

1976 A corroded Egyptian bronze: Cleaning and discoveries. *Studies in Conservation* 21:101–14.

KEEPAX, C. A.

1989 Corrosion "preserved wood": Advances since 1975. In *Evidence Preserved in Corrosion Products: New Fields in Artifact Studies*, 15–20. R. Janaway and B. Scott, comps. Leeds: The United Kingdom Institute for Conservation.

LAWTON, T.

1987–88 The Imperial legacy revisited: Bronze vessels from the Qing Palace collection. *Asian Art* 1:51–79.

NELSON, D. M.

1980 Bronze *Ming-Ch'i* vessels with painted decoration: A regional study in Han pictorialism. *Artibus Asiae,* 127–54.

POPE, J. A., R. J. GETTENS, J. CAHILL, AND N. BARNARD

1965 *The Freer Chinese Bronzes, Volume I: Catalogue*. Oriental Studies, no. 7, Smithsonian Publication 4706. Washington, D.C.: Smithsonian Institution, Freer Gallery of Art.

SMITH, A. W.

1978 Stable carbon and oxygen isotope ratios of malachite from the patinas of ancient bronze objects. *Archaeometry* 20:123–33.

THORP, R. L.

1980 Burial practices of Bronze Age China. In *The Great Bronze Age of China: An exhibition from the People's Republic of China*, comp. Wen Fong, 51–64. New York: The Metropolitan Museum of Art and Alfred A. Knopf.

1988 *Son of Heaven: Imperial Arts of China.* Seattle: Son of Heaven Press.

Biographies

Jane Bassett, associate conservator of sculpture and decorative arts at the J. Paul Getty Museum, received her bachelor of arts degree in art history from Stanford University and her master's degree in conservation from the Cooperstown Graduate Program in New York. From 1986 to 1991 she was objects conservator at the Pacific Regional Conservation Center at the Bishop Museum in Honolulu, Hawaii.

W. T. Chase is head conservator of the Department of Conservation and Scientific Research of the Arthur M. Sackler Gallery and the Freer Gallery of Art, Smithsonian Institution. He majored in conservation of art at Oberlin College and later studied at the Conservation Center of the Institute of Fine Arts at New York University. During his graduate studies, he took his student internship at the Freer Technical Laboratory under Rutherford J. Gettens. After completing his degree in 1966, he returned to the Freer, where he became head conservator in 1968. His primary research interest is technical study of ancient Chinese bronzes.

Tomography of Ancient Bronzes

STEPHEN D. BONADIES

The Cincinnati Art Museum has been fortunate to have had the technological resources of General Electric Aircraft Engines in Evendale, Ohio, close at hand. With General Electric's generous support and cooperation, the museum has been able to study a number of objects from its collection using industrial computed tomography (CT) to gain a better understanding of ancient methods of manufacture as well as the condition of these objects. Industrial CT is one tool in the repertoire of nondestructive evaluation techniques that is uniquely suited to the visual examination of the inner structures of an object. The images that result from an industrial CT examination are aesthetically appealing as well as intellectually tantalizing.

Industrial Computed Tomography

Industrial CT is an adaptation of computerized axial tomography (medical CAT-scanning) developed in the late 1960s and early 1970s (Dennis 1989). Since the 1970s, efforts have been directed toward the application of computed tomography in industry and research. The technique was first applied to the study of museum objects in the 1980s. Although medical scanners can provide high-quality images of ceramics, plastics, and aluminum as well as those of soft tissue and bone, these systems are limited in their ability to scan heavier metals, such as bronze and steel. Industrial CT systems, however, are capable of scanning large, dense objects, using higher-energy X-ray sources with higher-resolution detectors and imaging systems. Early applications of industrial CT included the inspection of large rocket motors, small precision castings, and other aircraft engine components.

The first CT studies of museum objects included those done by Miura (1980) and by Tout, Gilboy, and Clark (1980), involving wood and bronze sculptures, respectively. The latter study also noted the possible application to dendrochronology. While many of the initial studies relied on medical CT systems, their limitations for penetrating dense materials as well as scanning large objects were quickly encountered. The first use of an industrial CT system is noted by Miura and Fujii (1987), who examined a small gilt bronze in this way. Although hollow, the bronze's composition and thickness required the use of a high-energy (420 kV) source.

One problem that has consistently plagued computed tomographic examinations is the low quality of visual reproduction. Avril and Bonadies (1991) overcame this difficulty by using a high-resolution imaging system and photographing the scanned image directly off the monitor. Continued improvements in image-processing software and the use of direct digital-to-photographic printers will eventually eliminate this problem.

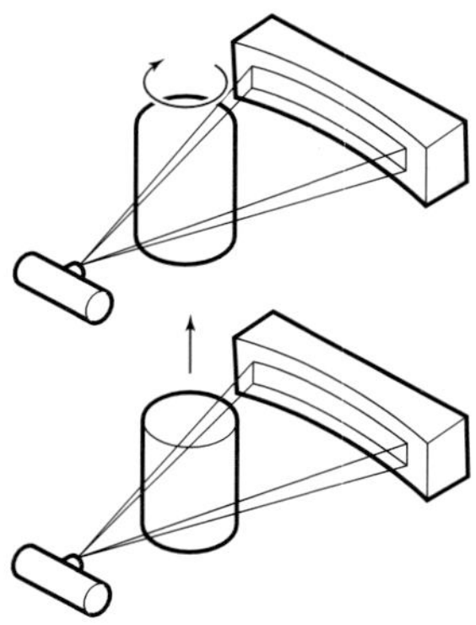

FIGURE 1a, b. *Schematic illustrating (a), top, the CT process, and (b), bottom, the digital radiographic process.*

The CT process begins with a narrow, fan-shaped X-ray beam—the thickness of which defines the thickness of the cross-sectional slice to be measured—interrogating the object under study (Fig. 1a). The attenuation of the beam as it passes through the object is directly related to the density and thickness of the material as well as to the energy of the X-ray beam. The object is rotated slowly so data can be collected from many different angles. The data are transmitted through a data-acquisition system to a high-speed computer, which generates a two-dimensional image by the application of a filtered back-projection algorithm. Within a matter of seconds, the image appears on a high-resolution video screen (1024 by 1024 pixels). Subsequent enhancement may be achieved through addition of color, by sharpening and enlarging, and by varying the brightness and contrast. A hard copy of the image may be obtained with a laser printer or by photography, and the image data can be stored on magnetic tape or optical disk for later retrieval. While an object 1.83 m (6 ft) in diameter by 0.91 m (3 ft) in height can be accommodated, the maximum scanning width is 0.61 m (2 ft).

Industrial CT systems can also be used to quickly scan an entire object to produce a digital radiograph (Fig. 1b). The digital radiographic image is similar to that obtained with conventional film radiography. Digital radiographic images are used for radiographic inspection of objects as well as to mark areas where the slices should be taken. While the resolution of the typical digital radiographic image is not as great as with film radiography, higher resolution can be achieved by repeating the digital radiograph scan with a slight shift of the detector. A significant factor limiting image sharpness is the detector resolution, which in turn is determined by spacing of individual detecting elements. Details smaller than this spacing cannot be accurately defined.

The advantages of CT over conventional radiography are numerous. Whereas radiography compresses the structural information from a three-dimensional object into a two-dimensional image, CT is uniquely able to inspect the internal structure of an object without interference from protruding parts or highly decorated surfaces. CT is extremely sensitive to variations in object density as well as to differences in material composition. Minute density differences can be visualized and measured. Unlike conventional film radiography, in which magnification frequently occurs due to the distance between object and film, a CT scan allows precise dimensional measurements of up to 50.8 microns (.002 inches). Cheng and Mishara (1988) utilized this capability to provide technical information on violin construction that included precise dimensions of the instrument's parts.

Perhaps the most exciting possibility presented by this technology is the ability to reconstruct an accurate three-dimensional image of an object. By taking hundreds of adjacent CT slices and then stacking them one on top of the other, a three-dimensional image can be created (General Electric's Digital Replica™ process) through computer-processing techniques. This computer model can then be "cut up" electronically to enable the study of previously inaccessible interior surfaces or the structures of enclosed hollow objects. Early medical uses of this technique were applied to the paleoanthropological study of fossil skulls (Conroy and Vannier 1984; Vannier et al. 1985).

The impetus for analyzing the museum's Chinese bronzes was the ongoing cataloging project of the permanent collections. In an attempt to gain as much technical information as possible, the museum decided to study these objects with X-radiography. The purpose of these analyses was to detect hidden damage and repairs in the vessels and to learn more about ancient casting techniques. Until fairly recently many museums studying ancient metal work have relied on conventional X-ray equipment (Bagley 1987). However, when museum staff approached officials at General Electric Aircraft Engines, they suggested that industrial CT be used instead. In addition to the Chinese bronzes, other objects examined with CT from the museum's collection included an Achaemenid silver rhyton, a Roman bronze bull, an Iranian portrait head, and an Egyptian mummy.

Ancient Chinese Bronze Ritual Vessels

The Chinese aristocracy of the Shang Dynasty (ca. 1550–1030 B.C.E.) commissioned sumptuous vessels cast in bronze for making ritual offerings of food and wine to their esteemed ancestors. The technical sophistication and extravagant consumption of raw materials required to produce these vessels attest to their importance as symbols of power for the ruling class. Enormous wealth was expended on the furnishing of tombs, in part because Chinese nobles believed that reverence for the needs of the ancestors in the afterlife would ensure the continued success of the clan on Earth. Indeed, such extravagance also served to legitimize the status of the ruling clans in Shang society.

The Cincinnati Art Museum's collection of Shang Dynasty Chinese ritual bronzes comprises superb examples dating from the thirteenth to eleventh centuries B.C.E. Most come from the area of the ancient Shang capital at Anyang (in modern Henan province) and represent the classic phase of Shang Dynasty bronze production.

Ancient Chinese Casting Techniques

During the Shang Dynasty the Chinese developed a multistage process for casting bronze in ceramic section molds. First, a model of the finished vessel was carefully built up in fine clay, then allowed to dry. The outer mold sections were produced by pressing moist primary loess around the model; this outer layer was then cut away in sections, fired, and reassembled to receive the molten metal (Wood 1989). Depending on the shape of the vessel being produced, the mold assembly needed one or more cores, also of loess, in addition to the outer mold parts. Metal spaces, or chaplets, were carefully placed between inner cores and outer mold parts to maintain the proper distance for the desired wall thickness. Vessels were cast upside down so that the legs or base served as inlets for pouring the alloy of copper, tin, and lead.

The lost-wax method, the more common technique for casting bronze in the ancient world, was used in the ancient Near East as early as the fourth millennium B.C.E., but was not employed in China until the sixth century B.C.E. (Thorp 1988).

Results of the Examination of Selected Bronzes

Digital radiographic and CT images were generated at 420 kV for each bronze examined by industrial CT, except for the Roman bronze bull, which required an even higher energy source. Repairs otherwise difficult to detect become obvious in the digital radiographic image because of differences in material density. With color enhancement, these differences can be made even more apparent. A yellow line through the digital radiographic image marks the location of the subsequent CT slices to be taken in order to provide more information such as wall thickness, core construction, and porosity. Slices taken for this particular analysis were 508 μm (20 mils) each.

Two ceremonial wine beakers of the *gu* shape illustrate basic casting techniques. The vessels were cast in a four-part mold with two inner cores, one to form the circular base, the other to form the wine container. The foot core was supported during the pour with cross-shaped projections. Remnant voids (approximately 1.0 by 1.6 cm in one *gu*, 1.4 by 1.9 cm in the other; opening thickness 0.16 cm in each) from the spacers were cleverly employed in the decoration of the vessels, appearing as openwork crosses in the completed vessels. Centering the inner cores was crucial for producing a thin-walled, symmetrical vessel; misalignment of mold parts could result in casting failure. CT scans reveal how successfully Chinese artisans achieved a thin wall of uniform thickness under conditions that allowed little tolerance for error (Fig. 2).

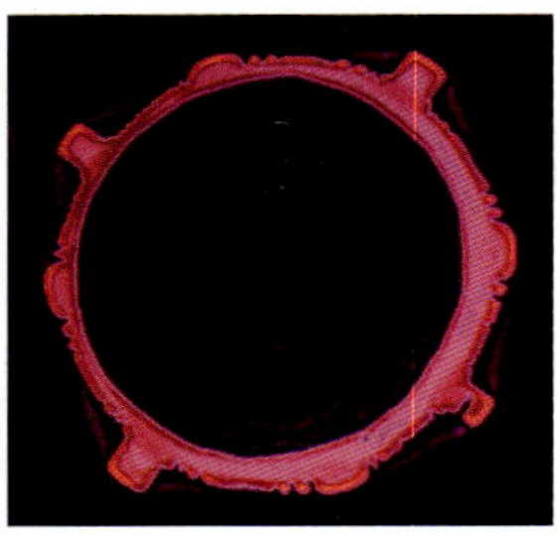

FIGURE 2. *CT slice of Chinese Shang Dynasty* gu *wine beaker. Section taken through circular base shows uneven wall thickness due to misalignment of core with respect to mold.*

Small tripod vessels with slender legs, such as the *jue*, were cast upside down in a single pour. The digital radiographs and CT images of the *jue* illustrate the porosity problems ancient artisans frequently encountered while working with molten metal and reveal concentrations of trapped air bubbles in the rounded bottom of the vessel. One of the fortuitous qualities of the primary loess used by the Chinese for section molds is its ability to absorb trapped gases that could otherwise pit the surface of a casting. Although the digital radiographic image reveals some porosity problems (Fig. 3), the CT scans specifically locate and reveal the extent of trapped air bubbles with respect to the vessel walls (Fig. 4). This dramatically illustrates how close the casting came to failure.

Study of the *jia*, another tripod wine-warming vessel, revealed the most startling condition problems of the bronzes analyzed. In the digital radiographic image (Fig. 5), numerous repairs—on the handle, in the core cast legs, and along the rim—appear in bright white due to the probable use of a lead-tin solder. Two of the legs have major repairs. One leg was reassembled using a modern screw, indicating that the object was probably repaired shortly before the museum purchased it in 1948.

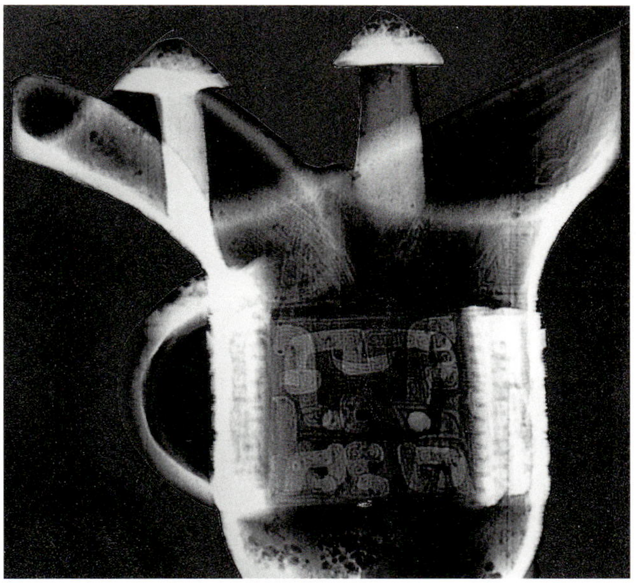

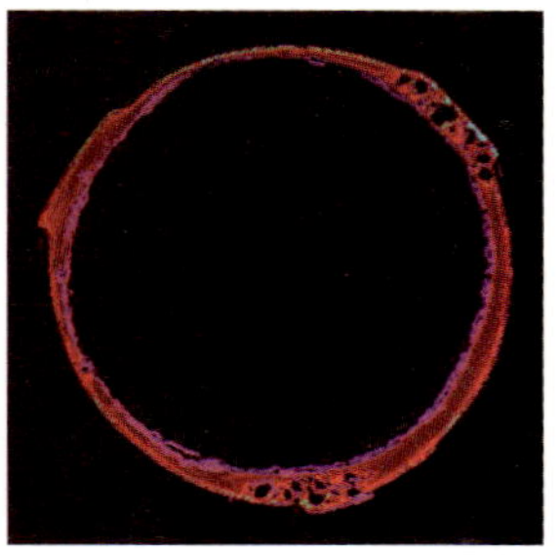

FIGURE 3. *Digital radiograph of Chinese Shang Dynasty* jue *tripod vessel, near right, showing trapped air in knobs and at rounded bottom of vessel.*

FIGURE 4. *CT slice of* jue, *far right, reveals extent of porosity, as well as variation in wall thickness.*

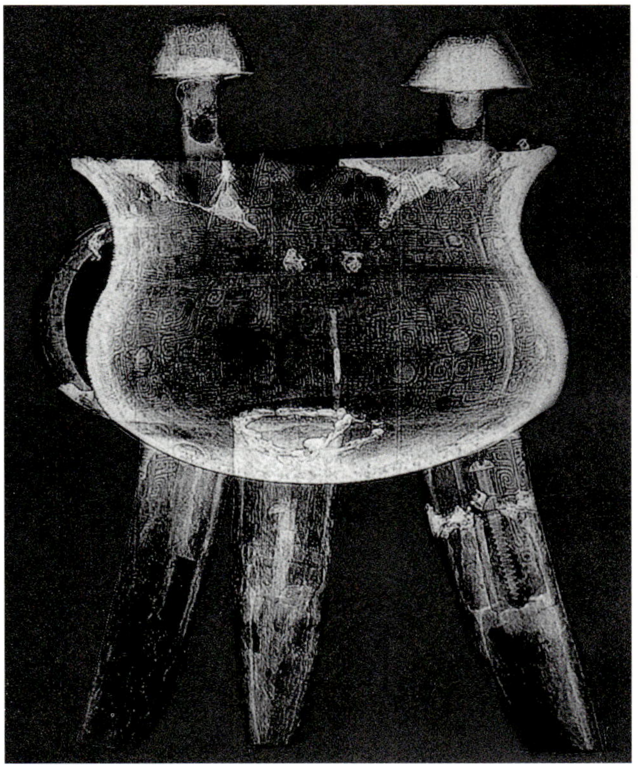

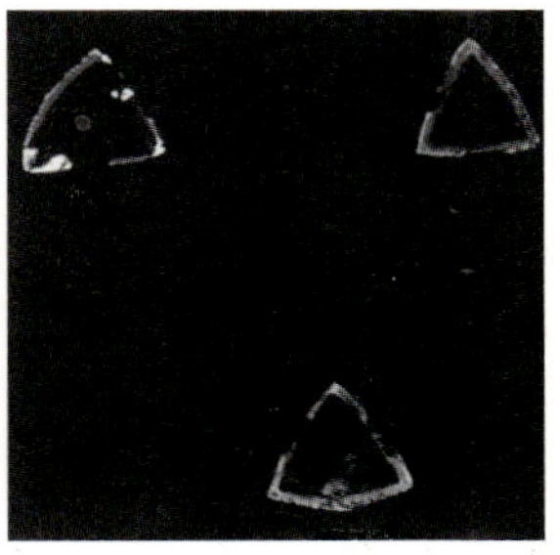

FIGURE 5. *Digital radiograph of Chinese Shang Dynasty* jia *wine-warming vessel, near right, showing core-cast legs, numerous repairs, and reassembly of legs.*

FIGURE 6. *CT slice taken through legs of the* jia, *far right, reveals location of screw and pin with respect to interior walls of the legs. Pin is in lower leg, whereas screw is in upper left leg.*

Another leg has a less obvious break and was reassembled using a metal pin. This repair appears older and may have been performed in antiquity, since no evidence of the break is visible on the surface; it is completely concealed by patination accumulated during its extended burial. CT scans taken through the legs show the exact location of the screw and pin with respect to the interior walls of the legs (Fig. 6). An additional CT scan, taken where the legs join the vessel body, clearly shows the reattachment of one leg.

A *guang*-type lidded wine vessel in the form of composite fantastic animals represents the most complex casting of the bronzes analyzed. Digital radiographic images reveal only one condition problem: a hairline fracture in the center of the lid. A close-up digital radiograph of the *guang* handle reveals that it was cast around a clay core. Four CT scans taken through the handle depict core shape and wall thickness in those locations as well as confirm the presence of tenons extending from the body for the handle's attachment (Fig. 7). Visual evidence of metal overflow onto the surface decoration of the body confirms that the handle was cast on as a separate unit (Bagley 1987).

To fully characterize the configuration of the clay core and determine the nature of the join between the handle and body of the *guang*, a three-dimensional model was constructed by taking CT slices horizontally through the entire vessel at 508 μm (20 mil) increments. The data were then processed to form a three-dimensional reconstruction of the *guang*. By electronically slicing the computer model in half lengthwise and rotating it on the screen, the interior of the handle was exposed, thus revealing the shape of the clay core inside.

Among the manipulations possible with three-dimensional replication (Fig. 8) is the ability to create a synthetic CT in any direction through the vessel using data accumulated in the previous scanning. This technique also proved useful in the analysis of the *guang*. Synthetic CTs were created vertically through the handle. The resulting images, when combined with horizontal CT and digital radiographic views,

FIGURE 7. *Digital radiograph of Chinese Shang Dynasty* guang *lidded wine vessel, below left, shows tenons extending from the body. Yellow lines indicate where subsequent CT scans will be taken.*

FIGURE 8. *Industrial CT workstation for examination and further manipulation of the three-dimensional reconstruction, below right.*

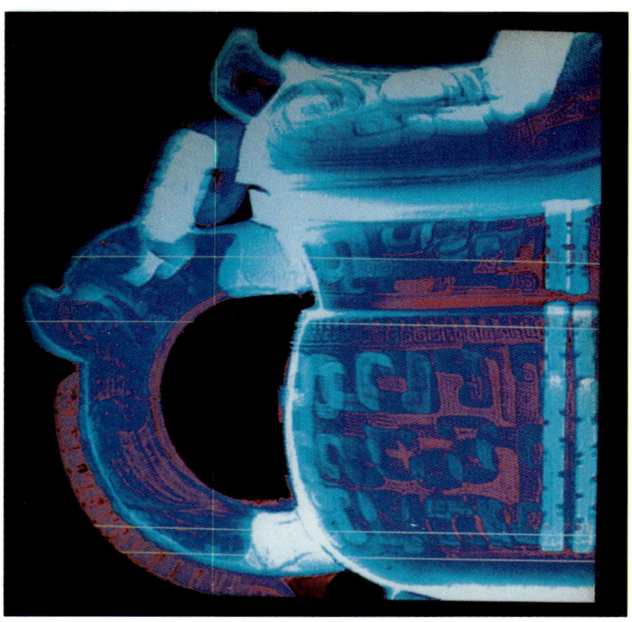

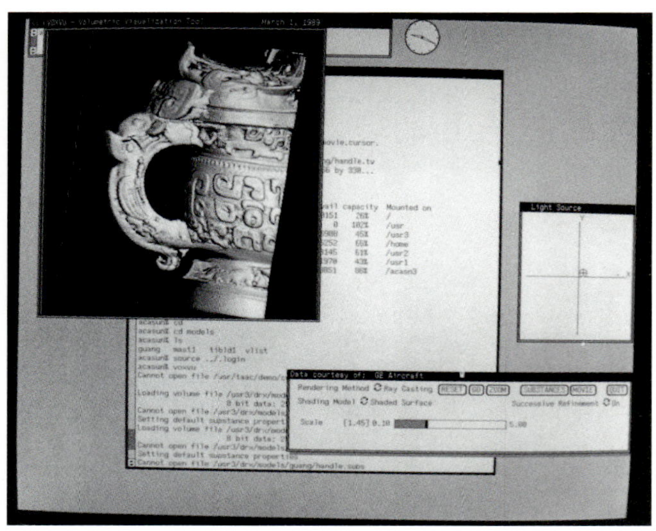

provided a full characterization of the join and confirmed suspicions that the handle was indeed cast over tenons extending from the previously formed body.

A unique Roman bronze bull, dating from the first century B.C.E. to the early first century C.E., proved interesting to study for two reasons. First, it is the largest known example of a cast-bronze representation of an Apis bull; and second, the object had previously been studied with conventional film radiography a number of years ago. Thus, direct comparisons between film radiography and industrial computed tomography could be made.

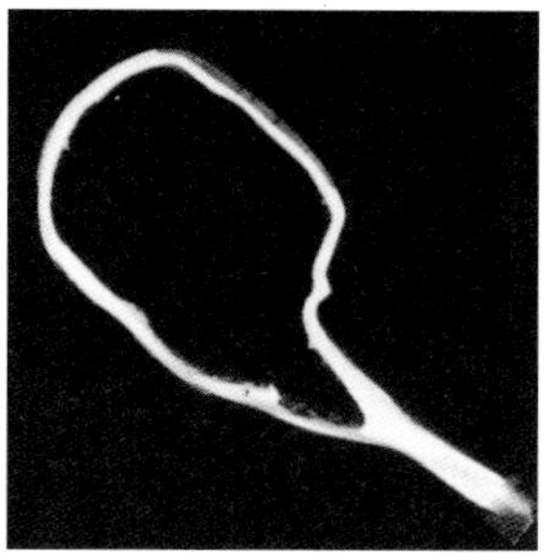

FIGURE 9. *CT slice of Roman bronze bull, first century B.C.E. to first century C.E., showing extent and specific depth of plaster restorations.*

As noted earlier, the 420 kV source was unable to penetrate the bull, resulting in electron scatter that effectively obliterated any details in the CT image. This was probably due to the thickness of the metal as well as to its composition (Cu 86%, Pb 4%, Sn 9%). Instead, a 2 million volt (2000 kV) source was successfully used to penetrate the object. Since there was still a slight problem with electron scatter in the CT, the image quality was not as fine as images made previously. Even though problems were encountered, however, useful information was obtained about the extent of surface restorations. While these restorations are obvious in the film radiograph, the CT slices additionally depicted their depth (Fig. 9).

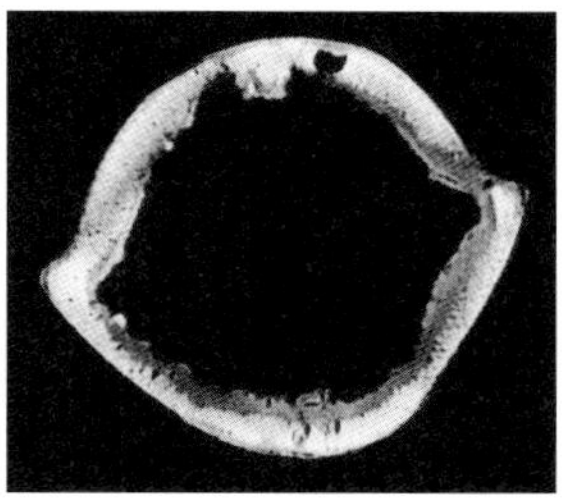

FIGURE 10. *CT slice of Bronze Age Iranian portrait head, revealing bands of differing densities.*

An Iranian portrait head (Cu 96.67%, As 1.77%, Fe 0.76%, Ni 0.74%) from the second millennium B.C.E. is one of the very few examples extant from the Bronze Age. This piece and a cast portrait of similar style and date from the collection of The Metropolitan Museum of Art are the earliest known examples of Iranian portraiture in any medium. The two objects allegedly came from the same find.

As part of a comparative study, the museum's portrait head was scanned. The CT slices revealed both extensive porosity of the metal and obvious bands of differing metallic density and composition within the wall (Fig. 10). Once the companion piece from The Metropolitan Museum of Art is examined, investigators hope to shed light on the early development of bronze casting in Iran.

The Future of Industrial Computed Tomography

An industrial CT workstation has been developed by General Electric Aircraft Engines that enables researchers to view and manipulate the three-dimensional reconstruction at a computer terminal. The model can be rotated, sliced, and viewed from any angle, and actual or synthetic CT images can be accessed individually to study details of an object. Examination of the reconstructed image allows one to view interior surfaces and to dissect joins. In the past such an inspection would only have been possible with the destruction of the object. Future developments will most likely be directed toward improving the speed at which the reconstruction is obtained by using a three-dimensional X-ray system.

Through the application of today's sophisticated nondestructive evaluation techniques, the combination of modern technology with art history advances the knowledge of ancient metallurgy, providing safe examination methods to aid in the preservation of many such works of art.

Acknowledgments

The author is grateful to the staff of General Electric Aircraft Engines, Quality Technology Center, for its continued support.

References

AVRIL, E., AND S. D. BONADIES

1991 Non-destructive analysis of ancient Chinese bronzes utilizing industrial computed tomography. *Materials Issues in Art and Archaeology*, vol. 185. Pittsburgh: Materials Research Society.

BAGLEY, R. W.

1987 Shang Ritual Bronzes in the Arthur M. Sackler Collections. *Ancient Chinese Bronzes in the Arthur M. Sackler Collections*, vol. 1. Washington D.C. and Cambridge: Harvard University Press.

CHENG, Y.-T., AND J. MISHARA

1988 A computerized axial tomographic study of museum objects. *Materials Issues in Art and Archaeology*, vol. 123. Pittsburgh: Materials Research Society.

CONROY, G. C., AND M. W. VANNIER

1984 Non-invasive three-dimensional computer imaging of matrix-filled fossil skulls by high-resolution computer tomography. *Science* 226(4673):456–58.

DENNIS, M. J.

1989 Non-destructive evaluation and quality control. *The Metals Handbook*, vol. 17, 9th ed. Metals Park: ASM International.

LEVEQUE, M. A.

1987 An approach to the conservation of Egyptian mummies; the mummy of Lady Nesmutaatneru. In *Recent Advances in the Conservation and Analysis of Artifacts, Jubilee Conservation Conference, London, 6–10 July 1987*. London: University of London.

MIURA, S.

1980 Investigation of a wooden sculpture of Buddha. *Science for Conservation* 19:9–14.

MIURA, S., AND M. FUJII

1987 Investigation of a gilt bronze statue by a high energy X-ray scanner. *Science for Conservation* 32:40–46.

REIMERS, P., AND J. RIEDERER

1984 Die Anwendung der Computertomographie zur Untersuchung kulturgeschichlicher Objekte. *Berliner Beiträge zur Archäometrie* 9:171–90.

RIEDERER, J.

1986 Neue Durchstrahlungstechniken zur Untersuchung von Kunstwerken im Rathgen-Forschungslabor. *Berliner Museen* 3:8–9.

1988 Ausgrabungen nach Röntgenbild. *Archäologie in Deutschland* 1:24–27.

TAGUCHI, E., I. NAGASAWA, S. YABUUCHI, AND M. TAGUCHI

1984 Investigation of a wooden sculpture using X-ray computed tomography. *Scientific Papers on Japanese Antiques and Art Crafts* 29:43–50.

THORP, R. L.

1988 *Son of Heaven: Imperial Arts of China*. Seattle: Son of Heaven Press.

TOUT, R. E., W. B. GILBOY, AND A. J. CLARK

1980 The use of computerized X-ray tomography for the non-destructive examination of archaeological objects. In *Proceedings of the 18th International Symposium on Archaeometry and Archaeological Prospection, Bonn, 14–17 March 1978*. Köln: Rheinland Verlag.

TYERS, I.

1985 Tree ring dating by X-ray. *The London Archaeologist* 5(4):87–88.

UNGER, A.

1988 Zerstörungsfreie Zustandanalyse an Kulturgut as Holzmittles Röntgen-Computertomographie. *Conservation-restoration of leather and wood; Training of restorers; Sixth International Restorer Seminar, 1987 Veszprém, Hungary*. István Éri and Gábriella Sárközy, eds. Budapest: National Center of Museums.

UNGER, A., AND J. PERLEBERG

1987 X-ray computer tomography (XCT) in wood conservation. In *ICOM Committee for Conservation: 8th Triennial Meeting, Sydney, Australia, 6–11 September 1987*. Marina del Rey, Calif: The Getty Conservation Institute.

VANDIVER, P., W. A. ELLINGSON, T. K. ROBINSON, J. J. LOBICK, AND F. H. SEGUIN

1988 Radiographic imaging technologies for archaeological ceramics. *Expedition* 29(3):55–63.

VANNIER, M. W., G. C. CONROY, J. L. MARSH, AND R. H. KNAPP

1985 Three-dimensional cranial surface reconstructions using high-resolution computed tomography. *American Journal of Physical Anthropology* 67(4):299–311.

WOOD, N.

1989 Ceramic puzzles from China's Bronze Age. *New Scientist* 18(2):50–53.

Biography

Stephen D. Bonadies, chief conservator at the Cincinnati Art Museum, received his initial conservation training at the Cooperstown Graduate Program in New York. He gained additional experience as part of a team of conservators sent to northeastern Italy under the auspices of the Friuli Italian Arts and Monuments Committee after a series of devastating earthquakes there in 1976. After receiving his master's degree in 1979, he became a Mellon Fellow at the Philadelphia Museum of Art. In 1980 Bonadies returned to the Art Conservation Program at Cooperstown to teach conservation science. He was appointed to the staff of the Cincinnati Art Museum in 1981.

Chinese Bronzes: Casting, Finishing, Patination, and Corrosion

W. T. CHASE

Why is the topic of the present conditions of Chinese bronzes of interest? From a historical point of view, conservators would like to know what the bronzes looked like when they were in use. Study of historical texts has proven to be difficult and fraught with problems, even in the case of modern outdoor sculptures (Weil 1977; Weisser 1992). Research is much more difficult regarding ancient Chinese texts, where the names of the metals have changed over centuries, and the interpretation of a 2,000-year-old language presents problems. It would be best to have some confirmation and indication of possible surface appearances from technical studies.

Trying to figure out what the bronzes looked like when they were made, and attempting to draw convincing inferences from the physical evidence that remains, is a stimulating intellectual exercise. To quote from a source in English rather than one in Chinese (Bowen [1959] 1989:45; Chase 1991:36):

> To re-create, even for an instant, what is laid low, dishevelled, or altogether gone into thin air is exciting. The study of anything that has disappeared is a call on faith. What is totally vanished raises peculiar questions.

Also of interest are the processes used to make and finish the bronzes, both from the view of the history of technology and from a more utilitarian perspective. Some of the bronzes are amazingly corrosion resistant, and knowing how to reproduce them would be very useful. The quality of some of the castings and their finishing is very high; casting and finishing methods used by the ancient Chinese can be adapted for modern use.

The study of corroded ancient Chinese bronzes is also of great interest to the study of long-term corrosion phenomena. Learning more about the processes of deterioration can enable conservators to control these processes in modern applications and to assess the deteriorated metal objects in museum collections. This paper focuses on the manufacture of ancient Chinese bronzes and the processes of their deterioration and corrosion. (For a discussion of the cleaning and stabilization of Chinese bronzes, see Bassett and Chase, herein.)

Terminology

The Chinese bronzes of concern here are ancient vessels, weapons, and mirrors from the beginnings of bronze metallurgy in China, dating from about 1500 B.C.E. up through the T'ang Dynasty. Most of these are corroded, following Gettens's (1969:171) use of the terms *patina* and *corrosion products*:

> Often the collectors' interest in ancient bronzes is quickened by the colored corrosion products or patina of their surfaces. The term patina . . . was used originally to describe the green, blue and red corrosion products of copper and bronze, but by extension is now used to describe any surface on wood, ivory, or marble or other materials resulting from age and exposure. Chemists use the terms "corrosion" and "corrosion products" to describe the salts and oxides that form during burial or long atmospheric exposure, while mineralogists prefer to call them "mineral alteration products." Patina, however, is a good term even if it implies artistic enhancement by corrosion, which is not always the case. Patina . . . connotes not merely a specific color but also a continuous surface layer of measurable thickness. . . .

For convenience and because of its wide acceptance by collectors, we will often employ the term patina; but in purely technical discussions, we will speak as a chemist of corrosion and corrosion products.

In terms of the changes caused by time and their amelioration by conservators and archaeologists, one should also consider terminology from the field of conservation of paintings. Based on the work of Sheldon Keck at New York University, the concepts of original condition, present condition, and actual condition may be adduced. Original condition is a simple concept: the condition of the object when it was new. With bronzes one has a series of original conditions. It's useful to think about this in terms of particular objects. With a mirror or a black sword, there is an "as cast" condition, a finished (polished) condition, a patinated condition, and the condition after use. In fact, mirrors that were in use underwent constant repolishing (Chase and Franklin 1979:239). Bronzes in ritual use in China and Japan today are usually tarnished; even the mirrors in the temples are rarely bright, and bronze *ding* incense burners, etc., are usually less so.

The term present condition indicates what the object looks like at the present time. In some cases, bronzes are much restored or covered with compensating layers of paint. Whole sides or parts can be restored, designs can be enhanced, inscriptions can be added, and so on (Gettens 1969).

Actual condition means how much of the original is left and what its state of preservation is. The mirror from the University of Michigan discussed in this chapter is in a cleaned and repaired state with heavy surface corrosion, much of which has been removed from the decorated back to reveal design. A sample was taken from the rim next to a crack, and the area and crack were compensated before returning the mirror to its owner. Without the restoration, however, the actual state of the mirror is corroded and cracked.

The actual condition is determined by what has happened during the corrosion process. The corrosion process is affected by a number of factors, as follows:

1. The composition of the bronze
2. Microstructure
3. Surface finish, including (a) smoothness, (b) change from structure of substrate, and (c) corrosion-resistant surface coatings
4. Environment, including (a) moisture, (b) pH, (c) access to oxygen, and (d) other corrosive ions, especially Cl^- and $PO_4^{=}$
5. Mechanical and thermal damage
6. Events in recovery of archaeological material, including (a) change in environment and (b) physical damage
7. Later cleaning and stabilization measures
8. Present environment

Casting

Until the Han Dynasty, Chinese bronzes were cast. There is no evidence of hammering bronze sheet at all until the appearance of a riveted bronze sheet repair in the Man-cheng tombs of about 170 B.C.E. (Barnard and Sato 1975:73). While other metals (gold, silver, and later iron) were worked by hammering, bronze was universally cast.[1]

Mold Material and Construction

Usually the casting was done into ceramic piece-molds fashioned from the local earth, which in many areas was typical windblown loess or yellow earth (Gettens 1969:107–14; Freestone et al. 1969; Holmes and Harbottle 1991). Loess has a high content of small, sharp, fractured silica grains and—with some levigation or treatment to separate coarser from finer material—forms the basis for the ceramic technology to create the bronze molds. The inner side of the mold is the surface that makes contact with the molten bronze and defines the decoration and detail of the casting. The mold construction for a *fang ding* is shown in Figures 1 and 2 (Chase 1991:24–25).

Molds were made in pieces so that the inner side of the mold could be directly retouched by the foundryman. This practice, plus replication techniques that became more sophisticated as time went on, forms the basis for the striking detail and sharpness of the decoration on the bronzes. A great deal has been written about ancient Chinese casting processes. Of more interest here are the finishing and patination processes.[2] The fact that nearly all the bronzes were cast into ceramic piece-molds is sufficient information about their manufacture.

FIGURE 1. *Diagram of mold for casting a* fang ding. *The four outer mold pieces surround the model; above the model is the inner core, and below the model is the inter-leg core piece. After the exterior mold pieces were made and fired, they could be used to produce the core pieces, which were then thinned down to make the casting space and lightly fired to drive out moisture. The cores and the mold pieces were then reassembled (insert A) and inverted (insert B) so that the metal could be poured in through the opening for one leg. A cross-sectional view of insert B on line a is shown in Figure 2.*

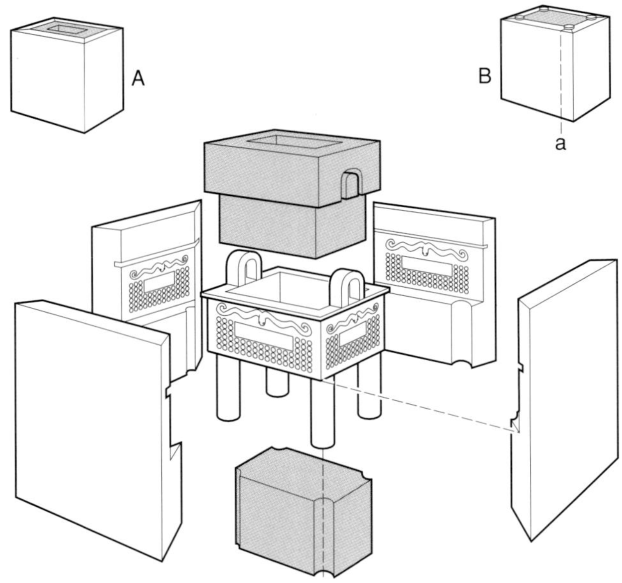

The question of mold temperature at the time of pouring is, however, of interest in the development of the microstructures. While one may suppose that the molds would have been preheated, as is common in modern practice, no evidence has been found of this in the Chinese objects studied. Sections through chaplets (the metal spacers which separate the inner core from the outer mold) usually show a chilling effect in which the molten metal impinges on the cold chaplet (Gettens 1969:132).

One modern technique very similar to that of the ancient Chinese is the Japanese *morokome* process shown in the film *The Enjo Foundry*.[3] Here the outer molds are decorated and fired to a high temperature. The core is then made from an impression taken from the outer mold and is lightly warmed until dry. The final assembly and placing of the chaplets is done just before casting with the molds totally cold. Professor Totsu of the Tokyo National University of Fine Arts and Music has been observed casting an iron teakettle by the *sōgata* (swept-mold) process, in which the iron chaplets were incorporated into the cold-mold assembly just before casting (Totsu, personal communication, Tokyo University, 1991). The chaplets were placed on the inner core of the mold, each on its own small pile of sand. The outer mold was put in place and some pressure exerted. Then the outer mold was removed. The sand had extruded out from under each of the chaplets, showing that each was in contact with the outer mold. Having determined that everything was fitting well, the sand was blown away and the mold reassembled and cast. The whole process of chaplet placement and checking took less than a minute; since the mold was cold, it could easily be handled.

Further discussion with Totsu (personal communication, 1991) leads to the conclusion that there are great similarities between chaplet use in ancient Chinese casting and modern Japanese art-foundry practice.

Another feature of the piece-mold casting technique, and one which is important to this discussion, is that *casting flash*, or mold flash, occurs at the joints where the molds butt up against one another. These are often called "mold marks," but casting flash is the more precise term. The casting flash indicates where the mold joints are, but it is often finished down. Careful examination of the condition of the mold flash

FIGURE 2a, b. *Casting the* fang ding. *The cross section of the mold assembly (a) shows how the pieces fit together. Chaplets (or spacers) are used to make sure that the casting space does not shift or close during pouring. The chaplets in the bottom of the vessel need to be larger, to counteract the upward hydrostatic pressure on the inner core when the metal is poured in. The bronze is poured in through one leg (b). The other three legs serve as air vents.*

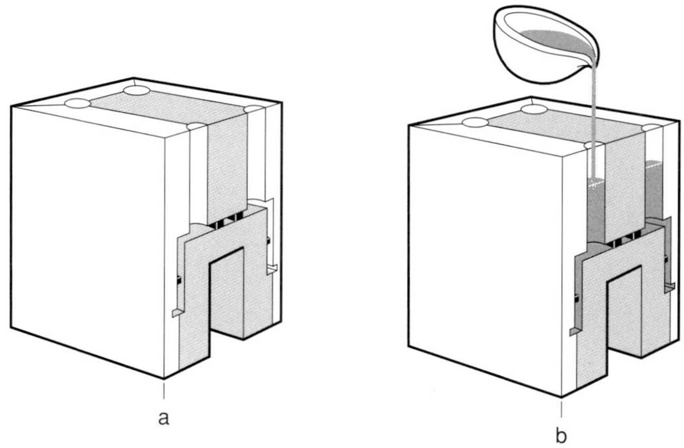

can show where finishing took place. Some mold flash is unavoidable in the piece-mold casting method, and it can be very helpful in technological detective work, as will be discussed further.

Composition

Since the ancient Chinese do not seem to have worked or hammered cast bronzes (in fact, many of the Chinese bronze compositions are difficult or impossible to work by hammering), the only method available to them for changing the properties of the bronze was altering its composition. The *Kaogongqi* section of the *Zhou Li* (The Rites of Zhou), dating from the Zhou or Han Dynasty (ca. 400–100 B.C.E.) mentions six different compositions and their uses, ranging from 14.3% tin for vessels and bells up to 33.3% tin for mirrors. Lead is not mentioned. Every author on Chinese metallurgy has his or her interpretation of this enigmatic text (Chase 1983:108; Hua 1985:511; Chase and Chen 1992). Regardless of its exact meaning in terms of percentages, this primary Chinese source does clearly imply that the way to adjust the properties of bronzes (hardness, ability to hold an edge, reflectance) is to change the composition.

Chemical analyses of the bronzes also demonstrate clearly that this approach was used (Chase and Ziebold 1978; Chase 1983:102ff). Vessels and bells do form a distribution around 15% tin, with lead acting as a diluent. Weapons (except for "parade pieces") are low in lead. Mirrors center around 27% tin and 5% lead. (The reasons for this will be discussed later.) Belt-hooks, a type of object not mentioned in the *Zhou Li,* were clearly made of a high-copper composition when they were intended for fire-gilding (Oddy 1985:70).

Differences in composition show up as differences in the color of the finished alloy. The relationship between color and composition will be discussed after considering the finishing.

Metallographic Structure

The usual structures of Chinese bronze vessels, bells, and many weapons (composition around 15% Sn) show columnar crystals. The chill zones are usually not apparent,

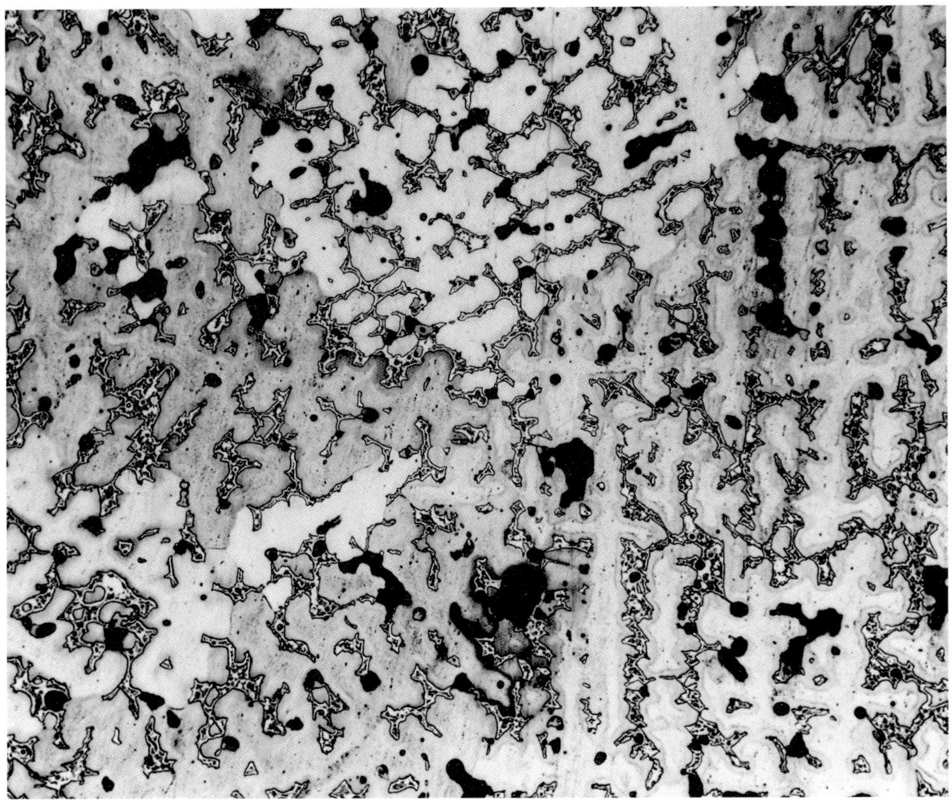

FIGURE 3. *Microstructure of a typical Chinese bronze, ×260, from a cast Chinese bronze dagger.*

confirming the lack of a chilling effect of the ceramic mold. Solidification proceeds according to the phase diagram (Johnson 1973:299; Saunders and Miodownik 1990:279), but the transformations shown below 350 °C do not take place. The low-temperature reactions proceed very slowly, and the normal rate of cooling, even in ceramic molds, prevents the transformation from alpha plus delta to alpha plus epsilon (Gettens 1969:122ff; Scott 1991:25). A typical Chinese bronze structure is shown in Figure 3. The usual alpha plus alpha-delta eutectoid with delta fringe can be seen. These structures have been tested, using both the scanning electron microscope and the electron microbeam probe, and they show considerable segregation of tin, with the highest concentration of tin in the delta fringe surrounding the alpha-delta eutectoid. Larger, thicker bronzes have a more equilibrated structure, and this will be illustrated later in this chapter with micrographs of the bell from the tomb of the Marquis of Cai.

Lead does not really affect the solidification structures; since it does not go into solution in copper, it remains in globules, which can be small and well distributed in low-lead bronzes or large and irregular in highly leaded bronzes. Some of the Chinese bronzes contain more than 50% lead. These have usually been solidified rapidly (Chase 1979:96); the highly leaded alloys were used for casting objects such as coins, which could be cast thin (for rapid solidification) and which did not require good physical properties, such as high tensile strength. In lower proportions, the lead adversely affects the physical properties of the final casting (Chase 1983:105), but also makes the melt less viscous and facilitates casting. Experiments by Li Gunsheng of the Archaeometallurgy Group of the University of Science and

Technology, Beijing, show that the fluidity of the molten bronze is greatest at 13% lead. This is exactly the proportion that was used to cast many of the large and thin-walled Han Dynasty bronze drums of south China (Wu and Li 1988:17).

The as-cast surface of an ancient Chinese bronze must have looked exactly like that of a modern bronze cast into fine sand or a loess mold (I have done both and could see no difference). The metal surface is shiny with a very fine roughness, almost like a finely shot-peened surface. On mirrors (a type of bronze often cast with extremely fine detail), the high points of the surface are often finished, while the low parts are simply left in the as-cast condition. The as-cast parts have the same sort of sheen as a good aluminum paint; they look like finely divided, silvery metal. This may come not only from the nature of the mold surface but also from the ends of individual crystals. One can see a fine pattern of raised dots under high magnification.

The ancient Chinese probably used a mold dressing to make the metal flow well into the mold and reduce the surface tension of the molten metal against the mold surface. Carbon is a traditional mold dressing and can be applied by smoking the mold using a smudgy fire (such as that of a candle with a badly trimmed wick) or by applying the carbon in a finely divided state in water. Chinese ink is used at the Enjo foundry. Remains of mold dressing can often be seen as a layer of black material between the foot core and the bronze in vessels where some core has been left in place.

Finishing

Many Chinese bronzes now exhibit a nice, shiny surface, often composed of tin oxide. Some of the mirrors are either black or silvery in appearance, with very shiny surfaces; mirrors that have a contrast of very shiny and not-so-shiny surfaces on the back have already been considered. In all cases (except those in which the bronzes have been refinished since excavation, and are thus easily detectable), the shiny surface is clear evidence of very good surface-finishing techniques in antiquity. Corrosion will often preserve a well-finished surface, and good surface finishing may also alter or slow the progress of corrosion. Corrosion will not, however, make a nice, smooth, shiny surface where one did not exist before; it can turn a nice, shiny surface into a rough, pitted one, but not the other way around. The nice, shiny surfaces we now see are not only good examples of beautiful patinas from corrosion but also a tribute to the skill of the ancient Chinese craftsmen in finishing the bronzes.

Finishing marks can often be seen on Chinese bronzes, but those that are clearly visible are often in areas not normally seen by the user. Rougher finishing marks can be viewed on the flat or concave bottoms of *ding* vessels. The bottoms of these vessels often have large areas of casting flash where the interleg core piece butts up to the molds for the vessel sides. The flash is often roughly finished off; these areas show deep abrasive scratches. Other areas can be quite smooth and show no finishing marks, even though there must have been mold flash there initially. On a Shang Dynasty *zun* in the Freer Gallery of Art (FGA55.1; Pope et al. 1967:79; Gettens

1969:117), the mold flash must have run down between the F-shaped decoration elements. There was also some mold slippage, so in addition to the mold flash there was a difference in levels between the two sides of the vessel. Finishing had to go so far that some of the F-shaped elements were totally abraded away. On some bronzes, notably a *chi* broadaxe in the Freer collection (FGA46.5; Brown et al. 1977:214), the metallographic structure can be seen, preserved in the corrosion products, right up to the cutting edge.

In fact, a lot of Chinese bronzes look just like heavily etched metallographic specimens (Gettens 1969:126), which, along with evidence of rougher scratch marks from finishing in less-visible areas of vessels, points to abrasive finishing. This technique was done with natural stones and grits, just as we do it today with abrasive papers, working down through decreasing sizes of grit. Each successive grit removes the scratches from the previous surface, until the desired degree of finish is achieved. It is a simple technique, though labor-intensive, which was presumably no impediment to the ancient Chinese.

Abrasive finishing must have originated with the lithic industries, particularly jade. Among the late-Neolithic jades from Jiangsu, near Shanghai, hundreds of jade ornaments were found in one tomb alone; their manufacture must have required a prodigious expenditure of effort in carving and polishing. The ancient Chinese also polished agates in the Neolithic period, as recent discoveries from Hefei attest. The surface polish on the jades was important not only for obvious aesthetic reasons, but also to show off fine so-called scratched designs (Bagley 1980:73). While there is no direct evidence from the ancient Chinese as to how long jade polishing took, we do have one interesting bit of information from the premetallic culture of New Guinea, where the author (Harrer [1963]1964:128, 142ff) attests to having seen jade axes ground and finished by primitive methods. One axe took three months to sharpen.

The results of abrasive finishing are clearly shown on a *dui* (FGA32.13) from the fifth century B.C.E. This piece is said to have been unearthed at Li-yü in northern Shansi province (Lawton 1982:32; Pope 1967:528), and it shares with other Li-yü bronzes a nicely preserved surface (Chase 1991:60). The mold mark occurred halfway between the handles, and seems to have been very well finished indeed (Fig. 4a). Were it not for the dislocation in the decoration band, this feature would not be visible at all. The scratches on the body below the decoration might also be a result of abrasive finishing, although they are fairly rough. Figure 4b shows an area on the body near one handle. The shininess of the metal is clearly visible, as are the finishing scratches running vertically parallel to the side of the handle.

The inside of the *dui* is a beautiful bronze color (Fig. 4c). As Gettens says, "On the inside the bronze surface has an untarnished and satinlike finish, as it must have had when it left the hands of its maker" (Pope 1967:531). The satin finish is the result of directional scratching that runs parallel to the major axis of the vessel on the bottom and around the vessel in the area just within the rim. The visual effect is to make the vessel appear to be stepped on the inside, although it is actually quite smooth. In some areas, the fine scratches appear under the edges of the corrosion in such a way as to indicate quite strongly that the shiny surface is not the product of some cleaning process. It is also of interest that the majority of the interior corrosion occurs in the bottom of the vessel, just where water from condensation would have run down. Drips and streaky corrosion, possibly from condensation, are fairly common occurrences on ancient Chinese bronzes, especially from the Li-yü and Hou-ma areas. In any case, this bronze may be a good example of what these vessels originally looked like, although the difference in color between the pewterlike outside (Fig. 4d) and the warm yellow-bronze inside gives one pause. Perhaps it was plated or pickled on the outside; this would provide an interesting subject for later study.

In terms of the evolution of finishing techniques, the earlier Shang (ca. 1100 B.C.E.) and perhaps pre-Shang (ca. 1500 B.C.E.) vessels appear to be rougher than the Zhou vessels (ca. 1000 B.C.E.), with the height of perfection being attained at the time of the late Eastern Zhou (ca. 350 B.C.E.) when color of bronze, inlay, gilding, cast iron, bronze mirrors, and other bronze objects all came into prominence (Chase and Franklin 1979:256). Production of marvelously finished, colored, or inlaid bronzes continues into the early Han period (ca. 150 B.C.E.), when reaction against such "conspicuous consumption" occurred (Gale 1931:14–15).

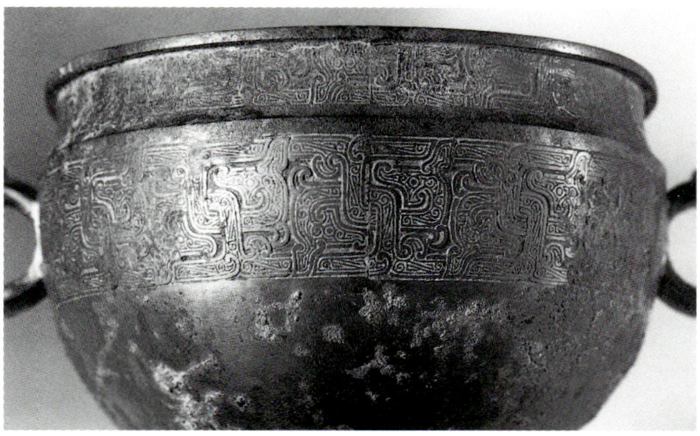

a

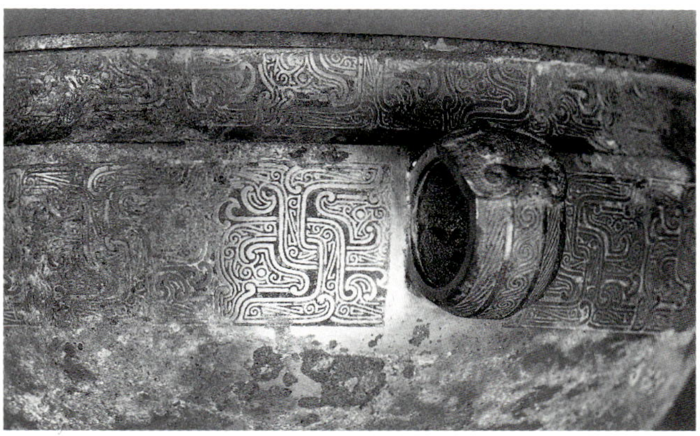

b

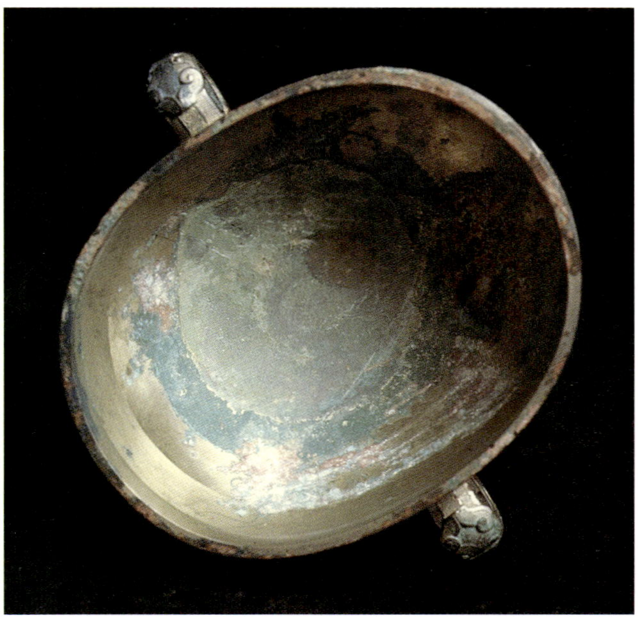

c

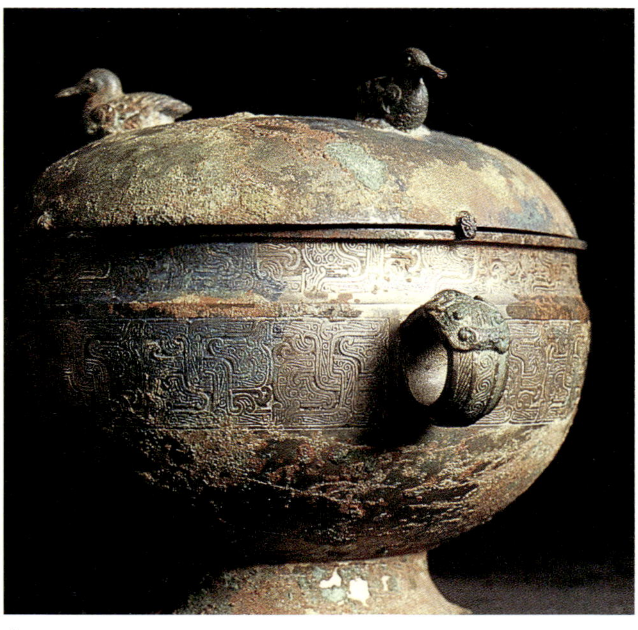

d

FIGURE 4a–d. *Four views of an Eastern Zhou (ca. 450 B.C.E.)* dui, *FGA32.13: (a) the vessel side, midway between the handles, shows the mold mark; (b) finishing marks run vertically next to the handle; (c) a color view of the interior shows the shiny yellow color of the metal as well as the effect of controlled surface scratching in the finishing; and (d) an overall view of the vessel shows the grayish color of the exterior surface.*

Appearance, Color, Tarnish, and Corrodibility

From the examples above, it is clear that at least some bronzes would have had the color of the finished metal alloy. In the finished (but not patinated) state, the color of the bronze depends solely upon its composition. (Heat-treated alloys do not seem to have been used in early periods in China, so the beta bronzes, heat-treated from 700 °C, do not enter into the picture, nor do other recrystallized bronzes.) The color of the finished metal was the way the ancient Chinese controlled their alloys. A short digression into color and appearance may be useful at this point and draws heavily on previous work (Johnston-Feller n.d.; Chase and Veloz 1989).

It is well known that the appearance of the metal depends on its degree of finish. Consider a flat piece of metal taken to a mirror finish, as in metallographic polishing: it will cleanly reflect an incident beam of light. The incident beam and the reflected beam will each form the same angle perpendicular to the surface. The reflected beam will not be scattered but will reflect off the surface as a perfect, clean image of the incident beam. The closer metal objects come to a perfect mirror finish, the more perfectly reflective and deeper they look.

The reflected beam does, however, change in color. The color change comes from the fact that the absorption index of the metal varies with wavelength. At the metal-air interface some wavelengths are reflected more than others. One can see that, in the case of a perfect mirror surface, there will be no diffuse reflectance. One must measure the color of the specularly reflected beam.

Most colorimeters are set up with geometries (such as 45°/0° illumination/viewing angles) and cannot include the specularly reflected beam. The work presented here was done on a Kollmorgen Color Eye, with an integrating sphere that can measure either total reflectance with specular component included (SCI), or diffuse reflectance with specular component excluded (SCE). The specular reflectance (S) can be determined by subtracting the diffuse reflectance from the total reflectance (S = SCI − SCE). For a perfect mirror, the ratio SCE/S is zero.

The colors of ten samples of bronzes supplied by various workshops (Chase et al. at the Freer Gallery, Rob Pond in Baltimore, and Harold J. Plenderleith at the British Museum) were measured. The compositions, tristimulus values, values in Munsell notation, Sn/Cu ratio, and Y_{SCE}/Y_S are shown in Table 1 (Johnston-Feller, n.d.). Y_{SCE}/Y_S is used for the SCE/S ratio mentioned above (see note to Table 1).

Color is directly related to tin content, as can be seen in Figure 5, where the Munsell hue is plotted against the Sn/Cu ratio. Two other studies of colorimetry of bronzes agree in the shape of the spectral-reflectance curves (Guiver et al. 1988) and in the dependence of color on the tin-copper ratio (Uchida et al. 1988), although the dependence is not so clearly demonstrated. Lead seems to have no effect on hue, as can be seen from the position of sample H, an extremely highly leaded bronze (Pb 32.5%). The lead does lower the chroma, although the relationship is not as clear. We hope to redo this work and extend it to include arsenical bronzes and brasses as well.

TABLE 1. *Composition and colorimetric properties of various bronze alloys (data from Johnston-Feller n.d.).*

			Total Reflectance					*Munsell Notation*		
Sample	*Cu/Sn/Pb*	*X*	*Y*	*Z*	*x*	*y*	*Hue*	*Value/Chroma*	*Sn/Cu*	Y_{SCE}/Y_S*
A	100/0/0	56.96	51.57	36.85	.3918	.3547	3.01YR	7.55/5.44	.00	0.277
B	90/7/3	69.34	67.73	54.31	.3623	.3539	7.73YR	8.47/3.56	.08	0.052
C_1	80/10/10	69.85	69.23	57.18	.3559	.3527	8.49YR	8.54/3.15	.13	0.064
C_2	80/10/10, old	65.33	64.49	51.89	.3595	.3549	8.47YR	8.30/3.35	.13	0.125
D	77.5/22.5/0	63.15	63.58	55.86	.3458	.3482	0.19Y	8.25/2.38	.29	0.028
E	76.5/12.6/9.6	63.19	63.63	56.48	.3447	.3471	0.09Y	8.25/2.31	.17	0.094
F	75/15/10	63.35	62.89	51.22	.3570	.3544	8.79YR	8.21/3.17	.20	0.077
G	70/27.5/2.5	61.21	62.19	62.86	.3286	.3339	0.91Y	8.17/1.22	.39	0.038
H	65/2.5/32.5	66.44	63.52	53.44	.3623	.3463	4.59YR	8.24/3.64	.04	0.126
I	71/26/3	63.18	63.99	63.12	.3320	.3363	0.28Y	8.27/1.45	.37	0.126
J	90/10/0	65.33	63.88	50.58	.3633	.3552	7.82YR	8.20/3.63	.11	0.020

*The ratio of the tristimulus value (Y) for the measurement made with the specular reflectance excluded, SCE (diffuse reflectance) relative to the specular reflectance (S), obtained by subtraction of the measurement made with the specular component excluded (SCE) from the measurement made with the specular component included (SCI). Thus, S = SCI − SCE, and the ratio indicates how near to a perfect mirror the sample appears to be; if perfectly mirrorlike, the ratio would be zero.

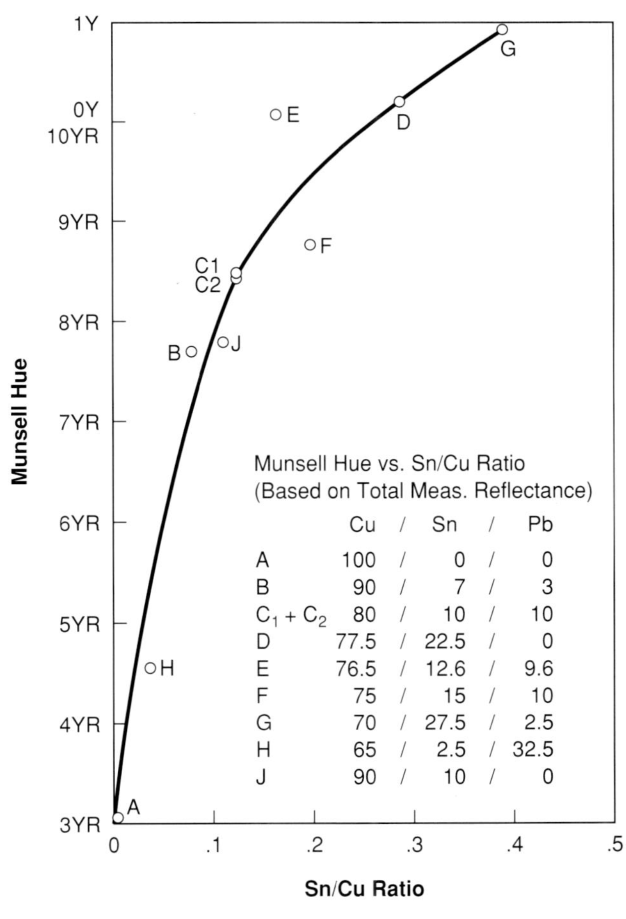

FIGURE 5. *Colors (Munsell hues) versus Sn/Cu ratio for various polished tin bronzes (Johnston-Feller n.d.).*

Most of these samples were polished at the same time—a week or so before the test—and then wrapped in tissue and stored in cardboard boxes for transport to Pittsburgh, where the tests were done. It is interesting to look at the plot of the Y_{SCE}/Y_S ratio against the Sn/Cu ratio (Fig. 6). While this is not a well-controlled experiment in terms of tarnishing rate, there does seem to be a relationship between the tin content and the Y_{SCE}/Y_S ratio, which can be used here as an index of the corrosion or surface tarnishing. Samples C_2, J, and I were repolished just before the test in Pittsburgh and have been excluded from the plot. The data are much more scattered than one would like, but it does appear that the higher-tin bronzes tarnish less rapidly. The pure copper (sample A) tarnishes much more rapidly than any of the others; this was the only change in surface appearance on the samples which was clearly visible to the naked eye. This method can detect growth of tarnish films in a very sensitive way. It is, in fact, so sensitive that it was used as the best indicator of the stability of organic coatings on copper alloys (Copper Development Association 1974:3).

Polarimetric Curves from Robbiola

The finding that tin bronzes are less corrodible with increasing tin content is also borne out by the pioneering polarization measurements on bronzes in aqueous conditions by Luc Robbiola (1990). In his doctoral thesis he reviews the current state of knowledge of the long-term corrosion of copper alloys, analyzes in detail the corrosion on bronzes from the French archaeological site of Fort-Herrouard, publishes some important fundamental corrosion experiments on bronze alloys, and propounds a new theory to explain archaeological corrosion (Robbiola et al. 1988). Looking first at the evidence for corrodibility, Robbiola was not able to find polarization curves for the copper-tin bronzes in the literature, so he performed the experiment, both by standard direct-current (dc) polarimetry and alternating-current (ac)

FIGURE 6. *Corrodibility as shown by Y_{SCE}/Y_S index for bronzes (date from Johnston-Feller 1991).*

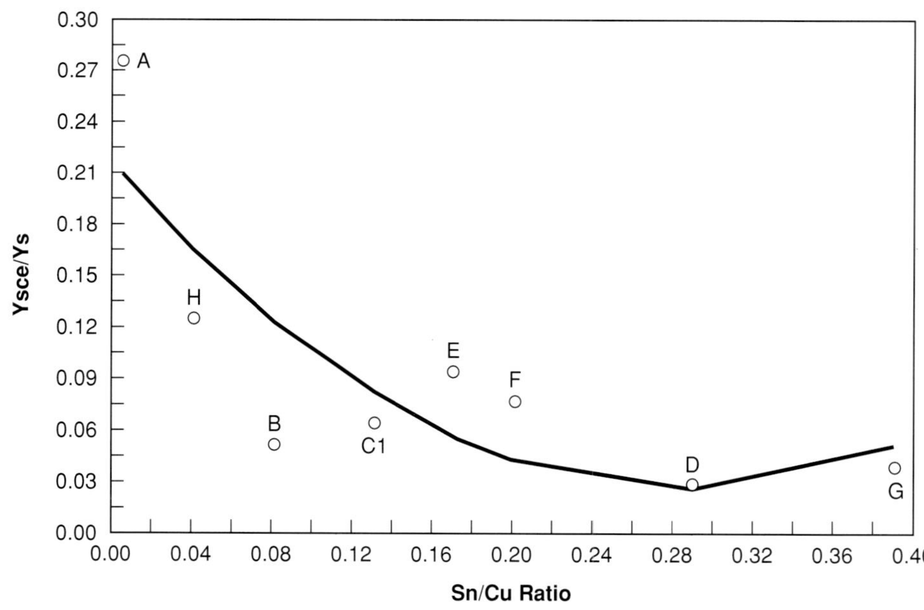

impedance measurements. Pure copper and tin were used, along with two bronzes with two different heat treatments, as shown in Table 2.

The polarization curves are very interesting. "The bronzes show an overvoltage for oxidation with regards to pure copper: the film previously formed on the surface is more resistant to corrosion than the film of cuprous oxide on pure copper" (Robbiola 1990:160). The heat treatments, though low enough for no visible changes to have taken place in the microstructure, do make a difference on the polarization curves, with the chill-cast bronzes being more resistant to corrosion. As one would expect, the presence or absence of air (dissolved oxygen in the solution) makes a large difference in the polarization curves. The presence of chloride has a destabilizing effect, especially on the cuprous oxide film. The corrodibility of the alloys also changes, with the chill-cast samples corroding more rapidly than the reheated ones.

The most important aspect of Robbiola's thesis is his emphasis on ion migration as the unifying theme in corrosion investigations. He states that conservators should not be concentrating on chemical reactions nor on environment, microstructure, or any of the conventional factors previously discussed, but rather on ion migration, transport, and movement. This is a very powerful concept, and it is probably correct in its emphasis.

Cuprite Marker Layers on Copper Objects

Many earlier studies confirm that the tarnish film in atmospheric corrosion of pure copper is Cu_2O, or cuprite. The other bronzes in the colorimetric study discussed here (in which tarnish was not detected visually) may be growing mixed or substituted copper and tin oxides. Those in Robbiola's polarimetric study are certainly growing films other than pure Cu_2O.

The phenomenon of a marker layer of cuprite, Cu_2O, or copper (I) oxide lying at the original surface of the antique metal is well known. A particularly good demonstration in this study came from fibulas and other small bronzes that were received as surplus material from an excavation in the Near East (Gettens 1970:72; Chase 1974:52). On these objects, a nice, compact dark cuprite lies where the original surface was. These are recrystallized 11% tin bronzes, and the marker layer appears to be cuprite. The phenomenon is also seen on Classical bronzes (Born 1985a),

TABLE 2. *Bronzes used for corrodibility studies by Robbiola (1990).*

	Cu	*Sn*	*Pb*	*S*	*Heat Treatments*	
Cu7Sn	91.5%	7.0	1.0	0.5	a:CuSn7 b:CuSn7	chill-cast reheated 1 hour @ 490 °C
Cu14Sn	84.5	14.0	1.0	0.5	1:CuSn14 2:CuSn14	chill-cast reheated 1 hour @ 490 °C

Egyptian bronzes (Eichhorn 1985), bronzes from the sea (Formigli 1990; although the layers here may be copper sulfides), Chinese bronzes, and many objects of pure copper. Cuprite marker layers may be used as evidence for the authenticity of the object (Eichhorn 1988). On modern metal, the phenomenon has been much studied because of the use of copper for electrical conductors.

All the evidence here suggests that the cuprite marker layer is initially formed by the oxidation of copper in (moist) air. Studies have shown that the apparent cuprite layers are not stoichiometrically Cu_2O, but actually $Cu_{2-x}O$, a lattice with holes. This material is a semiconductor (McNeil and Little 1991) and can act as a membrane for transport of ions across the layer, $O^=$ and Cl^- inward and Cu^+ (or Cu^{++}?) outward. While the evidence is not yet complete for the transport of Cl^- inward across the cuprite layer, one may well find that this is a convincing explanation for some of the phenomena observed, especially in bronzes with very even layers of cuprous chloride.

One should, however, look a little further to explain the transport of Cl^- ions deeply into the bronze substrate. This recalls the idea propounded by Smith (1975, 1959) concerning the transport of chloride being driven by differences in the interface-energy balance: If the Cl^- bearing material has a lower interfacial energy than that of cuprite or the metal, it can drive itself inward between the two just as a liquid metal (such as mercury or molten tin) will drive itself between the grains of cast aluminum, rendering the cast aluminum totally brittle in a very short time. This is consistent with the ideas of MacLeod (1981; see also Scott 1990).

It is important to carefully consider the formation of the cuprite layer. In many cases the layer is formed by initial tarnishing and may have even formed while the object was actively in use. It also forms in underground burial chambers, especially where the tomb has not caved in and the vessel has been sitting in moist air. In atmospheres containing sulfides (at about 100 ppm or above) the cuprite layer will not form. Discussions with Joe Payer (personal communication, Case Western Reserve University, Cleveland, 1991) indicate that if one puts an object with a cuprite layer into these atmospheres, the cuprite layer will be disrupted and eventually destroyed.

The model of corrosion that is proposed for pure coppers is as follows: first, the formation of a marker layer of cuprite on the surface of the copper; second, ionic transport through this membrane, with anions migrating inward and cations outward. The membrane remains intact in the absence of disrupting agencies (mechanical damage, sulfides, etc.) and serves to delineate the original surface of the object.

The question of intentional patination versus natural corrosion should also be raised here. Cuprite patinas are used (though rarely) on artistic objects. Two particular examples come to mind: a nineteenth-century Turkish water jug in the British Museum with a bright-red (almost cherry-red) patina, and the bronze maquettes in the Gorham Foundry in Providence, Rhode Island, which were finished around the turn of the century.[4] Formulas for reddish (probably cuprite) patinas are given in Hughes and Rowe (1983). Cuprite patina has also been used in the protection of an outdoor statue in Poland (Socha et al. 1988). If the cuprite patina is formed by chemical or electrochemical techniques, it may not be distinguishable from a natural corrosion product. The question of intentional patination versus corrosion by natural agencies is a recurrent one; this can be a difficult problem and must be dealt with on a case-by-case basis.

Marker Layers on Other Bronzes

The cuprite marker layer will probably form more easily (and thickly) on the higher-copper alloys in normal indoor atmospheres. In regard to the bronzes and leaded bronzes, one may logically assume that the single-phase structures (up to about 6% Sn) will exhibit the Cu_2O uniform marker-layer growth. Above this, where duplex structures are present, the films could be less coherent and probably consist of patches of Cu_2O and Cu_2O-SnO_2-mixed oxide over the alpha and delta areas, respectively. This may be a less corrosion-resistant structure. In looking at cross-sections of bronzes with both a duplex structure and a marker layer, however, one does not see any evidence of differences in the marker-layer formation over the alpha and delta phases where they lie on the surface of the object.

The experiments of Robbiola on polarimetry also show that coherent and corrosion-resisting films *are* forming on the higher tin alloys. Since these oxides are difficult to analyze by X-ray diffraction, and since they form very thin films, one does not yet have the quantitative data to predict the behavior of the films or even to see what their composition is. Perhaps a combination of X-ray fluorescence, Moessbauer spectrography, and colorimetry combined with electron optical techniques will provide the tools necessary for understanding these films. Serious questions exist about what is happening on the surfaces of these alloys and whether the oxide formation has components of sideways migration; one should not forget the migration of sulfide corrosion products across the surfaces of objects and even through holes in other materials.

When examining the 22% Sn hot-forged and quenched bronzes, or even the Japanese *sahari*, one finds that these are more corrosion resistant. Increased corrosion resistance may be coming from the structure of these alloys, which are mostly composed of fine needles of beta or beta prime (Goodway and Conklin 1987:8, 9; Scott 1991:95, 98); it is possible that a surface-oxide film of mixed oxides forms. This film is much more corrosion resistant than a film formed on the bronzes containing less tin.

This finding raises the concept of equipotentialization of the surface, meaning that corrosion is driven by differences on the surface that can be reflected in potential (voltage) differences when the object is immersed in an electrolyte (Gettens 1969:182). If one can equipotentialize a surface—for instance, by putting a perfect tin plating on it—the corrosion will have no differences in potential to attack, and the corrosion will slow or stop. This is also the concept of passivation of the surface by oxide films as discussed by Pourbaix (1966, 1973), and works for high-quality paint and lacquers as well. One now-classic method of equipotentialization is to make amorphous or glassy metals; in this technique the corrosion rate drops off dramatically.

In the case of the hot-forged and quenched high-tin bronzes, it seems that, in going from an alpha plus alpha-delta eutectoid structure to the beta needles, one is

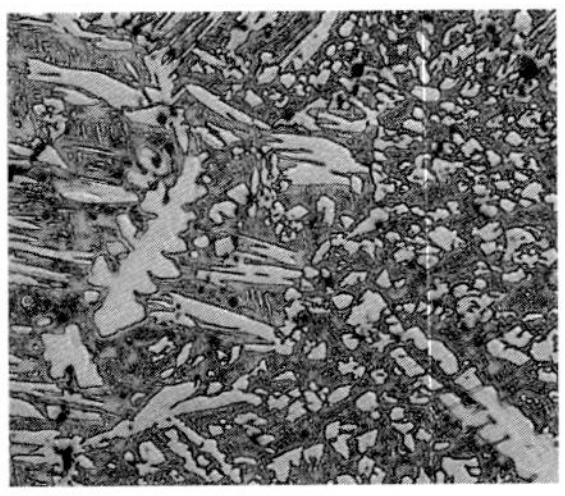

a

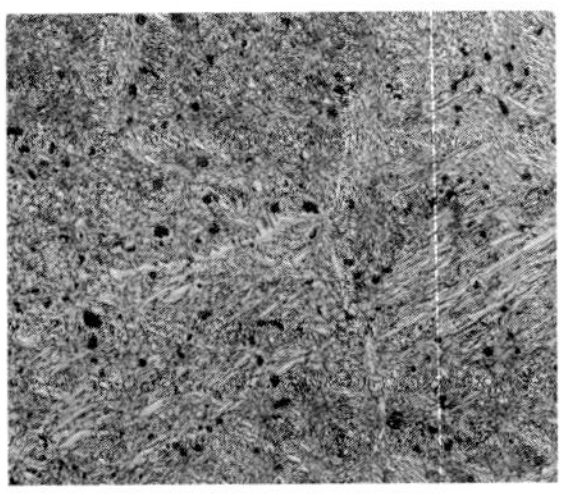

b

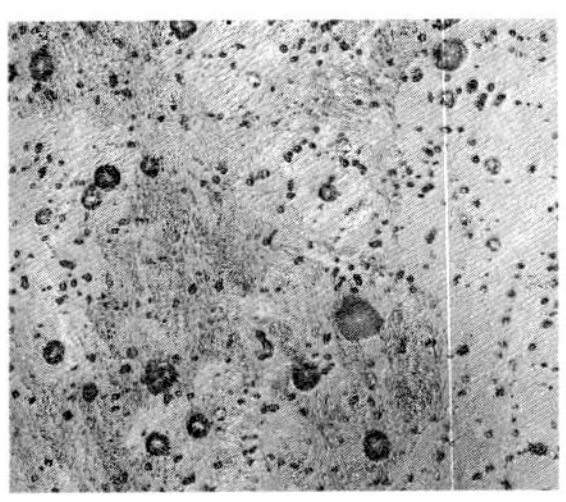

c

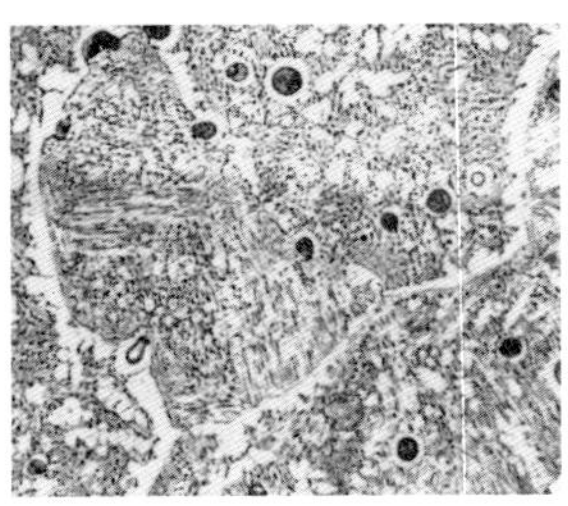

d

FIGURE 7a–d. *Microstructures of sand-cast bronzes with 22–26% Sn. All contain 3% lead, and all of the metallographs were taken in bright field at a magnification of ×320 after etching the sample with potassium dichromate. (Magnifications above are ×112.) Note remains of grain boundaries, particularly in c and d.*

taking a step toward equipotentialization; the corrosion rate dramatically slows. These alloys seem to form a natural, black-tin-oxide surface film that passivates the surface and inhibits further tarnish. It has been proposed that these surface films were put on intentionally, and some of them may have been. Others appear to have corroded naturally. A covered Korean high-tin bowl in the collection of the Honolulu Academy of Arts has the typical black appearance that could have been intentionally applied; there is, however, a rather distracting slanted line where the cover meets the bottom. This is probably evidence of corrosion during burial rather than artificial corrosion (intentional patination).

Turning to examples of alloys with an even higher tin content, the Chinese mirrors contain around 26% Sn and variable amounts of lead (about 1% to 6%). Twenty-six percent of tin is only 1.6% above the eutectoid transformation (24.6% Sn, 586 °C) where beta decomposes into alpha; it is also only 1% below the eutectoid transformation (27.0% Sn, 520 °C) where gamma decomposes into alpha and delta (Saunders and Miodownik 1990). The microstructure is extremely sensitive to small differences in the tin content, as can be seen in Figure 7. The same types of structures were seen by Mooradian and Gettens (1936) in the 1930s and called by the prevailing terminology *hypo-eutectoid*, *eutectoid*, and *hyper-eutectoid* structures (Chase and Franklin 1979:216ff3). The eutectoid can be extremely finely divided, with a composition of 26% Sn (from SEM and EPMA determinations). The tin differences may be beyond the sensitivity of many chemical techniques of analysis, but they are clearly visible in the microstructures shown in Figure 7a–d, as follows:

(a) 75% Cu, 22% Sn, 3% Pb; structure alpha + alpha/delta eutectoid. The alpha is large and light grey. Remains of a dendritic structure can be seen.
(b) 73% Cu, 24% Sn, 3% Pb; structure alpha/delta eutectoid + alpha. The alpha is much smaller and appears to have formed as a result of the eutectic phase transformation. Grain boundaries are partially outlined by alpha.
(c) 72% Cu, 25% Sn, 3% Pb; structure alpha/delta eutectoid. The structure of the individual alpha and delta is so fine that it is difficult to resolve it in the optical microscope. Lead (the dark dots) appears to be more prominent in this section.
(d) 71% Cu, 26% Sn, 3% Pb; structure delta + alpha/delta eutectoid. Delta can be seen at the edges of the old grain boundaries, as small dendrites forming around the lead globules, and within the eutectoid.

Remains of the grain boundaries from initial solidification can be seen on all four metallographs, most clearly on Figure 7d where the original grains are now delineated with a line of the white delta phase. At this magnification, three or four grains appear in the field, with five showing up in Figure 7c.

One wonders if the presence of the finely divided eutectoid contributes to the corrosion resistance of the Chinese mirrors. Work has been proceeding on the corrosion phenomena on these alloys, and a particularly stable form of the mixed oxide

($Sn_{1-x}Cu_xO_2$, where x = about 0.16) has been found to exist on some ancient Chinese mirrors (Wu et al., in press). In other words, these alloys are corrosion-resistant because of the formation under normal atmospheric conditions of a particularly stable form of the mixed oxide on the surface. It is possible, although not yet proven, that a similar stable oxide film could be produced by intentional patination techniques. Experiments are proceeding. One should note that these experiments *must* be done on the correct alloys. Even a few percentages of difference in tin will drastically alter the behavior of the alloy.

If one moves on to even higher tin alloys, speculum metal (32.5% Sn) is essentially pure delta and is also quite corrosion resistant, probably also due to formation of a stable oxide film.

Ion Migration: A Unifying Theme

At least some information now exists on the behavior of cuprite films in terms of ion migration, and one can, for the moment, postulate that there will be movement of cathodic ions outward and anodic ions (oxygen and chloride) inward. However, one cannot conclusively say much about the tin-oxide films and even less about the mixed-oxide films, which may be more common on bronzes than previously suspected. For the purposes of this paper, it seems reasonable to assert that ion migration through tin-oxide films will be much slower and less prevalent than through cuprite films.

Robbiola (1990) places a great deal of stress on considering ion migration as the unifying theme with which to examine corrosion processes. To state his qualitative model simply, he believes that corrosion starts by formation of a layer of tin oxide on the higher-tin bronzes, followed by corrosion controlled by either of two processes:

1. In cationic control, the cations diffusing to the surface control the rate of the process, which is generally slow and leads toward compact corrosion layers maintaining the object's shape. Thicker layers, especially of copper II compounds, may form on the outside.
2. Under anionic control, the process proceeds with more change of volume at the corrosion interface; thicker, less coherent corrosion layers may be the result, with the loss of the original shape. Chloride plays an important role in many instances of the second process.

The idea of concentrating attention on ion migration is a powerful one and may help to explain some phenomena hitherto not understood. To be fully applicable to Chinese bronzes, however, this approach also needs to be coupled with some understanding of the initial corrosion films.

One phenomenon seen fairly often on Chinese bronzes is the formation of fibrous malachite, as shown on a Shang Dynasty *gu* in the Freer Study Collection (Fig. 8). This looks just like a cuprite plus malachite analog of electrolytically deposited copper, the same long needlelike morphology that occurs when making copper electrotypes. The Cu^{+} ions migrate outward through the superficial cuprite film and are further oxidized to Cu^{++} when they get closer to the air, continuing to travel in a direction generally normal to the original surface. This may explain the formation of fibrous malachite as being similar in terms of ion transport to electrodeposition. In this case, though, the ions are moving outward from the surface and not inward toward it.

The second phenomenon—which needs a much more comprehensive look—is that of the corrosion of outdoor bronzes. The lack of cuprite layers due to the destruction of the cuprite by sulfides coupled with ion migration by dissolution of the copper in rainwater is illustrated by the condition of a bell erected outside of Washington's Union Station in 1981. The bell has a definitely silvery sheen. When observed in detail (Fig. 9), it is apparent that after ten years of outdoor exposure, the high-copper centers of the alpha dendrites are dissolving away and the products are being removed by the rain, leaving a finely etched crystalline surface. Other outdoor sculptures, in statuary bronze rather than in bell metal, sometimes show crystalline structures in their corroded surfaces, especially on the flatter planes where rainwash is not so rapid. The more rapidly flowing solutions tend to wash away a greater proportion of the corrosion products and promote a more general attack. It may be useful to go back and examine some models of corrosion of outdoor sculptures with ion migration (and solubility effects) in mind.

Chinese Bronze Corrosion: Type I

The normal type of corrosion seen on Chinese bronzes, or at least what we think of as normal, is the removal of the tin-rich phase to form tin oxide or mixed tin-oxide and cuprite corrosion products on the surface (sometimes with malachite or azurite being formed as well[5]), leaving behind the alpha phase of the alloy as the last to corrode. This sort of corrosion is well documented in Gettens (1969:171ff). It has been seen on many Chinese bronzes, including those from the original Anyang excavations (Carpenter 1933).

This sort of corrosion can be associated with a cuprite marker layer, although it would be desirable to go back and examine more samples before becoming dogmatic on this point. This corrosion type is often associated with chlorides. The question of amounts of chlorides in the burial environment versus the types of corrosion seen

FIGURE 8. *Fibrous malachite on the surface of a Shang Dynasty* gu, *below left, dark field, ×132.*

FIGURE 9. *Surface of a bronze bell, below right, Washington, D.C., Union Station Plaza, after outdoor exposure for ten years. The high-copper centers of the alpha dendrites are being etched away, leaving the high-tin phase behind.*

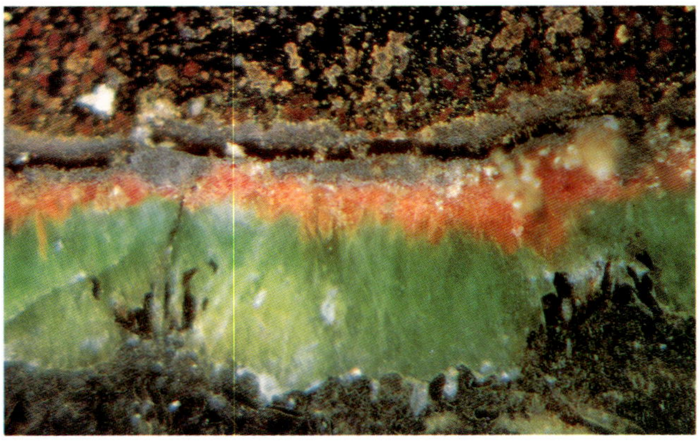

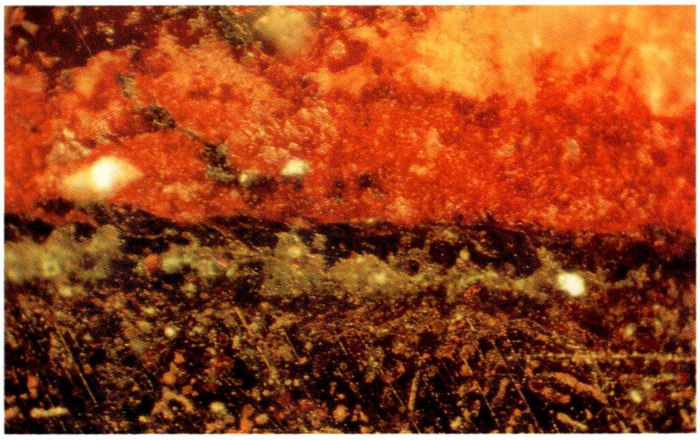

on the objects has not yet been satisfactorily resolved. However, from the examination of samples from the tomb of the marquis of Cai (Fan et al. 1991), along with others from both China and Japan, one concurs with the ideas of MacLeod (1981) that the electrochemistry will drive the Cl^- ions deeply into the bronze. Even a small amount of Cl^- in the burial environment will evince itself in Cl^- ions at the corroding front, deep within the intergranular corrosion cracks in the bronze. Here is an example of Robbiola's "anionic control," and these bronzes normally do not have beautiful tin-oxide patinas. One also wonders if these were patinated in antiquity, or if this sort of corrosion only occurs on bronzes that were simply naked bronze when they were buried.

Chinese Bronze Corrosion: Type II

The other sort of corrosion, often seen on later bronzes and mirrors, can be detected visually by a nice, smooth, so-called water patina of tin oxide. In cross section, one can see that the alpha phase of the alloy has dissolved and the delta remains in an uncorroded state. Type II corrosion corresponds to Robbiola's "cationic control" type of corrosion and can lead to very beautiful and stable bronzes. In some cases, at least, one suspects that these bronzes have been intentionally patinated in antiquity. This is certainly the case with the two-colored mirrors, where the color differences do not manifest themselves in a difference in surface level, but rather in a difference in the color of the surface itself. They are still a mystery (Chase and Franklin 1979:238; Chase 1991:75).

This sort of corrosion is believed to start with a layer of intentional patination, because subsequent corrosion (on cracks or pits) often proceeds as Type I corrosion, with the alpha remaining behind (Chase and Franklin 1979:248ff). One hypothesis is that the bronzes were put into the burial environment with some sort of protective patina layer, which determined the progress of corrosion. Later, when the bronzes broke or cracked due to soil pressure or movement and fresh surfaces were exposed, corrosion took place by more normal means. This hypothesis is strengthened by the fact that the heavier corrosion of Type I often occurs at sharp edges or changes in direction of the surface; corrosion often appears like that on chromed white-metal automotive trim parts, where edges are usually the first to go.

With these ideas in mind, the following case histories will examine some cross-sectional samples from Chinese bronzes to see if more sense can be made of their corrosion structures than heretofore.

Case Histories

Corroded Chinese Bronzes

Bell from the Tomb of the Marquis of Cai

The first example is an excavated Chinese bronze bell from the tomb of the Marquis of Cai, who was buried about 450 B.C.E. in the region of modern Shou Xian, near the Huai River in Anhui province (Institute of Archaeology 1956). The corroding conditions are rather extreme and the burial environment damp; from the evidence of bronze disease, chlorides are elevated, although there are no soil measurements to back this up. Another sample from the same object has already been examined and published in some detail (Wang et al. 1990; Fan et al. 1991) but without extensive optical microscopy.

The bronze is extremely corroded. Even though it is almost a half-inch thick, it can be broken easily. A cross section reveals deep corrosion (Fig. 10). This sample has more transgranular cracking than is normally seen in Chinese bronzes. The microstructure is that of a moderately high-tin bronze, without the coring and segregation in the delta phase (delta fringe) normally seen in sections from Chinese bronzes. Perhaps, due to its size and great thickness, this object naturally solidified more slowly and the metal was able to form a structure closer to equilibrium. A number of corrosion features can be seen—notably redeposited copper—and both alpha- and delta-removal corrosion (Types I and II) are present (Figs. 11, 12).

Intentional patination might explain the Type II corrosion and a cuprite marker layer might indicate the original surface of the object. All four of the bell's available sections (two each, from two fragments) were carefully inspected for these marker layers. Areas that look like original surface were found, but instead of resembling red cuprite, they were composed of two layers, one red and one black (Figs. 12, 13). No analysis of these layers has yet been performed with electron optical techniques; it would not be surprising to find that they are copper oxide and copper sulfide. The layers are badly broken up by subsequent corrosion, and it appears that the protective nature of the surface layers did not last too long, possibly because of the aggressive nature of the burial environment.[6]

Incidentally, black preservative layers were found not only on the bells but also on other bronzes from the tomb of the Marquis of Yi; the bronzes are now in the Museum in Wuhan. The surface layers were evidently reported as being sulfides. With this section of a contemporary piece tending to confirm black layers on Chu bronzes, it would be desirable to go back and open the question of marker layers again. A large amount of redeposited copper is also seen on the section from the Cai

FIGURE 10. *Sample from a bell from the Tomb of the Marquis of Cai, below left. Overall view, bright field, ×4.8, showing deep penetration of corrosion on outside and intergranular cracking running right through the section.*

FIGURE 11. *Same sample as Figure 10, below right, bright field, ×82, showing Type II corrosion at the upper left and Type I corrosion along the deep crack. The outer surface of the sample is upward.*

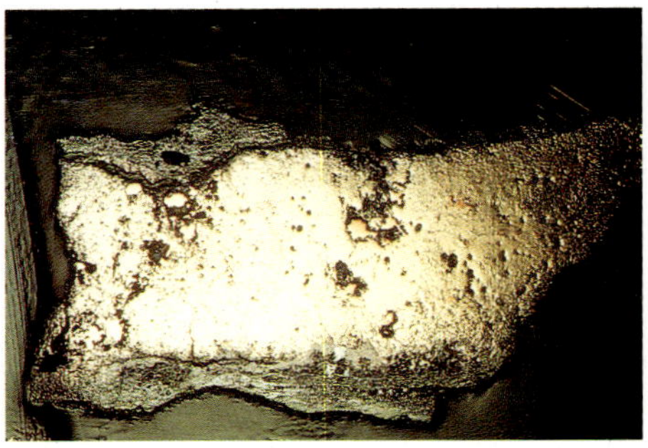

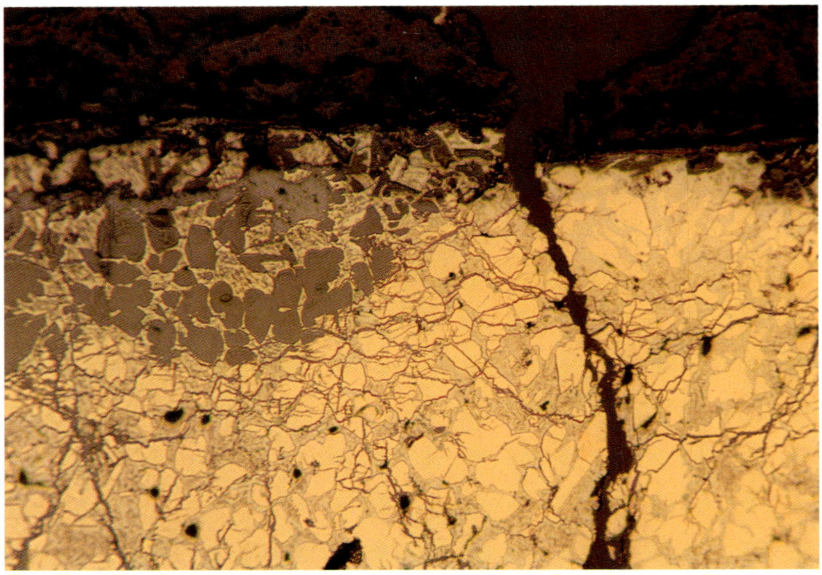

FIGURE 12. *Same sample as Figure 10, bright field, ×88, near right, showing Type II corrosion under a two-part marker layer.*

FIGURE 13. *Same sample as Figure 10, dark field, ×88, far right, showing the marker layer to be red on the top (outside) and black on the bottom.*

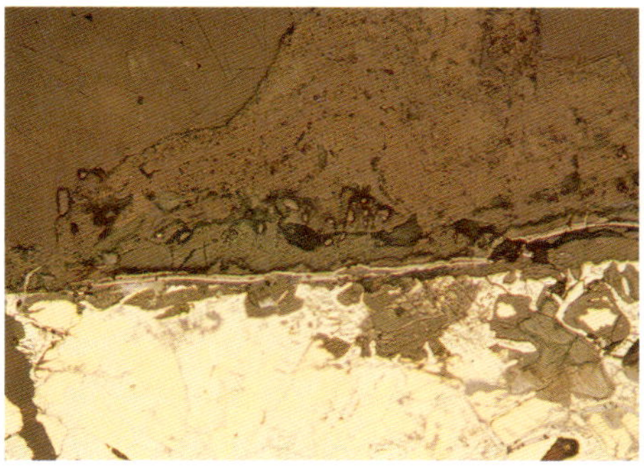

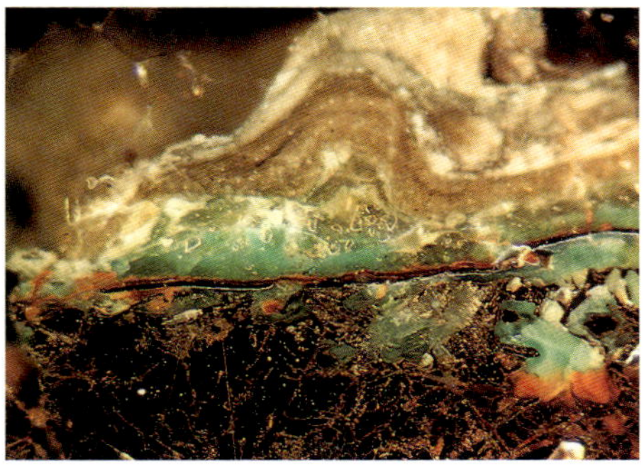

Hou bell. Redeposited copper will be considered below in the discussion of the Michigan mirror.

A Western Zhou *Gui*

The existence of what appeared to be a cuprite marker layer was discovered on a Western Zhou *gui* (FGA24.11) in the Freer Gallery of Art. A sample from the handle shows brazing metal, the metallography of which was well examined by Gettens (1969:135, 181). The Freer Technical Laboratory was interested in reexamining the sample to see if such a marker layer was indeed present. A black layer is clearly evident at the level of the original metal (Figs. 14, 15). The layer has only been identified visually thus far, and one cannot even guess as to whether it is oxide or sulfide. The appearance of the vessel where it has been cleaned, though, argues that it is a cuprite marker layer.

Mirror from the University of Michigan

This typical Sui Dynasty (ca. 600 C.E.) zodiacal animals mirror was so corroded that it was difficult to read the design (Fig. 16). The mirror was also cracked at the rim, and it was agreed with the University of Michigan that the Freer's conservators would clean the mirror if they could take a sample from the rim. After sampling, cleaning was done mechanically using fine-steel chisels, a Vibrotool with a sharp needle, and various brushes, including a small glass-bristle brush. The cleaning markedly improved the readability of the design (Fig. 17).

Although the cleaning was completed about ten years ago, the sample was not examined metallographically until recently. The sample, taken from the thick rim, reveals that the mirror is heavily corroded, with somewhat less corrosion on the reflective face (Fig. 18a, b). The metallographic structure is typical of Chinese mirrors, an as-cast structure of high-tin bronze, probably about 24% tin with some lead.

FIGURE 14. *Cross-sectional sample, below left, from a Western Zhou* gui, *FGA24.11, showing dark marker layer on outside (upper surface), dark field, ×3.7. The marker layer runs straight along the original surface under the overlying thick corrosion products. Area indicated by small rectangle is shown in Figure 15.*

FIGURE 15. *Detail, far right, of Figure 14, showing the dark marker layer under an overburden of red cuprite; dark field (Leitz Ultropak Objective), ×62. Area of magnification is indicated in Figure 14.*

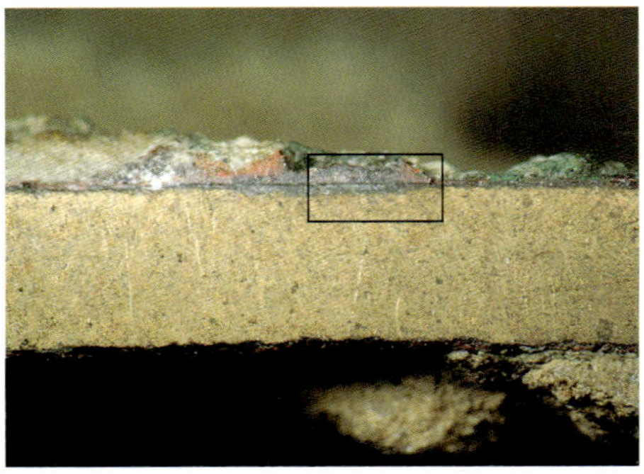

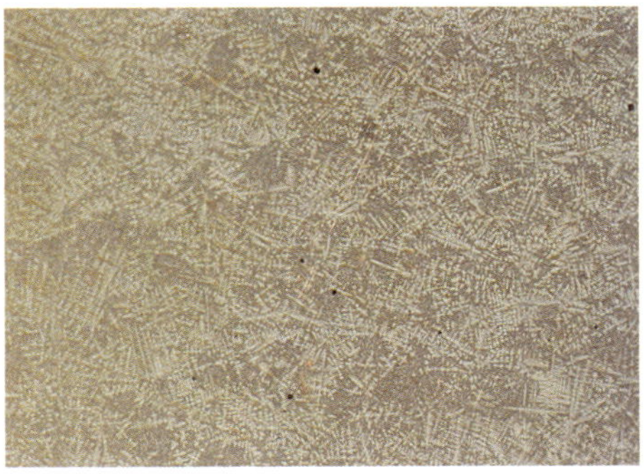

FIGURE 16. *Sui Dynasty Chinese zodiacal mirror from the University of Michigan, showing the decorated back before cleaning.*

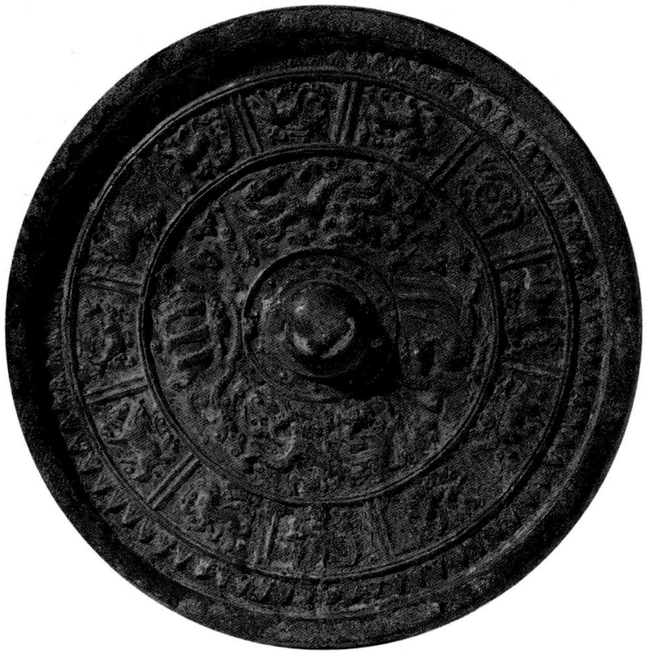

FIGURE 17. *Sui Dynasty mirror, shown after mechanical cleaning. The design can be read much more clearly.*

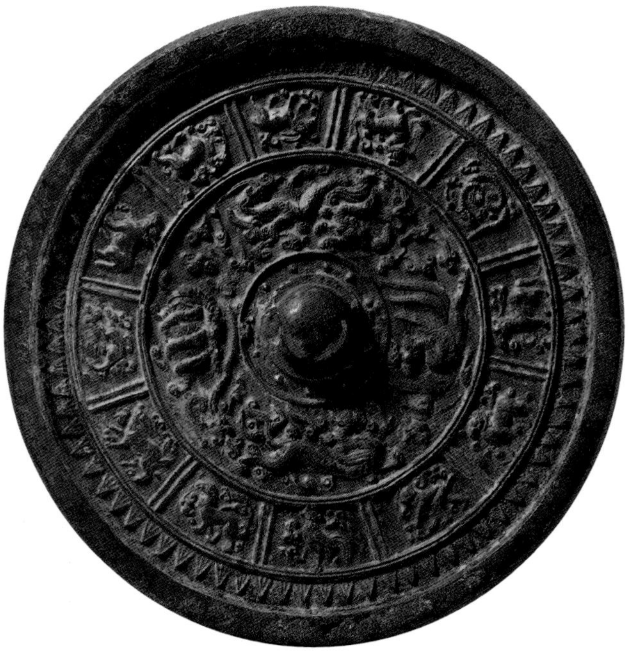

Surface layers can be seen to be preserving the original shape of the piece, with a totally corroded area and an interpenetration zone between the metal and the original surface (Fig. 19a, b). Much of the corrosion appears to be tin oxide, although a cuprite marker layer is seen at some spots on the surface.

The other notable factor on the cross section is the great proportion of redeposited copper in the corrosion zone (Fig. 20). Redeposited copper is present in holes left by corroding lead, throughout the structure as stringers, and also on the surface as a discontinuous surface layer lying at or near the original surface of the metal; in all cases, it is probably replacing earlier corrosion products of cuprite. Undisturbed inclusions of sulfides can be seen throughout the metal, the interpenetration zone, and even in contact with the original surface (Fig. 21). These, plus the ghost structures in the corroded zone, lead to the conclusion that there has not been

FIGURE 18a, b. *Cross section from the rim of the mirror (a) near right, in bright field, ×4.8; (b) far right, dark field, ×4.8. The thinner part of the mirror continues toward the right. The deep but fairly even penetration of corrosion can be seen, along with good delineation of the original surface.*

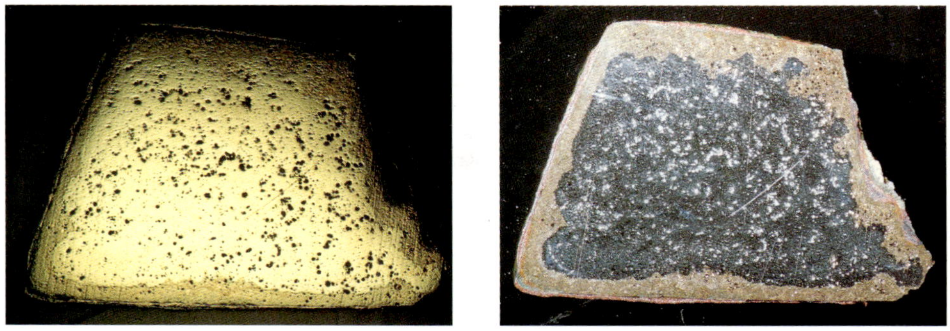

much (or any) dimensional change during corrosion. This mirror is, however, much more heavily corroded than the usual Chinese mirror in the Freer collections.

The presence of the large amount of redeposited copper in this mirror and in the bell from the tomb of the Marquis of Cai suggests a new explanation of the phenomenon. Earlier explanations of "a localized electrolytic process" (Gettens 1969:130), copper prills left over from casting (Barnard and Kennon 1986), and rapid change in the burial environment (Wan 1989) all seem rather unconvincing in light of the prevalence of the phenomenon and its occurrence throughout the sections where it is observed. One possibility may be that this mirror had only a light surface tarnishing when it was buried. Corrosion layers (perhaps oxides or other compounds) would have built up on the surface, and intergranular corrosion would have taken place as well. While there is little evidence of bronze disease on the surface, the underlying layers may contain some chloride. As was previously mentioned, MacLeod thinks that any Cl^- in the burial environment will be forced down toward the corrosion front by electrochemical forces.

Thus, there are two ongoing phenomena: increased Cl^-, which lowers the pH in the corrosion zone, and the build up of overlayers of corrosion products, which makes it more difficult for oxygen to penetrate. The oxygen in the underlayers also

FIGURE 19a, b. *The mirror section at a magnification of ×70 in (a) near right, bright field; and (b) far right, dark field, shows a corrosion overburden outside the original surface of cuprite, the original surface, heavy corrosion just within the original surface, an interpenetration zone with redeposited copper (mostly round structures), and remnant alpha phase from the original bronze (linear structures). A little uncorroded metal shows at the upper right.*

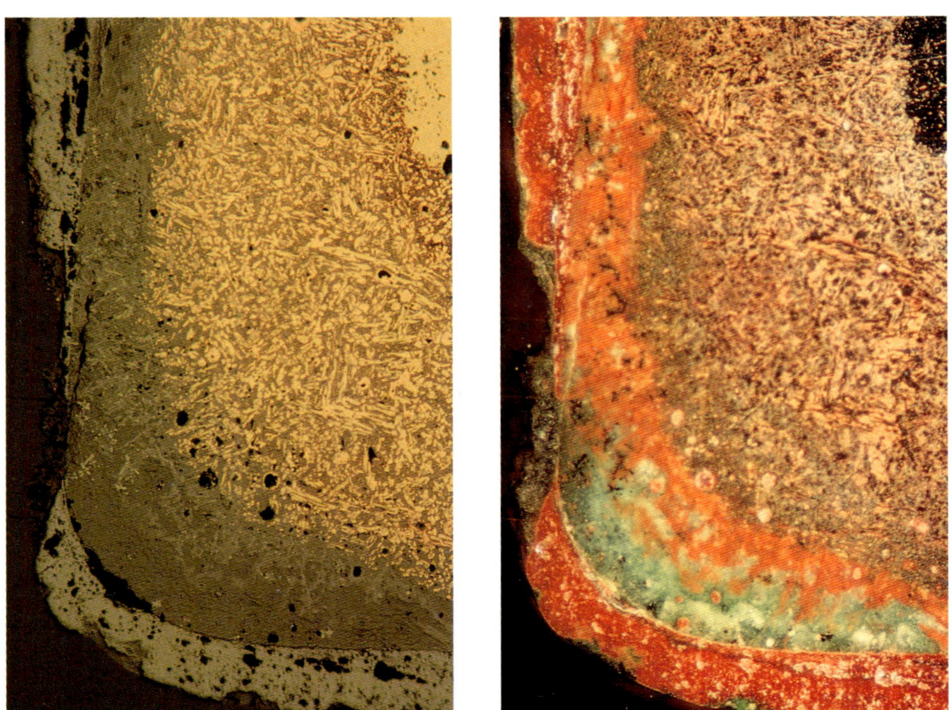

FIGURE 20. *Another area of the mirror, near right, bright field, ×65, shows large amounts of redeposited copper, both as round structures where redeposited copper is replacing cuprite formed in place of lead globules, and a linear area at the upper left, where redeposited copper is replacing cuprite in the outer layers of corrosion.*

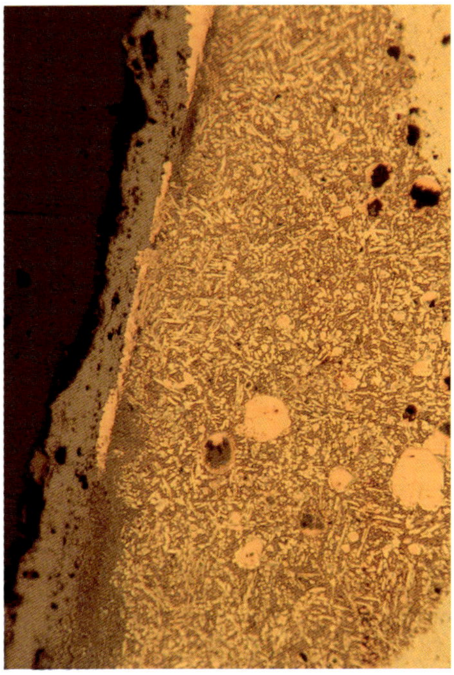

FIGURE 21. *The mirror section in dark field illumination, above right, at a magnification of ×223. The interpenetration zone is seen at left, the totally corroded zone at center, the sharp demarcation of the original surface at the center right, and cuprite overburden at the far right. Dendritic black-sulfide crystals can be seen in the first two zones, right up to the original surface.*

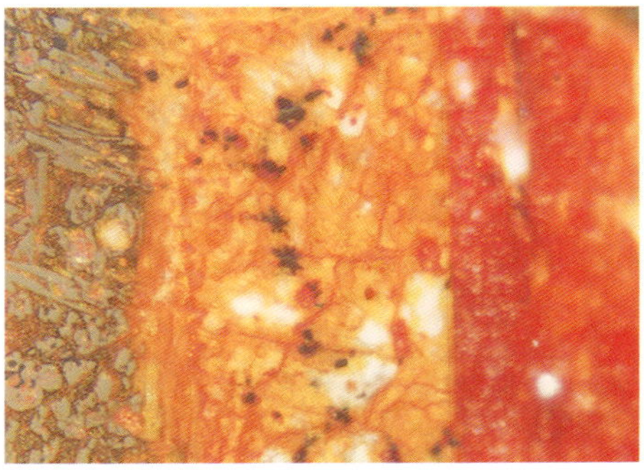

tends to be used up by oxidation processes, which take place on the surface of the sound metal, well within the corrosion zone. The layers away from the metal-corrosion interface could be depleted in oxygen.

If one looks at an Eh-pH diagram for copper in the presence of carbonate, chloride, and sulfur, one sees that at a pH of 7 and oxidation potential of 0.1 V, cuprite is the stable phase (Pourbaix 1973:267; Brookins 1988). If the pH or the oxidation potential drop, the locus moves into a region where copper metal is the stable phase. Taking into account the possible slowness of the growth of the external corrosion layers and the probable gradual intrusion of Cl^-, these processes may be very prolonged. The low rate could explain the nice, uniform transformations of cuprite to copper and also the common phenomenon in which the cuprite in the external part of one of the spheres has transformed to redeposited copper while the internal part remains as cuprite. The hypothesis is that the corrosion sequence is metal to cuprite to redeposited copper, the second step taking place gradually as a response to slowly changing oxidation and pH conditions within the corrosion layers, particularly to the reduction in available oxygen as the corrosion layers grow thicker over time.

It would have been worth the trouble to mount, polish, and observe this section before doing the mechanical cleaning; some areas may yield well to mechanical cleaning, but others, where the redeposited copper appears on the surface and is rooted in the inside, are impossible to clean mechanically. The piece may end up with a layer of redeposited copper showing, or with a breakthrough to the weaker, heavily corroded layer that underlies the redeposited copper. Mechanical cleaning should proceed slowly here, and one should not try to force a good outcome in areas that do not seem to clean well; in such areas it would be best to leave well enough alone. While the mirror looks quite good in its cleaned state, a look at the cross section would have indicated the inadvisability of proceeding so far in some spots and might have yielded a better result.

Intentionally Patinated Objects

Sections of intentionally patinated objects are difficult to find. The ones studied most by the author derive from the Freer study collection (Chase and Franklin 1979:245; Brown et al. 1977:197–207). These objects share the characteristics mentioned above: a well-preserved oxide surface on the outside, usually with Type II corrosion underneath; chloridic corrosion in areas of cracks or abrasions, or on the inside of the closed objects, with Type I corrosion proceeding in these areas; and a shiny and/or decorated surface.

At this point, there is no coherent explanation of the technique used to make intentionally patinated objects; an entire group of techniques may have been used. The possibility that pickling was used to promote Type II corrosion (easily done with ammonia) and that the tin-oxide layer was subsequently applied by a heating process or some other process is the best explanation at the moment. Further study of the question is planned. A more definitive explanation of corrosion phenomena and tarnishing phenomena above will greatly aid in this study.

Nevertheless, the question remains whether intentional patinas may have been even more prevalent on Chinese bronzes than was previously thought (Ma and Han 1988; Meeks 1986, 1988). These layers are subtle and easily lost in cleaning, and usually need to be seen on the actual object before they can be perceived to be legitimate subjects of inquiry. The difference in color on the Li-yü vessel mentioned above is a good case in point. The vessel has not yet been examined to see if there is an actual material difference between the inside and the outside. The bronzes from the tomb of Lady Meng Ji in Baoxiangsi in Henan province (Thorp 1988:52–55), which all have a peculiar, shiny, purplish-brown patina, are among many other objects that could use close investigation.

Seeing the red and black layers in the cross section from the bell from the tomb of the Marquis of Cai, coupled with the bronzes in the tomb of the Marquis of Yi, many of which are black, raises the question of whether blackened bronzes were more common than heretofore believed.

Finally, one may consider the bronzes from the tombs of Liu Sheng and Dou Wan (the tombs of the jade body suits with gold-wire attachments), Han Dynasty (ca. 175 B.C.E.), from Man-cheng, just to the northwest of Beijing. The tombs were cut into the mountain rock, and cast iron was poured in to seal them. The chambers were never filled in with earth, so these are examples of ancient bronzes that were not buried. The bronzes, as pictured in the archaeological report (Institute of Archaeology 1980) and as seen at various times on exhibition (Brown 1977:192–93), have a large number of different patinations: One sword has what looks like a transparent, shiny patina over deep-yellow, bronze-colored metal (Institute of Archaeology 1980:no. 1:5046, plate XLIV, 2). Another (which the author has not seen) appears to have silvery edges and a black center (Institute of Archaeology 1980:no. 1:5051, plate XLIV, 1). A *ge* dagger-ax has a surface pattern of black dots on a bronze-colored background; the mount is gilt bronze (Brown 1977:193; Institute of Archaeology 1980:plate XII). The famous tigers with carnelian eyes (Institute of

Archaeology 1980:plate XXVI) are inlaid with gold and silver, but the bronze color looks very much like a thick, dark-red cuprite patina. The mirrors are three different colors (Brown 1977). Many of the bronzes are dark and shiny, but the famous Boshanlu censer is a matte black, which sets off the gold inlay very well (Institute of Archaeology 1980:plate IX).

The best working hypothesis concerning these bronzes is that they had a variety of different patinas when they were put in the tombs. The closed environment in the rock-cut tombs has preserved the bronzes extremely well. The whole suite of bronzes and their surfaces are worth serious study.

Conclusion

Close study of ancient Chinese bronzes, the materials from which they are made, and the technology of their manufacture, coupled with study of the deterioration mechanisms they have undergone, has yielded some intriguing results. Probably the most interesting part of the whole study is that of patina and corrosion products. The patinas come from initial treatment, use, burial, and later events. Much information is encoded in the patinas of these objects, and it is important for conservators to try to read and understand it. There is also no point in trying to bring these bronzes back to their original condition, for their original condition has been altered totally by corrosion and, in some cases, subsequent treatment. Robbiola makes the additional point that removal of patina layers will probably be destabilizing, especially if the bronze has bronze disease. One should, at the very least, exercise a great deal of restraint in cleaning or altering the surface (Bassett and Chase, this volume).

Additional observations may be made here about conservation treatments of ancient bronzes. Sometimes conservation should go beyond just preserving the present condition; revealing design, removing old and possibly deceptive restoration, and cleaning down an ugly surface to a more attractive surface are all possible reasons for extensive treatment. These treatments are, however, irreversible. Conservators should act from a position of knowledge before beginning any of them. An understanding of the piece needs to come before the treatment of a bronze and, in many cases, the best understanding comes from metallographic study of cross sections.

Perhaps the implications are even more important for archaeologists. More information on the burial environment that bronze objects encounter would be very helpful indeed to those of us trying to interpret corrosion phenomena. The measurement of the soil moisture (often saturated, sometimes waterlogged), pH, Eh, Cl^- ion concentration, and carbonate concentration would be extremely helpful in trying to reconstruct the corrosion processes.[7] Measurement of cation concentrations in the soil (especially Cu, Sn, Pb, and Fe) in addition to the anions would be useful. Cross-sectional samples from excavated bronzes and their intensive study are also particularly vital to our understanding of long-term corrosion phenomena, and this work is now proceeding or beginning to proceed in various centers in China and Japan. Cleaning or chemical treatment in an indiscriminate fashion of all bronze finds from a site should be avoided until the finds have been carefully studied. At the least, some finds could be retained in an untreated state for study. If consolidation is necessary, reversible consolidants should be used. Bronze disease should be, at least initially, treated by keeping the objects dry and not by chemical means.

The maintenance of excellent and thorough records on what has been done to the objects in terms of recovery and treatments is also extremely important. In order to reconstruct what has happened or proceed to improve the condition of a deteriorating object, one really needs to know what has been done to it previously.

Study of the corrosion phenomena and the environments and initial treatments that have brought the object to its present condition is intellectually interesting and challenging, but more important, these phenomena enable us to appreciate the skill of ancient craftspeople and the beauty of what they made.

Notes

1. Recently, however, evidence of worked and annealed bronzes dating from the late Zhou period has come to light (He Tangkun 1993). We have also found evidence of worked and annealed bronze in a sample from the Jinsheng Tomb near Taiyuan in Shanxi (personal communication, Tao Shenggang, Shanxi Institute of Archaeology, 1992). Both of these are extremely thin bronzes, but they do illustrate that the process of working and annealing bronze was known before the Han Dynasty. It is, however, still fair to say that the vast majority of early Chinese bronzes were cast.
2. For example: Chase 1991; Bagley 1987; Bagley 1990; Keyser 1979; Han 1986; Huang 1991.
3. Produced by Bill Debold and Ed Yates in Matsue with a grant from The Desert Research Institute of Reno, Nevada.
4. When these maquettes were seen by the author in 1971 at Gorham, he was told that the foundry had lost the secret of this bright-red patina.
5. I will not attempt here to deal with the question of the formation of azurite versus malachite; it seems to be related strictly to local conditions and may well be due to local increases in CO_2 concentration from microbial action on decaying organic matter (McNeil and Little 1991).
6. It has recently come to my attention (personal communication, Zhou Baozhong, 1993) that all of the bronzes from the tomb of the Marquis of Cai were immersed in distilled water for a year prior to any treatment. Some of the swelling of the outer layers may have come from the prolonged corrosion, which did not, however, remove the chlorides deep in the structure.
7. Eh is difficult to measure; perhaps Eh measurements should be made before digging by a probe of some sort.

References

BAGLEY, R. W.

1980 The beginnings of the Bronze Age: The Erlitou culture period. In *The Great Bronze Age of China: An Exhibition from the People's Republic of China*, 67–94, Wen Fong, ed. New York: The Metropolitan Museum of Art and Alfred A. Knopf.

1987 *Shang Ritual Bronzes in the Arthur M. Sackler Collections*. Washington, D.C. and Cambridge, Mass.: The Arthur M. Sackler Foundation and the Arthur M. Sackler Museum.
1990 Shang ritual bronzes: Casting technique and vessel design. In *Archives of Asian Art* XLIII:6–20.

BARNARD, N., AND T. KENNON
1986 *Studies on the Free Copper in Chinese Bronzes*. Canberra: Privately printed.

BARNARD, N., AND T. SATO
1975 *Metallurgical Remains of Ancient China*. Tokyo: Nichiosha.

BORN, H.
1985a Korrosionsbilder auf ausgegrabenen Bronzen—Informationen für den Museumsbesucher. In *Archäologische Bronzen, antike Kunst, moderne Technik*, 86–96, Hermann Born, ed. Berlin: Dietrich Reimer Verlag.
1985b Polychromie auf prähistorischen und antiken Kleinbronzen. In *Archäologische Bronzen, antike Kunst, moderne Technik*, 71–83, Hermann Born, ed. Berlin: Dietrich Reimer Verlag.

BOWEN, E.
[1959] 1989 *A Time in Rome*. New York: Alfred A. Knopf, Penguin Books.

BROOKINS, D. G.
1988 *Eh-pH Diagrams for Geochemistry*. Berlin: Springer Verlag.

BROWN, B. F., H. C. BURNETT, W. T. CHASE, M. GOODWAY, J. KRUGER, AND M. POURBAIX, EDS.
1977 *Corrosion and Metal Artifacts—A Dialogue Between Conservators and Archaeologists and Corrosion Scientists*, National Bureau of Standards Special Publication 479. Washington, D.C.: National Bureau of Standards.

CARPENTER, H. C. H.
1933 Preliminary report on Chinese bronzes. In *Anyang fa jue bao gao*. Preliminary reports of excavations at An-yang, part 4. Peking, Nanking: Academia Sinica National Research Institute of History and Philology.

CHASE, W. T.
1974 *Bronze Disease and its Treatment*. Catalogue of an exhibition at the Thai National Museum. Bangkok: Thai National Department of Antiquities.
1979 Solid samples from metallic antiquities and their examination. In *Proceedings of the Second International Symposium on the Conservation and Restoration of Cultural Property*, 73–109. Tokyo: Tokyo National Research Institute for Cultural Property.
1983 Bronze casting in China: A short technical history. In *The Great Bronze Age of China: A Symposium*, 100–23, George Kuwayama, ed. Los Angeles: Los Angeles County Museum of Art.
1991 *Ancient Chinese Bronze Art: Casting the Precious Sacral Vessel*. New York: China House Gallery, China Institute in America.

CHASE, W. T., AND C. YUYUN

1992 Structures of alloys with the compositions of the "Six Formulas." *Science of Conservation and Archaeometry* (Wenwu Baohu ye Kaogu Kexue) 3(2):24–36.

CHASE, W. T., AND U. M. FRANKLIN

1979 Early Chinese black mirrors and pattern-etched weapons. *Ars Orientalis* 11:215–58.

CHASE, W. T., AND N. F. VELOZ

1989 Airbrasive cleaning of statuary and other structures: A century of technical examination of blasting procedures. *Technology and Conservation* 10(1):18–28.

CHASE, W. T., AND T. O. ZIEBOLD

1978 Ternary Representations of Ancient Chinese Bronze Compositions. In *Archaeological Chemistry IV*, chap. 18, Giles Carter, ed. Advances in Chemistry Series, no. 171. Washington, D.C.: American Chemical Society.

COPPER DEVELOPMENT ASSOCIATION

1974 *Application Data Sheet (copper, brass, bronze): The Properties of Clear Organic Coatings on Copper and Copper Alloys*. New York: Copper Development Association, Inc.

EICHHORN, P.

1985 Bergung, Restaurierung und Konservierung archäologischer Gegenstnde aus Bronze. In *Archäologische Bronzen, antike Kunst, moderne Technik*, 148–67, Hermann Born, ed. Berlin: Dietrich Reimer Verlag.

1988 Cuprit auf Bronze—ein Echtheitsbeweis? In *Arbeitsbltter für Restauratoren, Gruppe 2, Bronze* Heft 2:235–39.

FAN C., W. CHANGSUI, W. SHENJUN, Z. FANSUN, W. T. CHASE, Z. JINGGUO, AND H. ZIQIANG

1991 Research on powdery corrosion and conservation of bronze objects [in Chinese]. In *Keji Kaogu Luncong; Proceedings of the second National Conference on Science and Technology in Archaeology*, 100–10, Hu Hsuiyu, ed. Hefet, Anhui, PROC: University of Science and Technology Press.

FORMIGLI, E.

1985 Die Restaurierung einer griechischen Grossbronze aus dem Meer von Riace/Italien. In *Archäologische Bronzen, antike Kunst, moderne Technik*, 168–74, Hermann Born, ed. Berlin: Dietrich Reimer Verlag.

FREESTONE, I. C., N. WOOD, AND J. RAWSON

1989 Shang Dynasty casting moulds from North China. In *Cross-Craft and Cross-Cultural Interactions in Ceramics*, 253–74, Patrick E. McGovern, ed. Columbus, Ohio: American Ceramics Society.

GALE, E. M.

1931 *Discourses on Salt and Iron: A Debate on State Control of Commerce and Industry in Ancient China*. Sinica Leidensia Vol. 2. Leyden: E. J. Brill.

GARRELS, R. M., AND C. L. CHRIST

1965 *Solutions, Minerals, and Equilibria*. Harper's Geoscience Series. New York: Harper and Row.

GETTENS, R. J.

1969 *The Freer Chinese Bronzes, vol. II: Technical Studies*. Oriental Studies, no. 7, Smithsonian Publication 4706. Washington, D.C.: Smithsonian Institution, Freer Gallery of Art.

1970 Patina, Noble and Vile. *Art and Technology: A Symposium on Classical Bronzes*, 57–72, Suzannah Doeringer, ed. Cambridge: MIT Press.

GOODWAY, M., AND H. C. CONKLIN

1987 Quenched high-tin bronzes from the Philippines. *Archaeomaterials* 2(1):21–27.

GUIVER, T., N. KENNON, AND H. RUBIN

1988 A technical study of Chinese bronze mirrors. In *Conference: Ancient Chinese and Southeast Asian Bronze Culture*, Noel Barnard, ed.

HAN RUBIN

1986 Metalcasting in ancient China. *Castings* July/August:33–39.

HARRER, H.

[1963] 1964 *I Come from the Stone Age*. Originally published as *Ich komme aus der Steinzeit* by Verlag Ullstein GmbH. London: Rupert Hart-Davis.

HE TANGKUN

1993 Laboratory analysis of bronzes with incised patterns [in Chinese]. *Kaogu* (Archaeology) 5:465–68.

HOLMES, L. L., AND G. HARBOTTLE

1991 Provenance study of cores from Chinese bronze vessels. *Archaeomaterials* 5(2):165–84.

HUA JUE MING

1985 *Translation of R. F. Tylecote, A History of Metallurgy; with a section on Ancient Chinese Metallurgy* (II-Zhongguo Gudai Jinshu). Beijing: Science and Technology Document Publishing Company, Kehue Jishu Wenxian Chubanshe.

HUANG LONG

1991 A brief study on bronze casting techniques in ancient China [in Chinese]. *Wenwu Baohu Yu Kaogu Kexue* (Sciences of Conservation and Archaeology) 3(1):31–43.

HUGHES, R., AND M. ROWE

1983 *The Colouring, Bronzing and Patination of Metals*. New York and London: Van Nostrand Reinhold and the Crafts Council.

INSTITUTE OF ARCHAEOLOGY, ACADEMIA SINICA

1956 *The Excavation of the Tomb of the Marquis of Cai at Shou Xian* [in Chinese]. Archaeological Monograph, Series B, vol. 5. Peking: Cultural Relics Publishing House.

INSTITUTE OF ARCHAEOLOGY, CASS AND THE HOPEI COMMITTEE FOR THE PRESERVATION OF ARCHAEOLOGICAL MONUMENTS (CPAM), HOPEI PROVINCE

1980 *Excavation of the Han Tombs at Man-ch'eng* [in Chinese]. Beijing: Cultural Relics Publishing House.

JOHNSON, R. E.

1973 Cu-Sn (copper-tin). In *Metals Handbook*. ASM Handbook Committee, 299–358. Metals Park, Ohio: American Society for Metals.

JOHNSTON-FELLER, R.

n.d. *The Science of Colorimetry and Spectrophotometry in the Examination of Museum Objects*. Santa Monica: The J. Paul Getty Trust. In press.

KEYSER, B. W.

1979 Decor replication in two late Chou bronze chien. *Ars Orientalis* 11:127–62.

LAWTON, T.

1982 *Chinese Art of the Warring States Period: Change and Continuity, 480–422* B.C.E. Washington, D.C.: Smithsonian Institution Press for the Freer Gallery of Art.

MA ZHAOZHENG, AND HAN RUBIN

1988 Studies on the chromium oxide surface film of bronze arrowheads unearthed from the terra-cotta warriors' tomb of Qin Dynasty [in Chinese]. *Daxue Huaxue* 3(3):57–60.

MACLEOD, I. D.

1983 Stabilization of corroded aluminum. *Studies in Conservation* 28:1–7.

MCNEIL, M. B., AND B. J. LITTLE

1991 Corrosion products and mechanisms in long-term corrosion of copper. In *Scientific Basis for Nuclear Waste Management XIV*, T. Abrajano, Jr., and L. H. Johnson, eds. Materials Research Society Symposium proceedings (212):311–16. Pittsburgh: Materials Research Society.

MCNEIL, M. B., D. W. MOHR, AND B. J. LITTLE

1991 Correlation of laboratory results with observations on long-term corrosion of iron and copper alloys. In *Materials Issues in Art and Archaeology II*, Pamela B. Vandiver, James Druzik, and George S. Wheeler, eds. Materials Research Society Symposium proceedings (185):753–59. Pittsburgh: Materials Research Society.

MEEKS, N. D.

1986 Tin-rich surfaces on bronze: Some experimental and archaeological considerations. *Archaeometry* 28(2):133–62.

1988 Surface Studies of Roman bronze mirrors, comparative high-tin bronze Dark Age material and black Chinese mirrors. *Proceedings of the 26th International Archaeometry Symposium*, 124–27. R. M. Farquhar, R. G. V. Hancock, and L. A. Pavlish, eds. Toronto, Canada: University of Toronto, Archaeometry Laboratory.

MOORADIAN, V., AND R. J. GETTENS

1936 A metallographic and microscopic study of ancient Chinese bronze mirrors. Manuscript in Freer Gallery of Art Bronze Archives, Washington, D.C.

ODDY, W. A.

1985 Vergoldungen auf prähistorischen und klassischen Bronzen. In *Archäologische Bronzen, antike Kunst, moderne Technik*, 64–71. Hermann Born, ed. Berlin: Dietrich Reimer Verlag.

POPE, J. A., R. J. GETTENS, J. CAHILL, AND N. BARNARD

1967 *The Freer Chinese Bronzes, Volume I: Catalogue*. Oriental Studies, no. 7, Smithsonian Publication 4706. Washington, D.C.: Smithsonian Institution, Freer Gallery of Art.

POURBAIX, M.

1966 *Atlas of Electrochemical Equilibria in Aqueous Conditions*. NACE edition, 1976. Houston, Texas, and Brussels: CEBELCOR.

1973 *Lectures on Electrochemical Corrosion*. New York and London: Plenum Press.

ROBBIOLA, L.

1990 Characterisation de l'alteration de bronzes archeologiques enfouis a partir d'un corpus d'objets de l'age du bronze. Mechanismes de corrosion. Ph.D. diss., L'Universite de Paris, Paris.

ROBBIOLA, L., I. QUEIXALOS, L. P. HURTEL, M. PERNOT, AND C. VOLFOVSKY

1988 Etude de la corrosion des bronzes du Fort-Harrouard: altération externe et mécanisme d'altération stratifiée. *Studies in Conservation* 33(4):205–15.

SAUNDERS, N., AND A. P. MIODOWNIK

1990 The Cu-Sn (copper-tin) system. *Bulletin of Alloy Phase Diagrams* 11(3):278–87.

SCOTT, D. A.

1990 Bronze disease: A review of some chemical problems and the role of relative humidity. *Journal of the American Institute for Conservation* 29(2):193–206.

1991 *Metallography and Microstructure of Ancient and Historic Metals*. Los Angeles: The Getty Conservation Institute, and the J. Paul Getty Museum.

SMITH, C. S.

1959 Corrosion of an ancient bronze and an ancient silver alloy. In *Physical Metallurgy of Stress Corrosion Fracture*, 293–94, Thor N. Rhodin, ed. Metallurgical Society Conferences, Vol. 4. New York and London: Interscience Publishers.

1975 Letter to author, 10 April. Discussion read at R. J. Gettens Memorial Seminar, 21 March 1975.

SOCHA, J., M. LESIAK, S. SAFARZYNSKI, AND K. LESIAK

1980 Oxide coating in the conservation of metal monuments: The column of King Sigismundud III Waza in Warsaw. *Studies in Conservation* 25(1):14–18.

THORP, R. L.

1988 *Son of Heaven: Imperial Arts of China*. Seattle: Son of Heaven Press.

UCHIDA, T., T. ISHINO, AND H. YONEDA

1988 Physical property and color tone of excavated bronze mirror (II). In *Kobunkazai no Kagaku*. Kobunkazai Kagaku Kenkyukai (Association of Scientific Research on Historic and Artistic Works of Japan). Tokyo: Kokuritsu Bunkazai Kenkyujo.

WAN CHIAPAO

1989 On free copper particles in ancient Chinese bronze [in Chinese]. *Wenwu Baohu Yu Kaogu Kexue* (Sciences of Conservation and Archaeology) 1(2):32–36.

WANG CHANGSUI, FAN CHONGZHENG, WANG SHENJUN, ZHANG MAOSEN, ZHANG JINGGUO, AND W. T. CHASE

1990 Research on powdery corrosion of the ancient bell from the Tomb of Cai Hou. *Zhongguo Kexue* (Chinese Science), part B 6 (1990.6):639–44.

WEIL, P. D.

1977 The conservation of outdoor bronze sculpture: A review of modern theory and practice. In *Corrosion and Metal Artifacts: A Dialog between Conservators and Archaeologists and Corrosion Scientists*, B. F. Brown, H. C. Burnett, W. T. Chase, M. Goodway, J. Kruger, and M. Pourbaix, eds. National Bureau of Standards Special Publication 479. Washington, D.C.: National Bureau of Standards.

WEISSER, T., ED.

1992 *The Conservation of Bronze Sculpture in the Outdoor Environment: A Dialogue Among Conservators, Curators, Environmental Scientists and Corrosion Engineers. "Dialogue '89."* Houston, Texas: NACE.

WU KUNYI AND LI XIHUI

1988 Effect of lead on the properties of bronze drum [in Chinese]. *Preprint of Ancient South and Southeast Asian Bronze Drums and Cultures.* Kunming, Yunnan: People's Republic of China.

WU YOUSHI, WANG CHANGSUI, FAN CHONGZHENG, WANG SHENGJUN, LI ZICHAO, AND W. THOMAS CHASE

n.d. Approach to corrosion-resisting mechanism of "Hei Qi Gu" (black lacquer mirrors). In press.

Biography

W. T. Chase is head conservator of the Department of Conservation and Scientific Research of the Arthur M. Sackler Gallery and the Freer Gallery of Art, Smithsonian Institution. He majored in conservation of art at Oberlin College and later studied at the Conservation Center of the Institute of Fine Arts at New York University. During his graduate studies, he took his student internship at the Freer Technical Laboratory under Rutherford J. Gettens. After completing his degree in 1966, he returned to the Freer, where he became head conservator in 1968. His primary research interest is technical study of ancient Chinese bronzes.

The Corrosion of Bronze Monuments in Polluted Urban Sites: A Report on the Stability of Copper Mineral Species at Different pH Levels

ANDREW LINS AND TRACY POWER

Copper oxide (cuprite) and basic copper sulfates are the predominant crystalline corrosion products found on outdoor bronze and copper monuments in the United States. In urban-industrial areas, copper sulfates are readily formed and result in the streaked blue or blue-green color evident on many such structures. Caused primarily by acid deposition (dew, fog, rain, particulates, or dry deposition), this discoloration indicates degradation of the metal surfaces, usually manifested as extensive surface pitting. While this discoloration does not typically portend structural damage to well-executed castings, the aesthetic alteration of the surfaces may be profound and irreversible. The partially soluble corrosion products usually formed do not protect the underlying surfaces from attack, as typical exposure conditions do not produce even film growth. In some circumstances, however, thick and closely adherent corrosion crusts may retard the rate of degradation (Mattsson 1982:19; Graedel 1987a:730; Costas 1980:106–15).[1]

Found in lesser quantities than cuprite, brochantite—$CuSO_4 \cdot 3Cu(OH)_2$ or $Cu_4SO_4(OH)_6$—has long been noted as a very common component of the corrosion films on commercial copper and bronze alloys in polluted urban-industrial environments (Table 1).

The presence of antlerite—$CuSO_4 \cdot 2Cu(OH)_2$ or $Cu_3SO_4(OH)_4$—in the corrosion films on bronze monuments has been noted several times in recent years. Earlier studies of copper alloy and copper corrosion in the atmosphere seldom cite this corrosion form. The works of art on which this latter, more sulfate-rich species has been reported include some well-known sculptural monuments: the Statue of Liberty by Eiffel in New York Harbor (Baboian and Cliver 1986:74, 80; Nassau, Gallagher et al. 1987:671, 675, 684), The Horses of San Marco in Venice (Marchesini and Baden 1979:200–10), Lorenzo Ghilberti's Baptistry Doors in Florence (Garbassi and Mello 1984:174), and the equestrian monument of Marcus Aurelius in Rome (Leotta and Marabelli 1984:81; Marabelli 1987:214). In Philadelphia, where the following work was undertaken, two of the city's most prominent and best-known outdoor

TABLE 1.[2] *Basic copper salts in green patina from various atmospheres, defined by the anions ranked with respect to content.*

			Type of Atmosphere				
Reporter	*Object*	*Time of exposure (years)*	*Country*	*Rural*	*Urban or industrial*	*Marine*	*Mixed urban-marine*
Vernon & Whitby	Copper roofs; copper conductor in marine atmosphere	12–300 13	U.K.	1. SO_4^{2-} 2. CO_3^{2-}	1. SO_4^{2-} 2. CO_3^{2-} 3. Cl^-	1. Cl^- 2. CO_3^{2-} 3. SO_4^{2-}	1. SO_4^{2-} 2. CO_3^{2-}, Cl^-
Vernon	Copper roof on church spire on the isle of Guernsey	33	U.K.			1. Cl^- 2. CO_3^{2-} 3. SO_4^{2-}	
Freeman, Jr.	Copper roofs	16–78	U.S.				1. SO_4^{2-} 2. CO_3^{2-}, Cl^-
Thompson, Tracy, & Freeman, Jr.	Copper panels from field test	20	U.S.	1. SO_4^{2-}, Cl^- 2. CO_3^{2-}	1. CO_3^{2-} 2. SO_4^{2-}	1. Cl^- 2. SO_4^{2} 3. CO_3^{2-}	
Aoyama	Copper conductor for railway		Japan	1. NO_3^- 2. CO_3^{2-}	1. SO_4^{2-} 2. CO_3^{2-}	1. SO_4^{2-} 2. CO_3^{2-}	
Mattsson & Holm	Copper-base materials from field test	7	Sweden	1. SO_4^{2-} 2. NO_3^- 3. CO_3^{2-}	1. SO_4^{2-} 2. CO_3^{2-}	1. Cl^- 2. SO_4^{2-}	
Scholes & Jacob	Copper-base materials from field test	16	U.K.		1. SO_4^{2-} 2. Cl^-, CO_3^{2-}	1. Cl^- 2. SO_4^{2-}	

sculptures, William Penn (Fig. 1) by A. M. Calder and the Swann Fountain by A. S. Calder (Alexander Calder's grandfather and father, respectively), show significant amounts of antlerite (>10%) in parts of their corrosion crusts (Lins 1990). Figures 2–4 depict examples of corrosion in the Penn sculpture. Differences of opinion have been expressed about whether the appearance of antlerite indicates the emergence of a more acidic, more soluble corrosion form produced by increasing levels of pollution and acidity in urban-industrial environments, perhaps presaging a more rapid deterioration of such works of art in the future (Nielson [1981]1984:78; Baboian and Cliver 1986:80; Nassau, Miller et al. 1987:716–17, 773).[3]

In urban regions, many variables determine the corrosion process at a particular site, among the most important of which are (1) atmospheric effects, primarily ambient temperature, relative humidity, and pollution levels; the amount, frequency, and duration of rain, wind, fog, mist, and freeze-thaw events, complicated by wind and aerosol erosion effects that influence the time of wetness on a given surface; organic deposition; photochemically induced reactions; and (2) the composition of the metal, the distribution of primary and secondary phases in the microstructure, defects, and porosity, especially due to casting and soldering in the case of bronzes, surface roughness, shape and position with respect to prevailing atmospheric factors, intentional patination, and coatings.[4] Accepting the impossibility of duplicating

FIGURE 1. *A. Milne Calder, bronze statue of William Penn, cast 1889–91, City Hall Tower, Philadelphia.*

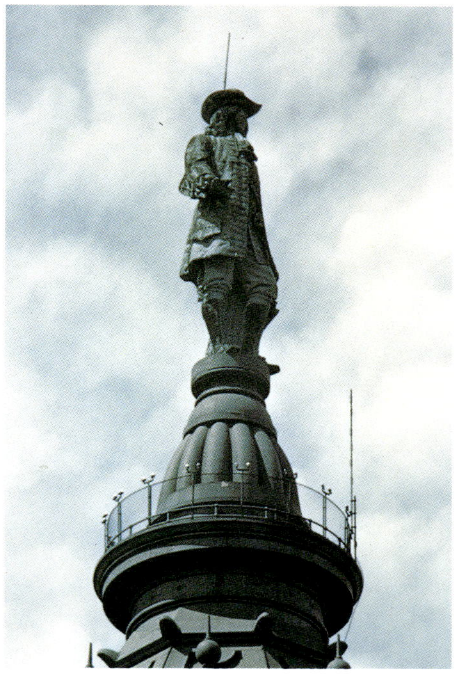

FIGURE 2. *BF ×96, near right, from the William Penn figure: a complex, multilayer corrosion crust can develop on bronzes after long exposure to urban-industrial atmospheres—in this case 97–98 years. The corrosion is approximately 200 microns thick and partly follows microstructural features, especially casting pores located near the surface.*

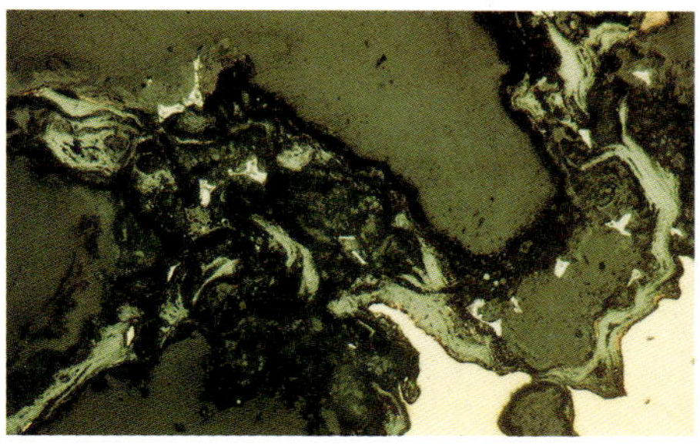

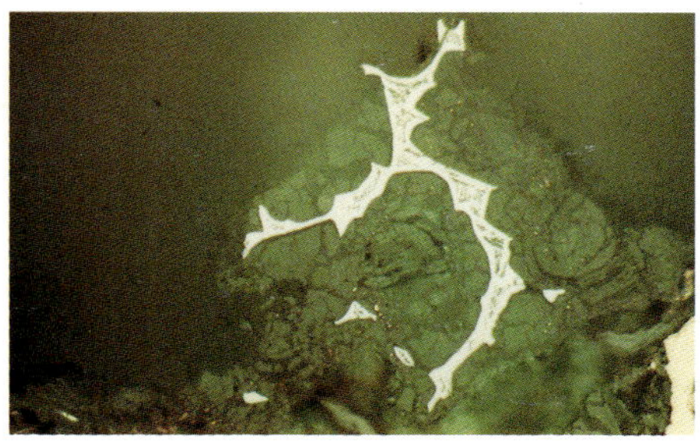

FIGURE 3. *BF ×240, above far right, also from the William Penn figure, illustrates that the delta phase is not as rapidly attacked as the alpha phase at this particular site.*

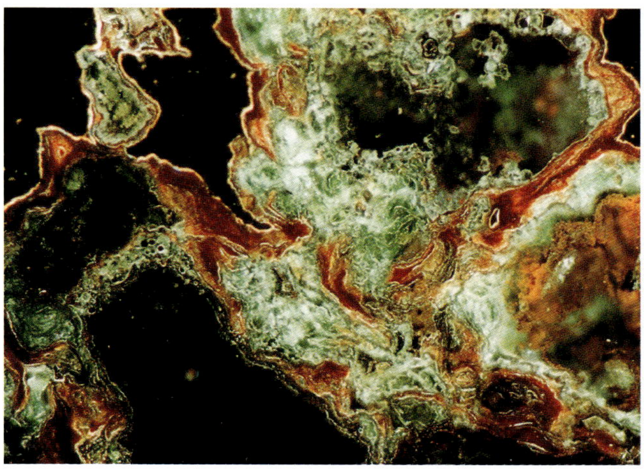

FIGURE 4. *BF ×310, above near right, shows the irregular, porous nature of the mineral formations within the crust.*

more than a few of these variables in the laboratory, the following experiments attempted to measure with simple means the relative stability of some of the principal species involved in the system $Cu\text{-}H_2O\text{-}H_2SO_4\text{-}O_{2(g)}/CO_{2(g)}$ at 20–22 °C. These analyses also attempted to assess the validity of antlerite as a pollution indicator, as well as to study more closely the predominant factors involved in the deterioration of bronze monuments, using concentrations of reactants not grossly different from those found in the environment.

Thermodynamics

Much of the work on the phase boundaries of the system Cu-SO_3-H_2O has been based on the thermodynamic properties of the substances involved.[5] The results of this research may be conveniently summarized in five graphs, in which pH and anion concentrations approach those found in the environment. The boundaries of antlerite stability are shaded for ease of visualization.

Based on Figures 5–9—derived primarily from thermodynamic values of E°, G°, H°, or u° at different pH values—one could predict that antlerite is formed under more acidic conditions in aqueous sulfate solutions than is brochantite, the most common sulfate corrosion product on copper and copper alloys in the atmosphere. If this theoretical thermodynamic situation is demonstrated to occur in a reproducible way under typical atmospheric conditions of moisture and pH and at a measurable rate with respect to a time scale of months or years, the ratio of antlerite to brochantite in a corrosion crust should serve as a measure of atmospheric aggressiveness.

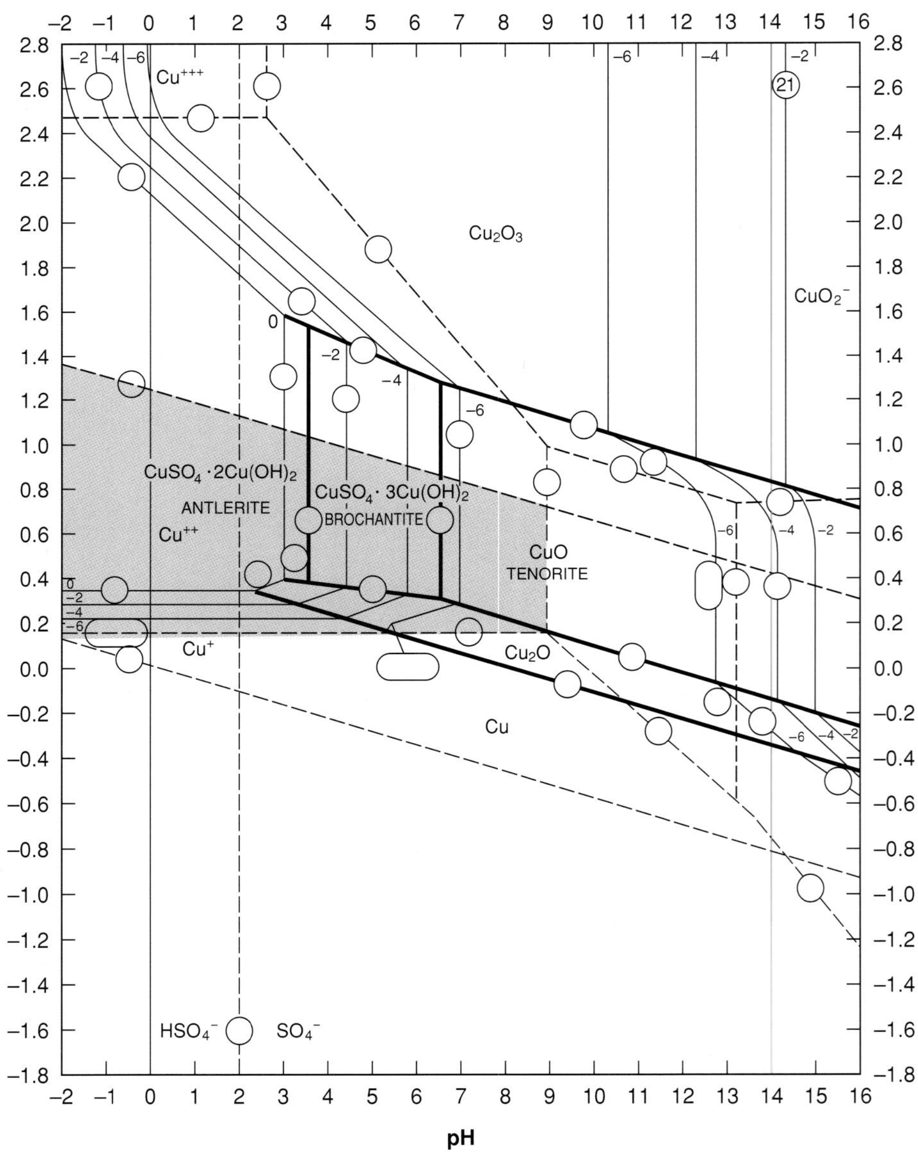

FIGURE 5. *This is a standard plot of E vs pH for Cu, SO_3, H_2O at 25 °C; for the specific condition that $10^{-3.24}$ M SO_3 (or 46 ppm) are dissolved per liter, from M. Pourbaix's work "to show the patina formed on copper in an atmosphere polluted by SO_2" (1977:13). Note that while the activity of Cu^{++}, H^+, and other charged species in solution vary, this plot is strictly confined to the stated SO_3 concentration. Changes in the activity of SO_3 would necessitate redrawing the graph.*

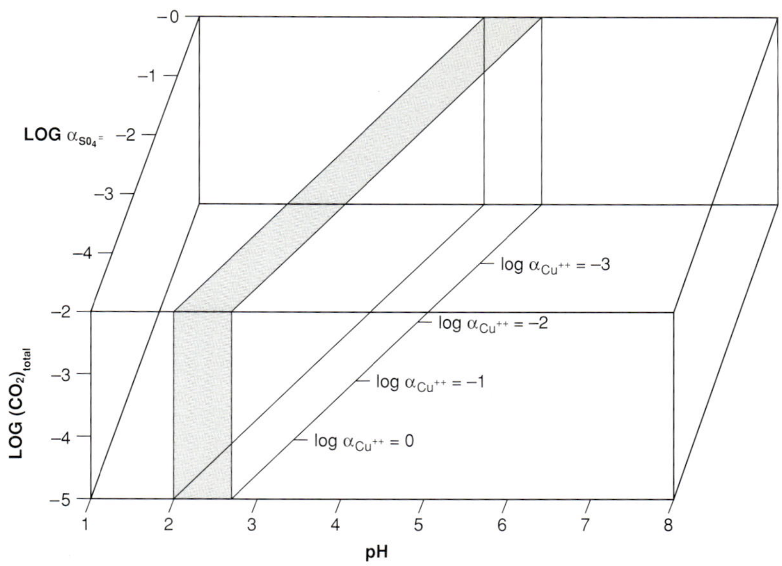

FIGURE 6. *This three-dimensional plot, and those below, show the stability of different copper minerals in water at 25 °C. At right are oxidized copper species found in rain or natural waters. The x axis shows pH, the y axis shows sulfate ion activity, and the z axis shows total* CO_2 *activity (including* $CO_{2(g)}$, H_2CO_3, HCO_3^-, *and* $CO_3^=$*) based on work of Silman and of Garrels (Garrels and Christ 1965:374–75). Zone of conditions favoring antlerite formation is shaded.*

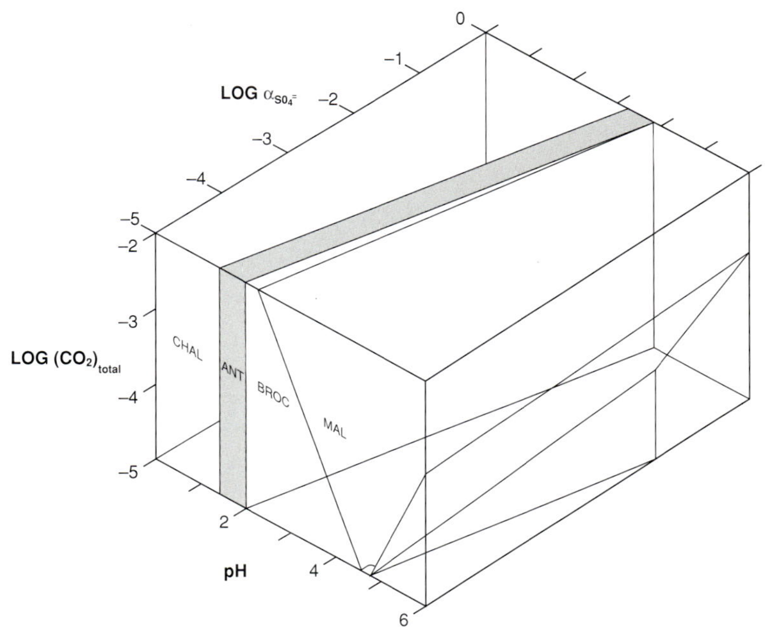

FIGURE 7. *Plot emphasizing the calculated lower limit of cupric ion activities* ($a_{ant\ v\ broch}$) *which would allow antlerite rather than brochantite to form, based on the work of Silman (Schmitt 1962:110–26).*[6]

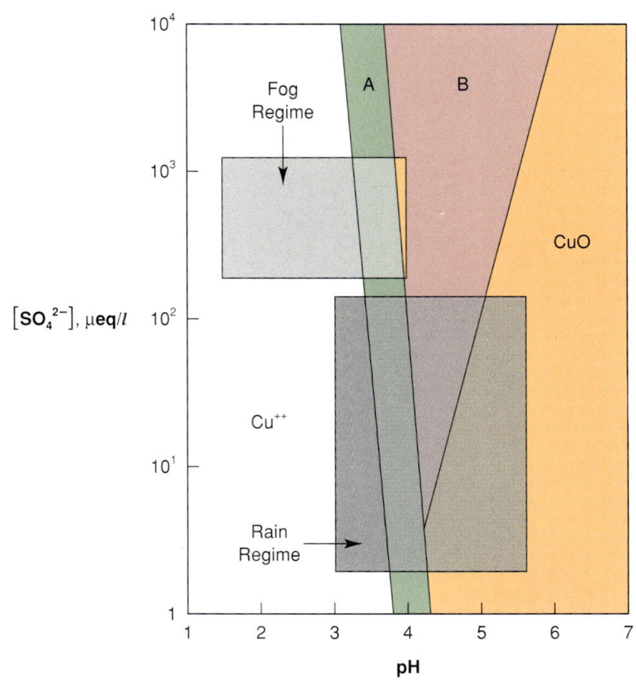

FIGURE 8. *Plot showing the solubility of different copper sulfate and oxide species in dilute sulfuric acid solutions. Data based on thermodynamic and other calculations at 25 °C, after Mattsson (1982:18) and Graedel (1987a:731). Zones for acid rain and fog are displayed, overlaps indicating which minerals are likely at particular pH levels and sulfate concentrations. The left side shows no stable mineral forms, only* Cu^{++} *in solution.*

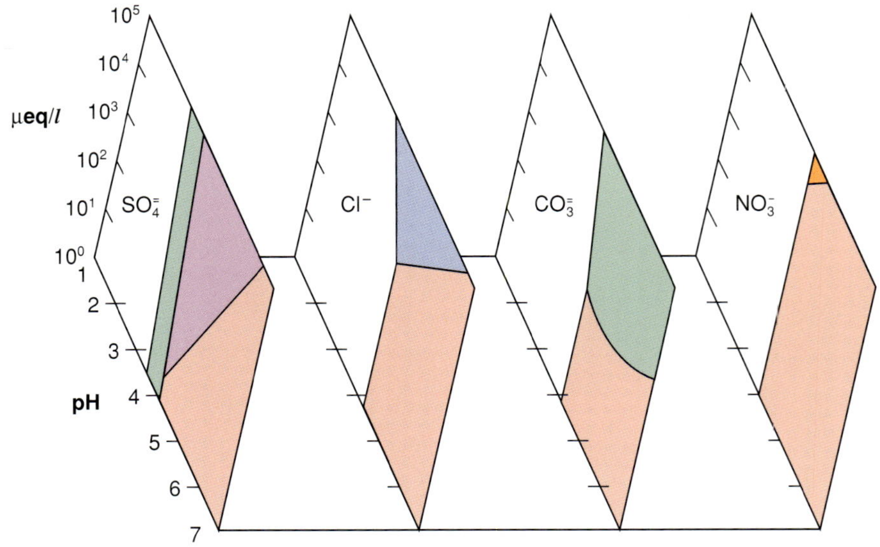

FIGURE 9. *Diagram comparing basic copper sulfate stability with that of basic copper carbonates, chlorides, and nitrates. It shows the relative insolubility of basic copper sulfates, and conversely, the solubility of the copper nitrates, as well as the high concentration of anion required for them to precipitate. The x axis is pH. The separate z axes give anion activity in microequivalents per liter. Each plot is separate and based on specific and different anion concentrations. The y axis is an arbitrary axis, necessary only for the sense of overlay in the diagram. The range of values on the z axis reflects concentrations that might be found in an average rain or fog-dew event, for each anion species.*[7]

The Behavior of Copper in Sulfuric Acid Solutions

Fundamental to any study of sulfate-film formation on copper substrates in the atmosphere are studies of Cu reactions in dilute H_2SO_4 and acidic salt solutions. These topics have been extensively investigated.[8] Space does not permit more than a brief summary of the results relevant to the present study:

In dilute sulfuric acid solutions, oxygen is necessary for copper dissolution or for cuprite formation to reach readily measured values. In very dilute sulfuric acid solutions and nearly pure water, cuprite formation is influenced by the concentration of gases dissolved in solution, enhanced in particular by oxygen. If oxygen is rigorously excluded from the reaction vessel, even at pH levels as low as pH 2, copper dissolution is difficult to measure reproducibly, and the growth of a cuprite layer is not demonstrated. In concentrated sulfuric acid solutions (>0.5 M), however, copper dissolves with or without oxygen present, but cuprite is not formed if the pH is kept low. During dissolution reactions with fixed amounts of reactants, the pH rises according to the general equation:

$$Cu + 2H^+ + {}^1/_2\, O_2 = Cu^{++} + H_2O$$

Two H^+ are consumed for each Cu^{++} that is put into solution. The equilibrium constant for this electrochemical reaction shifts to the left as more copper ions go into solution. At higher pH (>2.5), relatively small concentrations of Cu^{++} shut down the dissolution reaction. As the pH rises in unreplenished solutions, OH^- concentration increases and with it the likelihood of producing cuprite (Yoon 1971:38; Gregory and Riddiford 1960:950–56).[9]

In sulfuric acid solutions open to air at pH below 3.5, the initial rate of dissolution of copper is independent of pH if sufficient mixing is provided in the solution, as Yoon (1971:53) indicates.

In the absence of rapid stirring (typically above 1000 rpm for the copper disc electrodes employed in such analyses), the rate of dissolution is transport-controlled. In poorly stirred or quiescent H_2SO_4 solutions below pH 2.8, cuprite has not been observed to form. Above this pH, however, cuprite is formed, and the thickness of

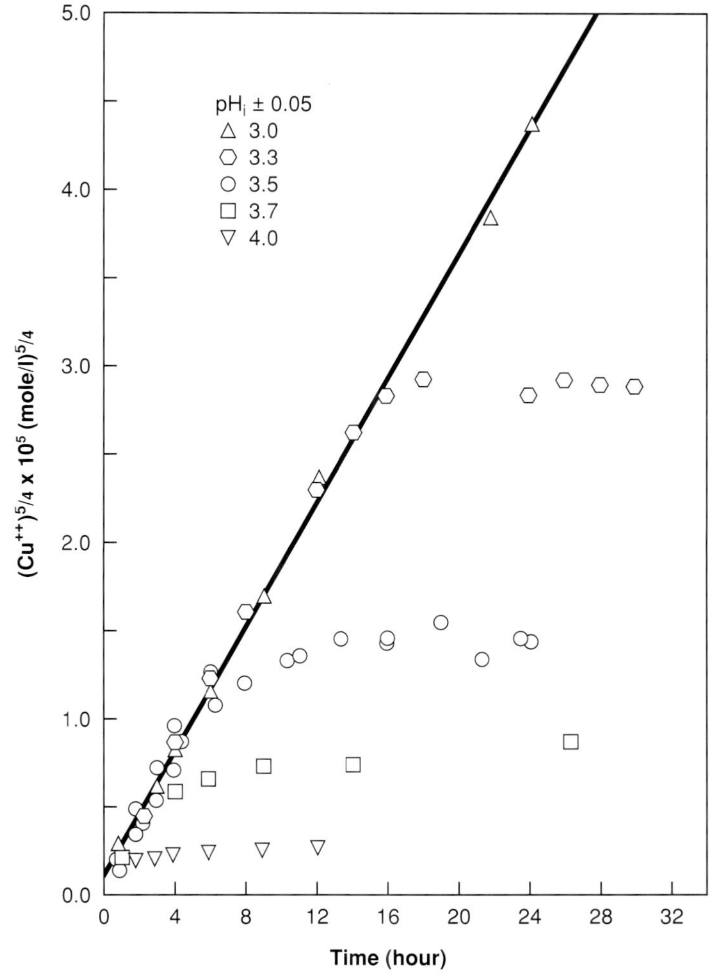

FIGURE 10. *Plot of Cu^{++} versus time for different initial pH values, after Yoon (1971:53).*

the layer formed depends on the relative quantities of solid and aqueous reactants present (Yoon 1971:47).

In sulfuric acid solutions containing $CuSO_4$, cuprite has not been observed to form below pH 3.1. The rate of formation of cuprite depends on the amount of exposed copper surface available for the reaction:

$$Cu + Cu^{++} + H_2O = Cu_2O + 2H^+$$

Near the surface of copper in such reactions, the pH decreases. The cuprite layer thus formed (above pH 3.1) grows at the copper-solution interface without extensive pitting or roughening of the substrate at the immediate interface, though etch pits are noted on adjacent exposed copper surfaces.

The following reaction is *not* favored in acid environments; Cu^+ is not stable in oxygen-rich solutions (Jenkins and Bertocci 1965:519, Scott and Miller 1966:884, 886, Miller and Lawless 1959:857):

$$2Cu^+ + 2OH^- = 2Cu_2O + H_2O$$

The presence of particular oxidants or depolarizers (HNO_3, $FeCl_3$, Cl^-, O_2, H_2O_2, CrO_4^-, $ClO_3^=$, and ClO_4^-) increases the rate of dissolution of copper in sulfuric acid solutions and solutions containing copper sulfate that are well stirred. In less well-stirred solutions, the initial dissolution of copper—often measured as the weight loss for copper coupons—also appears to increase as the concentration of the oxidant increases, typically up to a limiting value determined by the relative quantities of copper, sulfuric acid, and oxidant/depolarizer present. Following dissolution in these solutions, in the absence of stirring and resupply of H^+, corrosion films form. The composition of the films is determined by the components of the salt-acid solution. In some simple salt solutions (NaCl and Na_2SO_4), cuprite growth is not favored below a limiting pH level, for example, pH 5 for 0.1 N NaCl (Davenport et al. 1959:108–9; Miller and Lawless 1959:854; Ives and Rawson 1962d:462).[10]

Whether salts, oxidants, or depolarizers are present, the cuprite layer formed on copper in H_2SO_4 solutions is discontinuous. If formed rapidly or in poorly mixed solutions, the cuprite layer contains numerous voids and flaws that yield a nonprotective surface susceptible to further corrosion. Soluble aggressive anions (especially NO_3^-, Cl^-, $ClO_3^=$, and $SO_4^=$) trapped in the cuprite layer during film growth determine the susceptibility of the cuprite to subsequent dissolution. In bronzes and other copper alloys commonly used in monuments, secondary phase materials and their corrosion products (Sn, Zn, Pb, etc.) may increase the porosity of cuprite films.[11]

Gas-phase Reactions Between SO_2 and Copper Substrates

Also basic to any work on the mechanisms by which copper sulfate corrosion layers develop in the atmosphere are studies of the initial reactions between gas-phase SO_2, SO_3, and H_2SO_4 and Cu substrates. While the pioneering work by Vernon, Whitby, and associates in London in the late 1920s and 1930s described with great thoroughness the components of weathered patinas from copper roofs (Vernon and Whitby 1930:389–96; 1929:181–94; Vernon 1931:255, 582; 1932:153–61), laboratory work did not produce a close simulation of naturally weathered corrosion crusts. Very concentrated SO_2 atmospheres were used, up to and greater than 1% SO_2.

One important finding of their laboratory simulations was that the rate of corrosion induced by SO_2 dramatically increased as the RH rose to 65–75%, though rate increases above 75% RH were not as marked. Later work by Kosting (1937) and by Ericsson and Sydberger (1977) was also conducted under conditions much more concentrated with respect to gaseous sulfur oxides than are found in typical urban-industrial environments (Kosting 1937:18–19; Ericsson and Sydberger 1977:755–57; Nassau, Miller et al. 1987:703–19).[12] As a consequence, their findings of antlerite—the only listing for antlerite as a result of gas-phase reaction—do not relate to natural weathering phenomena, in view of the fact that the antlerite formed "directly" on the copper without the interposition of a cuprite layer and did not show a stratigraphy similar to what is found on monuments after decades or even centuries of exposure.[13]

Water in Thin Films

The presence of water on metal surfaces is essential for most corrosion reactions to develop at 20–25 °C. For most metal surfaces exposed outdoors, therefore, the rate of corrosion formation will be substantially affected by the RH. When a thin film of water is prevented from forming by very low RH, corrosion processes will become extremely slow despite the presence of relatively aggressive corrosion species in the air. On the other hand, as RH increases, the thickness of the water layer increases. The water layer becomes much thickened by dew, fog, mist, or rain; the chemical behavior of the water film, which includes absorbed and adsorbed gases, begins to approach the chemical behavior of bulk solutions. Some values for the weight and thickness of water films are shown in Table 2.

The processes of water-film formation are complicated during normal weathering by a myriad of factors, including those listed in Table 2: temperature and temperature change, RH and RH change, wind, ambient pollution levels, surface roughness and porosity, surface shape, and particulates, to list just a few (Schmidt 1970:2589[14]; Leidheiser, Jr. 1971:74). On already corroded surfaces, the presence of different minerals in the corrosion crust may enhance water absorption and retention. In the case of copper sulfates, there are a number of hygroscopic species which are likely to be present on corroding copper surfaces in urban-industrial atmospheres and which can absorb water from the air even at comparatively low RH levels (around 35%), thus promoting corrosion reactions. These include: chalcanthite ($CuSO_4 \cdot 5H_2O$), bonatite ($CuSO_4 \cdot 3H_2O$), $CuSO_4 \cdot H_2O$, and $CuSO_4$

In view of the many factors involved in copper corrosion and the many parameters to be taken into account in any study of atmospheric weathering, only a few of which are described here, it is not surprising that the study of thermodynamic stability regions and the kinetics of thin-film aqueous-phase reactions in the laboratory has not been combined, in particular for copper substrates with sulfate-rich films of little commercial value. Further thwarting any attempt to simulate outdoor exposure conditions is the fact that nonequilibrium reactions predominate in typical wetting events: there are never sufficient reactants (including water) nor time for true equilibrium to be established. The paucity of hydrogen and sulfate ions above pH 2.5 in sulfuric acid solutions is evident in Table 3. The majority of rain events result in only small volumes of rain per unit area of exposed copper or bronze surface, perhaps milliliters per square centimeter or less. In such hypothetical average rainstorms, the concentrations of hydrogen and sulfate ions that will be delivered to a square centimeter of exposed surface are expected to be only a small fraction of the values listed.

TABLE 2.[15] *Values of weight and thickness of water films on metal surfaces.*

Condition	*Deposited weight (g/cm^2)*	*Water thickness (nm)*	*Water monolayers*
Clean copper, 60% RH, 20 °C	0.0055	5.5	15
Clean copper, 90% RH, 20 °C	0.0095	9.5	27
Copper with deposited particles, $RH_{critical}$	0.01	10	30
Copper at 100% RH	1.0	10^3	3×10^3
Copper covered by dew	10	10^4	3×10^4
Copper wet from rain	10^2	10^5	3×10^5

TABLE 3. *Calculated concentrations of H^+, HSO_4^-, and $SO_4^=$ per liter in dilute sulfuric acid solutions from pH 0.9 to 7.0 at 25 °C, based on $K_2 = 1.20 \times 10^{-2}$ for $HSO_4^- = H^+ + SO_4^=$ (CRC Handbook of Chemistry and Physics 1989).*

pH	$[H^+]$ / liter	$[SO_4^=]$ / liter	$[HSO_4^-]$ / liter	Ratio $SO_4^=$ / HSO_4^-
0.9	126 mM	10.5 mM	105 mM	0.1
1.5	31.6 mM	7.0 mM	17.6 mM	0.40
1.9	12.6 mM	4.2 mM	4.2 mM	1.00
2.00	10.0 mM	3.60 mM	2.80 mM	1.26
2.50	3.16 mM	1.40 mM	0.35 mM	3.98
2.75	1.78 mM	0.83 mM	0.12 mM	7.08
3.00	1.00 mM	0.48 mM	0.038 mM	12.6
3.25	562 μM	275 μM	12.3 μM	22.4
3.50	316 μM	156 μM	3.9 μM	39.8
3.75	178 μM	88.4 μM	1.2 μM	70.8
4.00	100 μM	49.8 μM	0.395 μM	126
5.00	10 μM	4.99 μM	0.04 μM	1260
6.00	1 μM	499.9 nM	0.04 nM	12600
7.00	0.1 μM	50 nM	$<10^{-4}$ nM	126000

Added to the list of variables that complicate laboratory simulation of weathering reactions are solid-phase phenomena involving crystallization, hydration, resolidification, aggregation, and precipitation of mixed phases.

The following work concentrates on the primary phases detected in sulfate-rich films on copper substrates—principally copper, cuprite, brochantite, antlerite, posnjakite, chalcanthite, bonatite, and hydrated copper sulfates—as a first approach to understanding the sequence and mechanisms of mineralization.

Experimental Program

Initial Aging

Initial aging experiments were conducted in which corrosion rates were measured for 90Cu10Sn bronze exposed to sulfuric acid solutions, and 85Cu5Zn5Pb5Sn coupons exposed to acidic gas environments, at levels close to those given in the literature for urban-industrial environments with rain and fog to determine how closely those concentrations could be approached in subsequent experiments. A primary motivation was to avoid concentrations that would produce results skewed so far from conditions of natural weathering that they became meaningless in terms of investigating the processes involved.[16]

The most pertinent results may be briefly summarized as follows: (1) on coupons exposed in and over dilute sulfuric acid solutions (pH 2.8–5.0)—with and without oxidant (H_2O_2) but with sufficient acid to cause attack—the copper dissolution rate was very low over a period of six months, as was the rate of corrosion-film formation; and (2) in atmospheres with constant RH ≥75% containing HNO_3, 1 ppm $SO_{2(g)}$ (81× the level in typical Philadelphia air[17]), rapid corrosion did not develop on the 85-5-5-5 bronze coupons over a period of seven months, despite condensate being maintained at pH 2, 3, 4, and 5. Only cuprite was detected by XRD analysis (Philips 1840 diffractometer), except in the SO_2 gas-phase experiment with condensate at pH 2, where a very small amount of posnjakite was found initially.[18]

Phase Change

Analysis of phase change was conducted by wet-and-dry cycling using XRD. Based on observations from the initial simulated weathering experiments, it is clear that cyclic wetting and drying enormously increases the rate of corrosion formation on copper substrates, even when the concentration of reactants are not pushed many orders of magnitude above ambient levels.

Samples of antlerite[19] and brochantite[20] were sorted and scraped under the binocular microscope by scalpel and fine needle to eliminate unwanted impurities; the minerals thus cleaned were >95–98% pure, though optical inhomogeneities were evident under the microscope. The samples were analyzed elementally by atomic absorption spectrometry and XRF analyses and emission spectroscopy. Semiquantitative XRD powder analysis was undertaken on the materials, using the following 2θ lines: cuprite 42.3°, brochantite 35.6°, antlerite 18.3° and 24.8°.

Samples of bronze (85-5-5-5), copper, cuprite, antlerite, and brochantite, and a mixture of bronze and cuprite (equimolar with respect to Cu) were wetted for two different periods of time (500 and 10,000 hours) at regular intervals with well-controlled volumes (250 and 500 µL, respectively) of: (1) H_2SO_4 solutions of pH 2.50, 3.00, 3.50, 4.00, and 4.50; and (2) synthetic rain solutions[21] of pH 3 and 4 blanks that were wetted with deionized water in equivalent volumes. The samples were allowed to dry between wettings. The results of the cycling (drying/rewetting with temperatures between 15 and 40 °C) were compared to unexposed samples and the blanks, using XRD analysis to determine the extent of phase transformation or conversion. All samples were run in duplicate.

Dissolution

Dissolution studies were based on the authors' observations that the well-crystallized copper sulfate minerals antlerite and brochantite appear to have different stability in dilute sulfuric acid than is described in the corrosion literature (Graedel 1987b:760). Studies of the stability of the minerals—and of cuprite and copper for comparison—in sulfuric acid were undertaken, including some simple kinetic analyses of the relative rates of dissolution of the four substances in dilute sulfuric acid and synthetic-rain solutions based on Environmental Protection Agency data for the Philadelphia region.

Copper (99.999% Aesar), cuprous oxide (99.95%, Alfa), brochantite, and antlerite were separately exposed in three different procedures:

1. To 500 ml deionized water—open to the air—into which 0.002 N, 0.01 N, 0.05 N, and 0.1 N H_2SO_4 (ACS grade) were admitted in order to

reach different pH levels (from 6.20 to 2.50) with slow stirring (100–300 rpm). Samples were withdrawn at intervals for atomic-absorption analysis; each initial solution contained 100 μM/L solid.

2. By introducing the solids into: (a) H_2SO_4 solutions at different pH levels [2.50, 2.75, 3.00, 3.50, 4.00, 4.50, and 5.00]; (b) synthetic-rain solutions over the same range of pH, adjusted by addition of sulfuric and nitric acid [0.2 N H_2SO_4 : 0.1 N HNO_3], and (c) H_2SO_4 solutions over the same range of pH, to which was added 1 ml of ionic strength adjustor [0.5 N $NaNO_3$: 0.5 N Na_2SO_4] per 100 ml of solution. The solutions were continuously monitored for Cu^{++} by ISE measurements.
3. By following the increase of Cu^{++} in solution over time (1,800 minutes) for the above four solids exposed to (a) a solution of pH 3 sulfuric acid, and (b) a solution of synthetic rain at pH 3. The solutions were continuously monitored for Cu^{++} by ISE measurements.

All solutions were maintained at 20–23 °C, and all pH measurements were corrected to 20 °C.[22]

Precipitation

Precipitation studies were initiated by the lack of evidence for antlerite formation or stability in the preceding work. The goal of these studies was to form antlerite at room temperatures (20–23 °C).

In the first precipitation study, varying amounts of anhydrous $Cu(OH)_2$ (Aldrich no. 28,978-7, technical grade, Cu content 57%) and anhydrous $CuSO_4$ (Aldrich no. 20,917-1, 98%) were mixed dry. Over these were introduced 100 ml H_2SO_4 solutions of varying pH (2, 3, 4, and 5). At intervals of one month and seven weeks any new phase material was collected and analyzed by XRD. A constant volume of solution (100 ±3 ml) was maintained over the two months by adding deionized water. In the second precipitation study, varying amounts of Cu (Alfa, 99.95%) and anhydrous $CuSO_4$ (as above) were mixed in sulfuric acid solutions of pH 2, 3, 4, and 5 initially, and the new phase development monitored by XRD. The pH levels were measured over the course of each precipitation study.[23]

Quantitation studies were also carried out for cuprite:antlerite and cuprite: brochantite mixtures totaling 50 μM in the ratios of 90/10, 75/25, and 50/50. Ternary mixtures of cuprite:antlerite:brochantite (50/25/25 and 50/10/40) were also prepared and analyzed quantitatively, as were cuprite:covellite and cuprite:chalcocite mixtures.

Experimental Results

Wet-dry Cycling

The results of the wet-dry cycling experiments are given by the accompanying XRD plots of counts per second versus °2θ (Figs. 11–14). Briefly, the data indicate that for the longer exposure period (10,000 hours), all samples were converted at pH 2.5 to

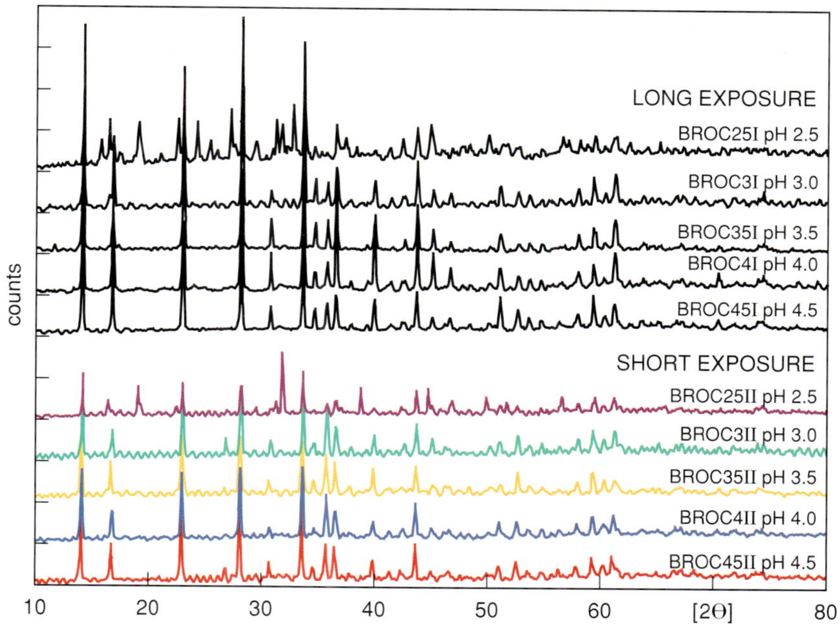

FIGURE 11. *Overlay of normalized XRD patterns for brochantite, showing changes at lower pH levels (2.5 and 3.0) in the long exposures (top) and less developed changes for the short exposures (bottom). The y axis gives normalized counts for the samples, overlaid for comparison.*

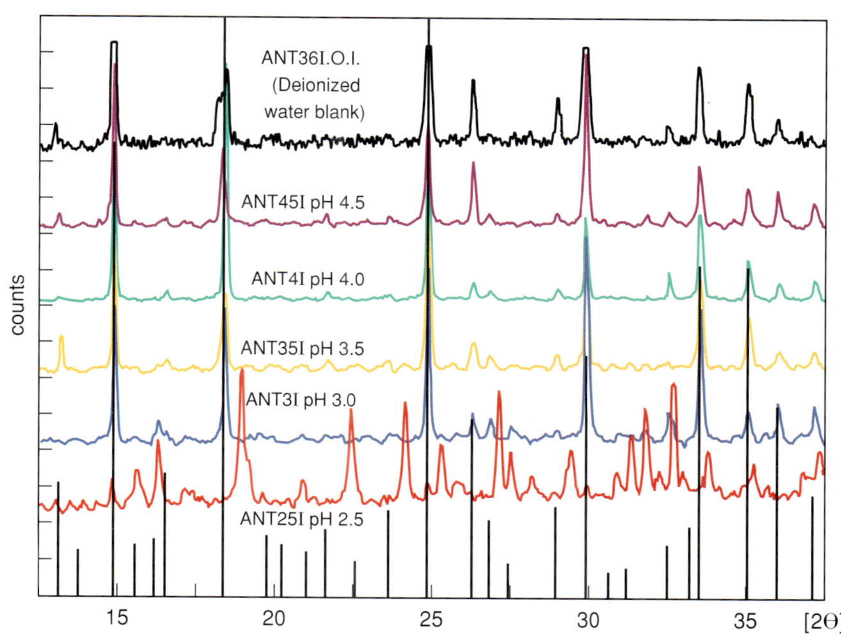

FIGURE 12. *Overlay of normalized XRD pattern for antlerite, on long exposure. The lines indicate the peak locations and relative heights for the unexposed antlerite "standard." The y axis gives normalized counts for the samples, overlaid for comparison.*

soluble copper sulfates—primarily chalcanthite and, to a lesser extent, bonatite. Over the shorter exposure period (500 hours), the low pH environment did not produce a conversion of brochantite to antlerite. Instead, antlerite, brochantite, copper, and cuprite all converted to more soluble copper sulfates. At pH 2.5, cuprite and copper formed chalcanthite rather than antlerite or brochantite.

At pH 3, antlerite and brochantite show much more conversion to soluble sulfate at the long exposure than at the short one; the peak heights for equivalent weights of sample are much reduced for these two species compared with higher pH levels (3.5) and show some alteration of the relative peak heights for each sample, suggesting that some dissolution and reprecipitation is occurring at the outside of the crystals and that a certain amount of noncrystalline or cryptocrystalline precipitation may be occurring. At pH 3, copper undergoes the only clearly observed alteration to antlerite

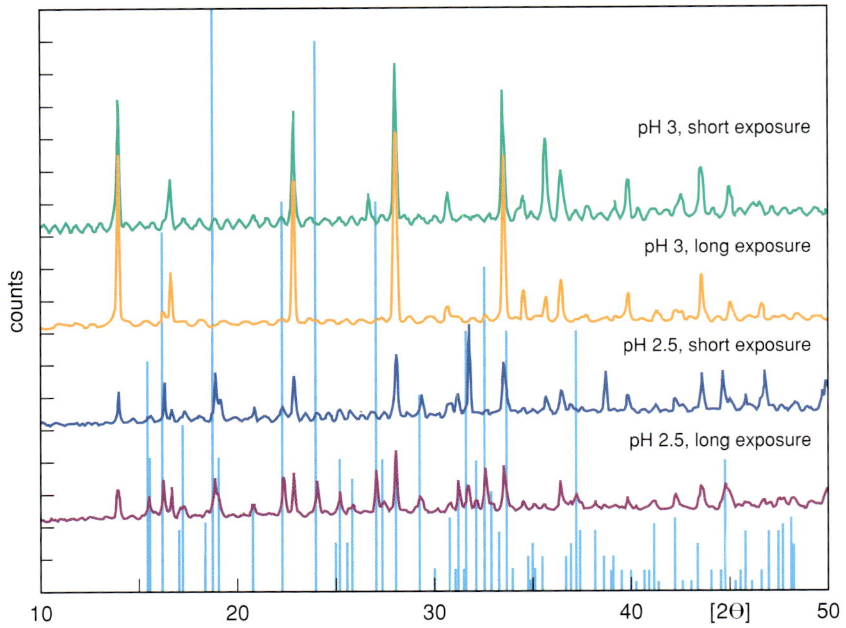

FIGURE 13. *Alteration of brochantite is shown over short and long exposure periods, at pH 2.5 and 3.0. The vertical lines give the position of the primary chalcanthite peaks over this range of 2θ angles. The counts are normalized for overlay on the y axis.*

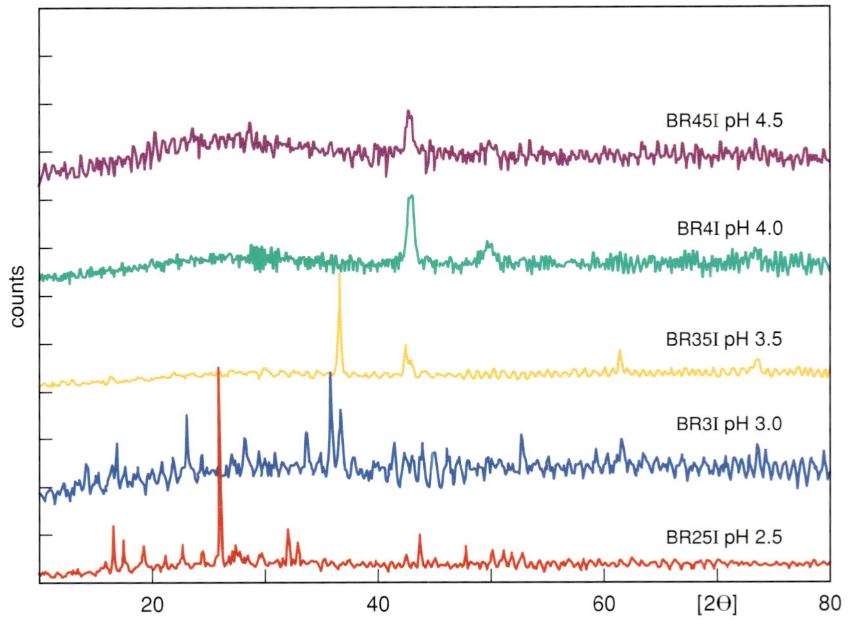

FIGURE 14. *Alteration of bronze is given for the long exposure period, illustrating the development of mostly cuprite at pH 3.5, mixed cuprite and brochantite at pH 3.0, and predominantly chalcanthite at pH 2.5. The counts are normalized for overlay on the y axis.*

in the course of all the experiments to date, brochantite is also produced, and cuprite underlies the sulfate layers. Curiously, both the bronze and the mixture of bronze and cuprite failed to develop antlerite under similar exposure conditions. Cuprite itself at pH 3 tended to develop into brochantite and later chalcanthite.

At pH 3.5, the rate of transformation is considerably slower than at the lower pH levels. The net (low) concentration of reactants available is in part responsible.

Table 4 sets out the theoretical yields for each run: The first figure is for a conversion dependent on H^+ only, the second for a conversion dependent on sulfate ion only. In run 1, where a total of 25 ml of synthetic acid rain solution (L/40) was delivered over the period from July 17 to October 25, 1990, in 250–500 μl portions (54×), the theoretical conversion or dissolution of the various samples—all at 50 μM—is given for each pH:

TABLE 4. *Theoretical yields for brochantite and antlerite samples at various pH levels, the first figure in the theoretical yield being based on H⁺ dissolution only and the second figure being based on $SO_4^{=}$ as the primary reactant in the dissolution process.*

	RUN 1		RUN 2	
pH level	*Available H^+ per 25 ml*	*Maximum theoretical yield*	*Available H^+ per 15 ml*	*Maximum theoretical yield*
2.5	3.2mM/40	100 / 88	3.2mM/66.7	93 / 46
3.0	1.0mM/40	50 / 25	1.0mM/66.7	30 / 16
3.5	0.3mM/40	16 / 8	0.3mM/66.7	9 / 5
4.0	0.1mM/40	5 / 2.5	0.1mM/66.7	3 / 1.5
4.5	0.03mM/40	1.6 / 0.8	0.03mM/66.7	0.9 / 0.5

In run 2, where a total of 15 ml (3L/200) of synthetic rain was delivered in 500 µl portions over the period from August 23 to October 25, 1990 (30X), the theoretical conversion or dissolution of antlerite, brochantite, bronze, copper, and a mixture of bronze and cuprite—all 50 µM initially—is given at each pH. In run 2, the cuprite sample was increased to 200 µM; the yield at each pH level should thus be divided by 4 in Table 4. The data in this table suggest that at higher pH, little H^+ is available to drive the conversion of these samples with the small volume delivered.

It would appear from the data that more reprecipitation or recrystallization occurs at pH 3.5 and to a lesser degree at pH 4 for antlerite and brochantite. This is in keeping with the dissolution data above, which show that both basic sulfates are soluble to a degree at these intermediate pH levels.

At pH 4.5, little change from the blanks and the standards is evident in either the long or the short run, suggesting that the changes are below the readily detectable limits for the XRD technique employed, probably about 3% for these particular materials.

Dissolution and Kinetic Studies

Results of dissolution and kinetic studies show that both antlerite and brochantite begin dissolving in very dilute acid solutions. The amount of Cu^{++} produced is a direct function of the amount of available H^+ in solution. For simplicity, some of the results are summarized in graph form as often as possible.[24]

Figures 15–24 indicate the direct dependence of antlerite and brochantite dissolution on H^+ concentration. At low pH (<3), the distinction between the rate of basic sulfate dissolution and of cuprite dissolution becomes indistinct under the given conditions of the experiment. At pH levels above 4, cuprite dissolution is notably slower than that of the basic sulfates.

Table 5 gives the results of a stability-dissolution study of antlerite, brochantite, cuprite, and copper in H_2SO_4 solutions. The pH, ppm Cu^{++} found in solution, and H^+ in mM/L added to reach each pH point are given for each substance. Each solution contained 100 µM/L of solid initially, except for copper, which contained 200 µM/L (weights were: antlerite 17.99 mg, brochantite 23.55 mg, and cuprite 7.10 mg). The initial volume of each solution was 500 ml (deionized water; the temperature was 20–22 °C). The solutions were stirred during each measurement period (5–10 m)

FIGURE 15. *Dissolution of brochantite showing dependence on available pH.*

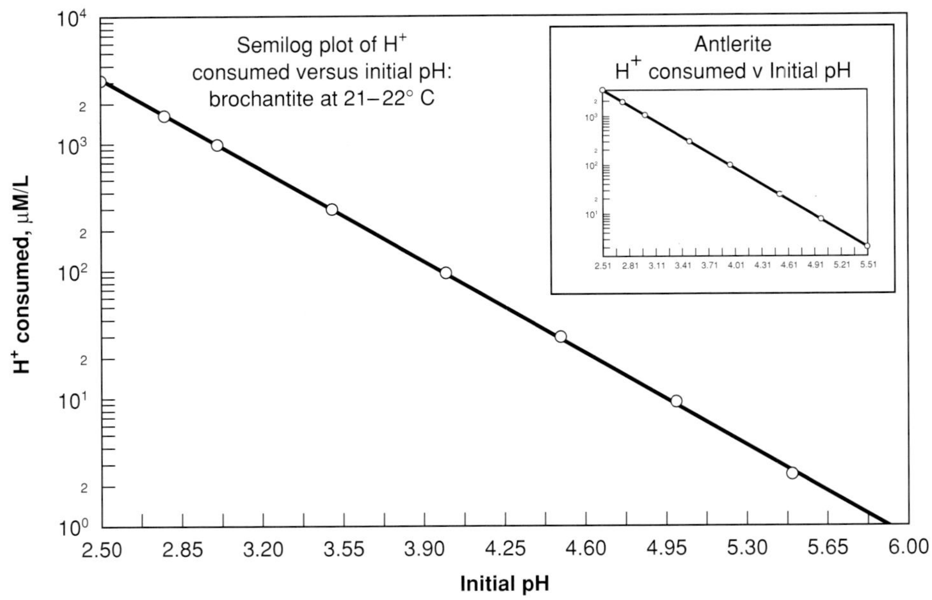

FIGURE 16. *Dissolution studies for antlerite showing linear relationship between Cu^{++} in solution and H^+ consumed.*

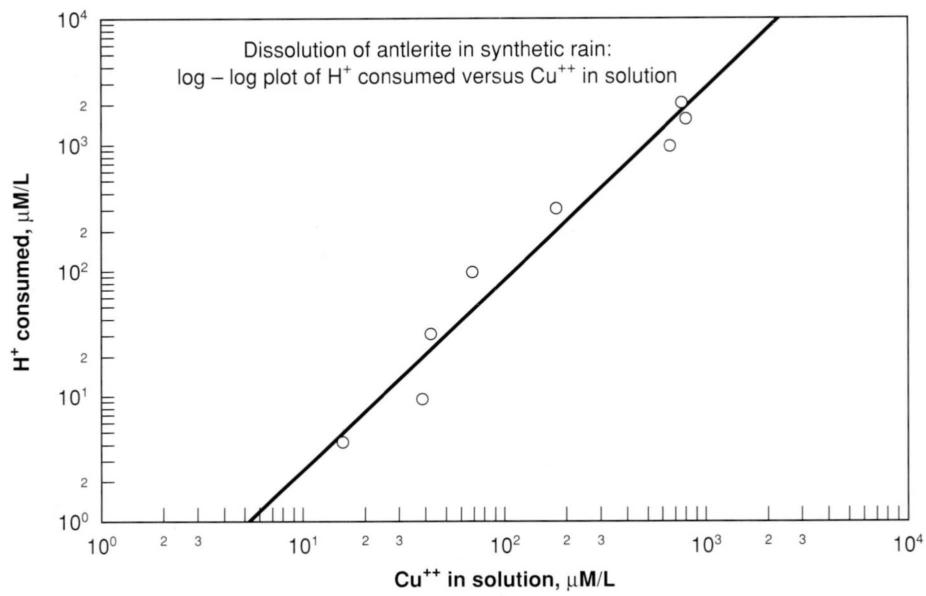

FIGURE 17. *Kinetic dissolution study in sulfuric acid.*

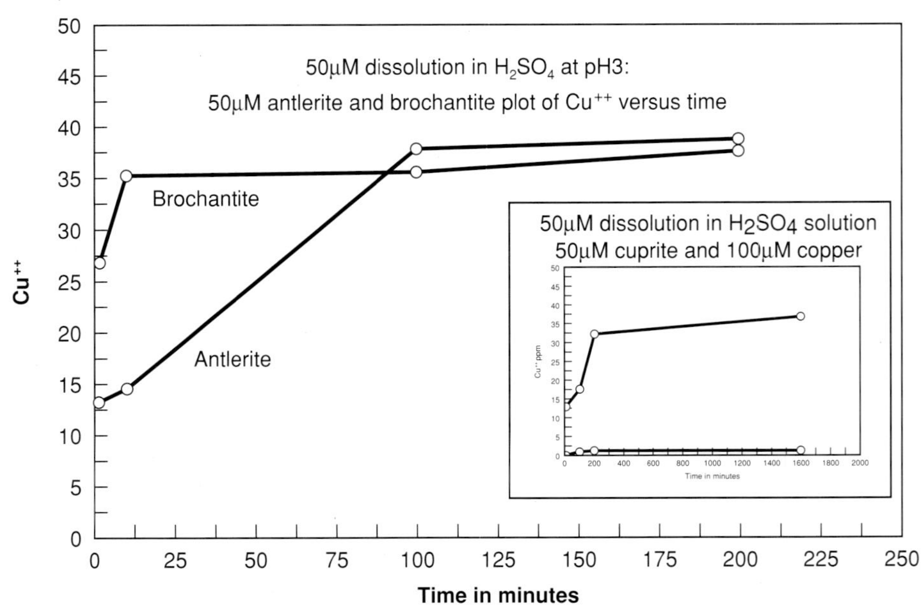

FIGURE 18. *Kinetic dissolution study in synthetic rain. Note that antlerite dissolves more slowly than brochantite.*

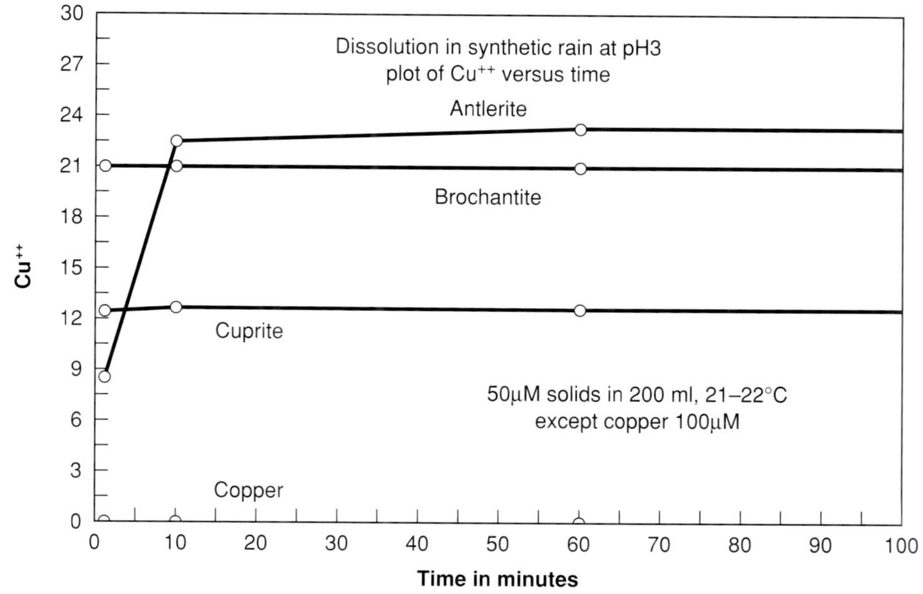

FIGURE 19. *Relative solubilities in H_2SO_4 solutions of different pH.*

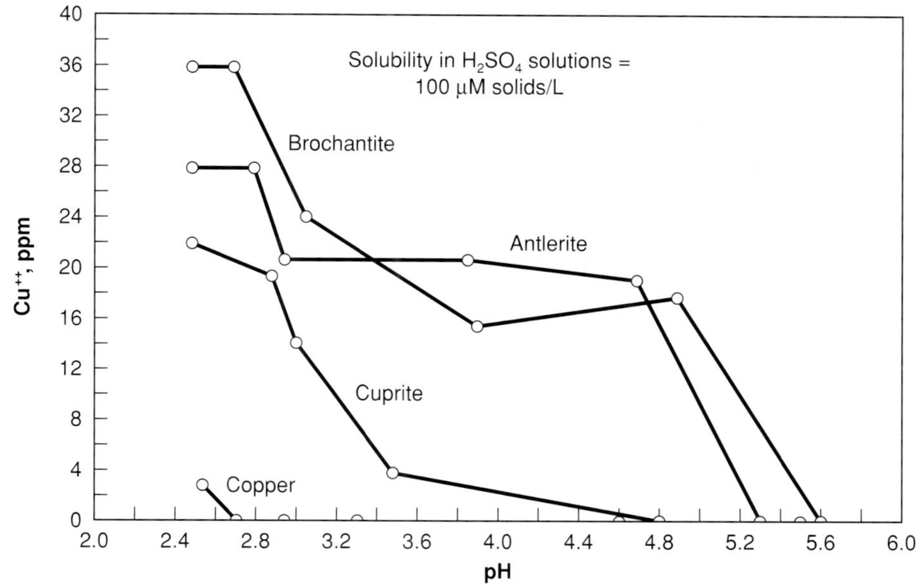

at 100–300 rpm. Regular volumes of sulfuric acid (0.002, 0.01, and 0.1 N) were introduced by burette into the solution. Following pH measurement, equivalent volumes of solution were extracted for atomic absorption analysis.

The kinetics of dissolution at pH 3 in H_2SO_4 solutions is shown in Table 6. The heading of each column indicates the time in minutes at which the pH and ppm Cu^{++} measurements for the given sample was made. The samples were 200 µM/L—$Cu_3SO_4(OH)_4$17.89 mg; $Cu_4SO_4(OH)_2$23.01 mg; and Cu_2O 7.55 mg, except for copper, which was 452 µM/L (7.23 mg). The volume of the solution in each case was 250 ml; the temperature was 20–22 °C.

Table 7 shows the kinetics of dissolution at pH 3 in synthetic-rain solutions. The heading of each column indicates the time in minutes at which the pH and ppm Cu^{++} measurements for the given sample was made. The samples were 250 µM/L—$Cu_3SO_4(OH)_4$ 18.34 mg; $Cu_4SO_4(OH)_2$ 23.35 and 20.02 mg; and Cu_2O 9.53 mg, except for copper, which was 500+ µM/L (7.53 mg). The volume of the solution in each case was 200 ml; the temperature was 20–22 °C.

TABLE 5. *Data from stability-dissolution study of sulfates in H_2SO_4 solutions, given as pH/ppm Cu^{++}/H^+ added in mM/L for each sample.*

Antlerite	*Brochantite*	*Cuprite*	*Copper*
5.30/nd/.02	5.60/1.0/.02	5.50/nd/.002	5.30/nd/.002
3.85/6.2/.12	3.90/4.6/.12	3.48/1.1/.088	3.30/nd/.088
2.95/6.2/.52	3.05/7.2/.52	3.00/4.2/.49	2.94/<<1/.49
2.80/8.4/1.02	2.70/10.8/1.02	2.88/5.8/.99	2.70/<1/.99
2.50/8.4/3.02	2.49/10.8/3.02	2.49/6.6/2.99	2.53/1.0/2.99

TABLE 6. *Kinetics of dissolution at pH 3 in H_2SO_4 solutions. For each sample pH/ppm Cu^{++} figures are given for the stated time.*

Sample	*1.2 min.*	*10 min.*	*100 min.*	*200 min.*
Antlerite	3.05/13.2	3.18/16.8	3.90/38.4	4.05/39.6
Brochantite	3.10/27.0	3.27/35.4	3.80/36.0	4.20/39.2
Cuprite	2.87/10.4	2.91/13.2	3.30/17.0	3.20/10.6
Copper	3.00/nd	3.05/nd	3.05/0.6	3.05/0.8

TABLE 7. *Kinetics of dissolution at pH 3 in synthetic-rain solutions. For each sample pH/ppm Cu^{++} figures are given for the stated time.*

Sample	*1 min.*	*10 min.*	*60 min.*	*1080 min.*
Antlerite	3.30/8.4	3.90/22.5	4.05/23.4	4.70/22.8
Brochantite	3.10/21.0	3.70/21.0	4.60/31.2	4.70/21.0
Cuprite	3.10/13.2	3.35/13.2	3.30/13.2	3.70/13.2
Copper	3.10/nd	3.10/nd	3.20/nd	20/nd

Precipitation Studies

The precipitation results were produced as follows: Mixtures of anhydrous $CuSO_4$ and $Cu(OH)_2$ at 20–23 °C were dissolved in acidified sulfuric acid solutions of different pH; 0.01 M ratios by weight, measured on a microbalance to 0.1 mg, solution volume of 100 ml. The maximum theoretical concentration of Cu^{++} species was 0.4 M/L. Solution volumes were maintained with deionized water to within 5 ml of initial volume for two months and covered with plastic film during standing. The capital letters in Tables 8 and 9 indicate new phases being formed: A=antlerite; B=brochantite; C=cuprite, L=langite—$Cu_4SO_4(OH)_6 \cdot H_2O$; P=posnjakite—$Cu_4SO_4(OH)_6 \cdot H_2O$; and R=ramsbeckite—$Cu_{15}(SO_4)_4(OH)_{12} \cdot 6H_2O$.

Experimental Procedures

During the course of laboratory work, features of aqueous mixtures containing hydrated $CuSO_4$ basic copper sulfates and cuprite-copper H_2SO_4 became more apparent and certainly influenced the results of the quite simple empirical tests described above.

The hygroscopicity and reactivity of the copper sulfates (anhydrous as well as antlerite and brochantite) contrasted strongly with the poor wettability of purified

TABLE 8. *Predominant new-phase growth and pH change of solutions are shown after one month and after seven weeks.*

Molar ratio initial pH 1 month/7 weeks	*pH 2*	*pH 3*	*pH 4*	*pH 5*
2 $CuSO_4$:1 $Cu(OH)_2$	B	B	B, ?L, P	B, ?L, P
pH change	4.22/4.18	4.19/4.17	4.28/4.23	4.24/4.19
1 sulfate:1 hydroxide	B	B	B	B
pH change	4.39/4.39	4.42/4.41	4.42/4.40	4.38/4.37
1 sulfate:2 hydroxide	B	B	B	B
pH change	4.50/4.42	4.51/4.48	4.55/4.53	4.48/4.45
1 sulfate:3 hydroxide	B ≥ P, L	B + P	P > B	P >> B
pH change	4.64/4/59	4.71/4.68	4.76/4.70	4.76/4.68

TABLE 9. *The predominant new-phase development after one month is shown for mixtures of copper and anhydrous cuprous sulfate.*

Molar ratio initial pH	*pH 2*	*pH 3*	*pH 4*	*pH 5*
3 Cu:1 $CuSO_4$	C	C	C	C
pH change				
2 copper:1 sulfate	C	C	C	C, ?B
pH change				
1 copper:1 sulfate	C	C >> B	C > B	C > B *?R
pH change				
1 copper:2 sulfate	C	C	C > B	C > B
pH change				
1 copper:3 sulfate	C	C	C > B	C
pH change				

cuprite, copper, or bronze. The rate of wetting and mixing of the finely powdered samples in the dissolution studies was affected by these differences. For the atomic-absorption and ISE measurements, data for the first 10 minutes of dissolution must have been determined in part by these differences, complicated by problems of extracting samples for atomic-absorption analysis while solids floated on the surface of the solution or adhered to the surface of the ISE electrode. Later measurements are less susceptible to variations produced by clumping and uneven distribution of solids, as more effective mixing had occurred and small clumps were broken by stirring rods.

The longer exposures of the drop experiments eventually produced films on the copper, cuprite, bronze, and bronze-cuprite samples that modified and improved the wettability of the samples: in short-term exposures the differences in hydrophilic-hydrophobic mixing were more apparent. Mounting the samples for XRD analysis with deionized water also revealed the differences between samples on the basis of wetting, which are significant in corrosion phenomena.

In the precipitation studies, the mobility of simple copper sulfates in water resulted in the easy separation of the copper sulfates from the newly formed basic copper sulfates, or cuprite, on drying. The simple sulfates moved to the evaporation fronts. Of interest to those collecting samples from outdoor monuments was the observation that thin film and very finely divided deposits of simple copper sulfates are often transparent and colorless, in contrast to the strong blue transparent examples of these compounds found in textbooks; in very thin films they may be easy to overlook during collection from a weathered surface.

As the various phases underwent dissolution or conversion in the drop and in the precipitation experiments, the size and shape of the remaining and newly formed solids became more varied with the passing of time. On slow drying, as one might predict, the crystalline structures that developed on some faces tended to grow into solution or into the air, while the faces that pointed toward the bottom and sides of the glass vessels were not easily distinguished and tended to grow together in the constrained space, perhaps analogous to their growth and recrystallization within a corrosion crust. The more rapid drying in the drop experiments produced some new-phase growth near the evaporation fronts; this was very finely divided and approached the appearance of naturally weathered solids (Hemming 1977:98–99, figs. 3, 4), in which erosion and particulate abrasion play significant roles in reducing and rounding crystalline forms.

The results of the dissolution studies indicate that antlerite may be slightly slower to dissolve near pH 3 than is brochantite, but it is not clear how this kinetic difference plays a significant role in the long-term development of a corrosion crust.

Conclusions

The kinetics of the acid dissolution of well-crystallized copper sulfates indicates that in low pH environments common to polluted urban-industrial environments, acidified copper-sulfate solutions are readily produced from the corrosion crust of bronze and copper substrates, in particular from brochantite and antlerite. These aggressive $CuSO_4$ solutions are likely to be responsible for part of the growth of the cuprite crust based on the reaction:

$$Cu + CuSO_4 + 2\,H^+ + {}^1/_2\,O_2 = Cu_2O + H_2SO_4$$

This finding is similar to the findings of Miller and Lawless (1959) and Robertson et al. (1958) in their tightly controlled investigations of systems in which $CuSO_4$ was present.

Dissolution of the copper-sulfate-rich materials on the crust exterior is initially accompanied by a rise in pH:

for antlerite:

$$CuSO_4 \cdot 2Cu(OH)_2 = 3Cu^{++} + SO_4^{=} + 4[OH]^-$$

or

$$CuSO_4 \cdot 2Cu(OH)_2 + 4[H]^+ = 3Cu^{++} + SO_4^{=} + 4H_2O$$

for brochantite:

$$CuSO_4 \cdot 3Cu(OH)_2 = 4Cu^{++} + SO_4^{=} + 6[OH]^-$$

$$CuSO_4 \cdot 3Cu(OH)_2 + 6[H]^+ = 4Cu^{++} + SO_4^{=} + 6H_2O$$

Certainly during rainstorms and other periods of substantial wetness, the surface copper-sulfate films modify the pH of the aqueous phase that reaches the metal-corrosion crust interface, raising the pH while supplying Cu^{++} and $SO_4^{=}$ ions. The modification of the pH of atmospherically deposited solutions in the crust is not a simple process. For instance, as the thin film of the aqueous phase dries out on the corroded metal surface, the pH is expected to fall. A 10:1 reduction in volume by evaporation of water should drop the pH one unit. Opposing this pH lowering, the activity of the Cu^{++} and $SO_4^{=}$ ions rapidly decreases as their concentration exceeds 0.05 M in concentrated solutions. The precipitation studies, moreover, indicate that after long wetting of solids typically found in corrosion crusts, the initial pH of the aqueous phase is overwhelmed by the copper sulfate material—solids and ions in solution—and reaches intermediate values, perhaps in the region of pH 4–5.

At the surface of weathered sulfate-rich crusts on copper alloys, it would appear that ionic dissolution reactions predominate, rather than electrochemical reactions such as the dissolution of cuprite or copper. The direct electrochemical dissolution of copper or cuprite, which is usually found near the metal interface, is hindered by the thickness and convoluted nature of the crust. At the metal-corrosion interface, however, a different series of reactions takes place, modified by the chloride ions which appear to concentrate often in this zone, and by the available copper-sulfate-rich solution that has permeated through the crust.

Past models for these phenomena have suggested that bulk aqueous-phase reactions occur throughout the corrosion films on copper alloys. The present work indicates that the outermost sulfate-rich layers will supply ions to the aqueous front from heavy rain or other condensation and then serve as sites for solidification as the aqueous phase dries out and evaporates at the surface.

This work also suggests that the reactions and transformations that occur during the relatively dry periods of exposure, when the main mass of pollutant anions and cations are deposited, are equally important. In these situations, the extensive regions of hydration that form around basic copper-sulfate crystals would include (in their outer, last-to-solidify faces) the very soluble chlorides and nitrates. These outer zones would then be the first to redissolve in the presence of additional water. While one would expect some of this material—which is mechanically soft and friable—to be eroded in rainstorms, hydrated and disordered crystalline forms would allow for the accumulation of anions on drying, increasing with time, and providing the copper-rich and anion-rich solutions that stabilize and build the cuprite layer.

It is clear from the wet-and-dry cycling experiments and from the precipitation experiments that the transition of brochantite + acid = antlerite does not easily occur in corrosion crusts by gross aqueous-phase reaction, even over periods of several months. Rather, it would appear that the transformation of brochantite to antlerite

$$3[CuSO_4 \cdot 3Cu(OH)_2] + H^+ + SO_4^{=} = 4[CuSO_4 \cdot 2Cu(OH)_2] + OH^- + H_2O$$

and

$$2[CuSO_4 \cdot 3Cu(OH)_2] + CuSO_4 = 3[CuSO_4 \cdot 2Cu(OH)_2]$$

is not kinetically favored at 20–23 °C. Similarly, the transformation of Cu_2O to antlerite does not appear to be favored near 20 °C. This transformation would be:

$$3Cu_2O + 2H_2SO_4 + 3H_2O + O_2 = 2[CuSO_4 \cdot 2Cu(OH)_2]$$

In the authors' work to the present, which has not strictly duplicated the full range of conditions described in typical Pourbaix-type diagrams at 20 °C, the only laboratory evidence for antlerite formation appears in the cycling experiments, where in low-pH solutions (pH 3) antlerite was found to grow in association with brochantite on copper substrates. In these samples, the presence of a cuprite layer over the copper is evident under the microscope. At lower pH (2.75 and 2.5), with the initial absence of Cu^{++} ions in solution, both chalcanthite and bonatite develop with time.[25] Using the precipitation experiments described, further aging experiments are currently underway to determine whether much longer periods (six months to one year or more) are necessary for the development of antlerite at 20 °C, as well as experiments at higher temperatures (35–40 °C), which appear to favor antlerite formation.

In summary, the zones of stability predicted for antlerite and brochantite at 20 °C by thermodynamic calculations for the $Cu-H_2SO_4-H_2O$ system have not been observed for well-crystallized mineral specimens. It is likely that weathered corrosion films in polluted atmospheres are even less well behaved, particularly with respect to the observable (real-time) events of dissolution and precipitation in a corrosion crust. In the analysis of corrosion processes in sulfate-rich environments, strict adherence to thermodynamic considerations appears to produce a misleading picture of the sequence and nature of the precipitation and dissolution of the corrosion layers, in which a number of complex hydrated species may exist. This complexity discourages the usefulness of antlerite as an indicator of corrosion aggressiveness, though its existence cannot be denied in corrosion films from monuments sited in polluted urban atmospheres (in temperate zones), nor can its high sulfate content be overlooked.

Acknowledgments

The authors gratefully acknowledge that part of this work was conducted with the support of the National Park Service (NPS Bronze Statues subcontract no. B104985-3) administered by Susan Sherwood and Jan Meakin, the latter at the University of Delaware. Also we want to especially thank Karen Fried for her tireless work in quantifying the XRD data; and Elizabeth Pirrotto, Nancy Heller, and Michelle Barger for their help in running samples.

Notes

1. See also Holm and Mattsson 1980:85–104.
2. After Mattsson 1982:16.
3. See also Mattsson 1982:17. No antlerite region is drawn in his E_H versus pH diagram. Also Leidheiser, Jr. (1971:3–24) omits antlerite as a stable corrosion species in his discussion of long-term atmospheric corrosion.

4. The role of dry deposition has until recently been understated in analyses of corrosion processes in the atmosphere (Dolske and Meakin 1991:A–13). The graph from Graedel 1987b:759 indicates that dry deposition may account for two-thirds of the sulfur dioxide that comes in contact with a metal surface. In urban-industrial environments, the surfaces of copper-sulfate corrosion crusts are hygroscopic, a property that probably facilitates SO_2 incorporation into the crust. In and near marine sites, chlorides play an important role in the rate and formation of corrosion crusts on copper-based metals. At marine sites, chloride deposition rates of 0.3–300 mg Cl^-/m^2/day have been cited by Mattsson (1982:10) in comparison with the following SO_2 rates in mg/m^2/day: rural 0–30, urban up to 100, and industrial up to 200. Some esoteric but less dominant factors are described by Cobb and Gross (1969:796–804) and Leidheiser (1971:71–75), including thermogalvanic effects.
5. The selective attack on the alpha phase in bronzes exposed to the atmosphere has been previously noted by Leoni (1977:245) and by Lins (1992). The phase boundaries of the system $Cu-SO_3-H_2O$ have been studied repeatedly in work by Tunell and Posnjak (1929:1ff), Pourbaix and his colleagues (1966:385–92; 1949:53–81; Bustorff and van Muylder 1964:607), Silman (1958), Garrels (1960:50–71), Yoon (1971), and more recently by Graedel et al. (1987), Woods and Garrels (1986a, 1986b), Williams et al. (1990), and Livingston (1991).
6. Silman indicated that within the parameters of the system he considered a stability zone for antlerite as only one log ($a_{Cu^{++}}$) unit wide. For example, the limiting activity for cupric ions at the boundary between antlerite and chalcanthite at point * (where log $SO_4^{=}$ = 24, log CO_2 = 25, and log $a_{Cu^{++}}$ = 0) is: (log $a_{Cu^{++}}$) + 1 = (0) + 1 = 1.
7. Values given for the range of H^+ and principal anions in rain, fog, and dew are numerous and show considerable variation (Graedel and Schwartz 1977:17–25).
8. Among the most relevant references for this study are: Bockris and Enyo 1962:1187; Elwakkad 1950:3563; Fenwick 1926:860; Giles and Bartlett 1961:266; Guthrow and Miller 1966:415; Halpern 1953:421; Hill 1953:345; Hurlen 1961a–c; Ives and Rawson 1962a:447; Ives and Rawson 1962b:452; Ives and Rawson 1962a–d; King and Weidenhammer 1936:602; Kruger 1959:847; 1961:503; Kruger and Calvert 1964:1038; Lal and Thirsk 1953:2638; Lambert and Trevoy 1958:18; Lu and Graydon 1954:153; Mattsson and Brockis 1959:1586; Miller and Lawless 1959:854; Näsänen and Tamminen 1949:1995; Petit 1965:291; Porterfield and Miller 1966:528; Robertson et al. 1958:569; Royer, Kleinberg, and Davidson 1956:115; Russel and White 1927:116; Scott and Miller 1966:883–86; Topham and Miller 1966:421; Tourky and Elwakkad 1948:740; Wadsworth and Wadia 1955:755; and Weeks and Hill 1956:203.
9. Kruger (1961) suggests that near neutrality CO_2 may allow soluble tenorite to be reduced to cuprite in very pure water solutions.
10. See also Yoon 1971:9, 41, 48–49; Gregory and Riddiford 1960:952; Cocleugh and Graydon 1962:1370–72; Bjorndahl and Nobe 1984:82–87; and Hurlen 1961c:1246.
11. This is the authors' observation, illustrated repeatedly in conservation work. See also Miller and Lawless 1959:859; Cathcart and Peterson 1968:595–97; Bambulis 1962:1130–34; Marchesini and Badan 1979:206; Marabelli 1992; Langennegger and Callahan 1972:252–53; and Finnegan et al. 1981:256–61. The authors found that thin films of cuprite on very short-term exposures retarded the rate of dezincification. It is

clear that near 20 °C the rate of cuprite growth is many orders of magnitude greater in solution than in air. See also Krishnamoorthy and Sicar 1969:734–36, and Roennquist and Fischmeister 1960:65–75.

12. See also City of Philadelphia Department of Public Health 1987.
13. Additional work has been carried out in this area in recent years. Among the most pertinent studies for the weathering of monuments has been the work carried out at the Bell Laboratories, summarized in part in Franey 1987:306–14.
14. Few differences were observed in a seven-year exposure test between samples ranging in hardness from 45 to 120 HV and of varying surface roughness. The time of wetness, determined by the angle of exposure, was more significant, the more horizontal and slow-drying coupons corroding faster than the vertical ones.
15. After Graedel 1987:728.
16. In one experiment, 1.0 cm × 7.5 cm coupons of 90 Cu 10Sn bronze were suspended in excess sulfuric acid solutions (pH 5), with and without oxidant in the form of hydrogen peroxide (5% by weight), at 20–23 °C. The solutions were replenished weekly; the spent solutions were saved for atomic absorption analysis. In addition, coupons of 85-5-5-5 bronze (analyzed by XRF, Kevex Analyst 8000 System, data shown in table below), both patinated and polished to 400 grit, were suspended in research-grade air (MG Scientific Ultra Zero Grade, less than 0.1 ppm THC) raised to 75% RH by bubbling through deionized water to which nitric acid at pH 2, 3, 4, and 5 were added. The exposure chamber, divided into four equivalent volumes of 20 × 20 × 20 cm, was fitted to a permeation tube (Kin-Tek) that delivered 1.0 ± 0.1 ppm sulfur dioxide gas under flow conditions of 2.4 L per minute at 20–20 °C. Initial composition of coupons suspended in acidic atmosphere:

XRF analysis	*Cu*	*Zn*	*Pb*	*Sn*	*Fe*	*Sb*	*Ag ppm*
as polished	85.42	5.58	4.35	5.12	0.14	0.11	443
patinated	85.28	5.49	4.57	5.15	0.14	0.13	525

The coupons were all examined directly by insertion into an X-ray diffractometer (Phillips 1840). Those immersed in acid solutions were examined after 25 days, 69 days, and 7 months of exposure. Those exposed to the acidic gas environment were examined after 500 hours and after an 8-month interval. The solutions were analyzed for copper ions by atomic absorption spectrometry (Perkin-Elmer 303). The calibration standards were prepared by dissolving 99.999% Cu in ACS-grade nitric acid and diluting with deionized water (>2MW resistivity, from Continental dual-bed cartridges with carbon filter) to 1, 2, 4, 8, 10, and 12 ppm for the atomic absorption work, with deionized and synthetic-rain blanks. The pH measurements were made with a glass electrode (Orion research-grade 910100) and the double-junction-reference electrode on a Fisher 910 meter and on a Fisher 800 meter.

17. Much of Vernon's work was undertaken at 500,000–1,000,000x the ambient level of SO_2 (12 ppb) in Philadelphia air (City of Philadelphia 1987). Graedel and Schwartz (1977:17–25) state that in 17% of 447 measured sites, the SO_2 concentration was found to average 0.020 ppm or higher.

18. After storage for one year (45–55% RH and 70–76 °F), basic copper nitrate had crystallized on the surface of the pH 2 sample, though this very soluble form was not detected as a crystalline phase immediately after exposure in the chamber.
19. Courtesy of Ward's Scientific Department.
20. Courtesy of the University of Delaware Department of Geology, Mineral Collection.
21. Taken by averaging the data for 31 rain events recorded in the Philadelphia area varying in pH from 3.40 to 5.91 with average pH of 3.84 and the following concentrations of ionic species in microequivalents per liter: H^+ 143, NH^{4+} 58, Ca^{++} 73, Mg^{++} 30, Na^+ 31, K^+ 11, Zn^{++} 8: $SO_4^=$ 219, NO_3^- 98, Cl^- 24. The synthetic-rain solution was composed of the following in micromole/l: zinc sulfate 4, calcium sulfate 36.5, ammonium sulfate 29, sodium sulfate 9, magnesium nitrate 15, potassium chloride 11. All salts were 3 N or purer. The pH was adjusted by adding a 1:1 volume mixture of 0.02 N sulfuric and 0.01 N nitric acids (ACS grade).
22. The solutions in Section 1 were analyzed for copper ions by atomic absorption (Perkin Elmer 303). Sections 2 and 3 used cupric ISE measurements for Cu^{++} assays (ISE Orion 94-29 with a double-junction Ag/AgCl reference electrode Orion 900200 on a Fisher 910 Isomet meter with digital readout to ±1 mV). The calibration standards were prepared by dissolving 99.999% Cu in ACS-grade nitric acid and diluting with deionized water (>2 MW resistivity, from Continmental dual-bed cartridges with carbon filter) to 1, 2, 4, 8, 10, and 12 ppm for the atomic-absorption work, with deionized blanks and synthetic-rain blanks. The standards were to 0.1 ppm, 1 ppm, 10 ppm, 100 ppm, and 1000 ppm for the ISE analysis; for ISE measurements, calibration curves including blanks were run before and after each measurement. The pH measurements were made with a glass electrode and the double-junction-reference electrode mentioned previously.
23. Prior to use, all solids were assayed by atomic-absorption and emission spectroscopy, optical microscopy, and XRD. The emission spectroscopy results (Baird Spectrograph, 3m path length with 15,000 lines per inch grating, alternating-current [ac] arc at 1,100 V and 3–4 A for 30 seconds, plate set for the range 230–375 nm) indicated that the Cu and Cu_2O were primarily copper with impurities at less than 10 ppm, while the principal contaminants for antlerite were Si and Al at or below 1% by weight, and for brochantite Ag (1–5%), Fe (0.1–1%) and Mg and Si at less than 0.1%. The atomic-absorption yields for Cu and Cu_2O were 95% or better following digestion in hot, concentrated HNO_3; the yield for the cleaned antlerite samples was 95 ±3% and for cleaned brochantite 96 ±3%. Copper (alpha copper, JCPDS no. 4-667) was identified in the cuprite sample at a level below 3%. The copper sample did not show an oxide layer in the analysis, indicating only that the layer was below the level of detection by the instrument. Prior to cleaning, the detectable contaminant phases in the antlerite were alpha quartz, beta-$CuAlO_2$, kyanite, alumina, pyrope ($Mg_3Al_3[SiO_4]_3$), and cuprite; the brochantite sample contained detectable amounts of cuprite, chalcopyrite, covellite, and stromeyerite.
24. An example of the raw data is included in Table 10.
25. Which phase was detected was probably an artifact of preparation caused by gentle warming of the samples (usually <40 °C) to drive off excess moisture prior to XRD analysis.

TABLE 10. *Raw data from dissolution and kinetic studies of antlerite.*

Species/pH	Date	Titration 1°	Titration 15m	Intermediate 8 m	Final 1 y	H^+ consumption 15 m	H^+ consumption 1 y	Cu^{++} μM/L 15 m
Ant 2.5	(5/90)	2.51	3.01	4.88	4.94	2.16 mM	3.15 mM	790
Ant 2.75	(5/90)	2.75	5.2	5.24	5.15	1.77 mM	1.77 mM	830
Ant 3	(5/90)	3.0	5.4	5.56	5.51	996 μM	997 μM	690
Ant 3.5	(5/90)	3.5	5.6	5.71	5.61	313 μM	313 μM	190
Ant 4	(8/1)	4.0	5.65	6.49	6.23	97 μM	99.5 μM	85
Ant 4.5	(8/1)	4.5	6.10	6.35	6.23	30.8 μM	31.0 μM	40
Ant 5	(8/1)	5.0	6.18	6.43	6.42	9.3 μM	9.6 μM	40
Ant 5.5	(8/1)	5.5	6.34	6.44	6.49	2.7 μM	2.8 μM	15
AntAR 2.5	(8/6)	2.50	3.01	4.82	4.81	2.18 mM	3.14 mM	535
AntAR 2.75	(8/6)	2.75	3.81	5.10	4.95	1.6 mM	1.77 mM	771
AntAR 3	(8/6)	3.00	5.08	5.30	5.05	992 μM	991 μM	425
AntAR 3.5	(8/6)	3.50	5.53	5.89	5.40	313 μM	312 μM	165
AntAR 4	(8/6)	4.00	5.84	6.40	5.47*	99 μM	97 μM	42
AntAR 4.5	(8/3)	4.50	6.20	6.27	6.10	31 μM	30.8 μM	37
AntAR 5	(8/3)	5.00	6.27	6.46	6.41	9.5 μM	9.6 μM	19
AntAR 5.5	(8/3)	5.32	6.28	6.37	6.38	4.3 μM	4.4 μM	14
AntISA 2.5	(8/3)	2.52	3.27	4.84	4.89	2.48 mM	3.02 mM	1071
AntISA 2.75	(8/3)	2.75	4.36	5.36	5.46	1.73 mM	1.77 mM	818
AntISA 3	(8/3)	3.00	5.13	5.54	5.52	998 μM	997 μm	661
AntISA 3.5	(8/2)	3.50	5.43	6.31	6.03	312 μM	315 μM	212
AntISA 4	(8/2)	4.00	5.87	6.16	6.03*	98.4 μM	99.2 μM	66
AntISA 4.5	(8/2)	4.50	6.07	6.42	6.43	30.7 μM	31.2 μM	34
AntISA 5	(8/2)	5.00	6.35	6.55	6.56	9.6 μM	9.6 μM	14
AntISA 5.5	(8/2)	5.50	6.44	6.37	6.65/.49	2.8 μM	2.8 μM	19

REFERENCES

BABOIAN, R., AND B. CLIVER

1986 Corrosion on the *Statue of Liberty*. *Materials Performance* 25(5):74, 80.

BAMBULIS, J.

1962 Dissolution of brass in sulfuric acid solutions. 1. 85/15 Brass. *Journal of the Electrochemical Society* 109(12):1130–34.

BJORNDAHL, W. D., AND K. NOBE

1984 Copper corrosion in chloride media: Effect of oxygen. *Corrosion* 40(2):82–87.

BOCKRIS, J. O'M., AND M. ENYO

1962 Mechanism of electrodeposition and dissolution processes of copper in aqueous solutions. *Transactions of the Faraday Society* 58:1187.

BUSTORFF, A., AND J. VAN MUYLDER

1964 Comportement électrochimique du cuivre en solutions sulfuriques: Diagrammes d'équilibres tension-pH du système $Cu-SO_4H_2-OH_2$, à 25 °C. *Electrochimica Acta* 9:607.

CATHCART, J. V., AND G. F. PETERSON

1968 Structural characteristics of the oxide on the (111) of copper. *Journal of the Electrochemical Society* 115(6):595–97.

CITY OF PHILADELPHIA DEPARTMENT OF PUBLIC HEALTH

1987 *Emission Inventory and Air Quality Data Report to the Air Pollution Control Board.* Philadelphia.

COBB, A. W., AND G. W. GROSS

1969 Interfacial electrical effects observed during the freezing of dilute electrolytes in water. *Journal of the Electrochemical Society* 116(6):796–804.

COLCLEUGH, D. W., AND W. F. GRAYDON

1962 The kinetics of hydrogen peroxide formation during the dissolution of polycrystalline copper. *Journal of Physical Chemistry* 66(7)1370–72.

COSTAS, L. P.

1980 Atmospheric corrosion of copper alloys exposed for 15 to 20 years. In *Atmospheric Corrosion of Metals*, 106–15. Technical Publication 767. Philadelphia: American Society for Testing and Materials (ASTM).

CRC HANDBOOK OF CHEMISTRY AND PHYSICS

1989 70th edition, D–165. Boca Raton: CRC Publishing Co.

DAVENPORT, W. H., V. F. NOLE, AND W. D. ROBERTSON

1959 An investigation of chemical variables affecting the formation of films on copper in aqueous solutions. *Journal of the Electrochemical Society* 106(12):108–9.

DOLSKE, D. A., AND J. D. MEAKIN

1991 Acid deposition impacts on historic bronze and marble statuary and monuments at Gettysburg, Pennsylvania. *Corrosion* 91 (NACE), Paper 152:A–13.

EDNEY, E. O.

Effects of acidic deposition on metals. In *State-of-Science/Technology, Report 19; Effects of Acidic Deposition on Materials*, eds. P. A. Baedecker, E. O. Edney, and P. J. Moran et al., 2–47. Draft for public review. Washington, D.C.: National Acid Precipitation Assessment Program.

EDWARDS, R. A., M. POLLARD, R. G. THOMAS, AND P. A. WILLIAMS

1990 Mineralogical aspects of the oxidation of copper alloy artefacts toward the establishment of authenticity. In *Surface Colouring & Plating of Metals*, June 14–16. London: British Museum Research Laboratory.

ELWAKKAD, S. E. S.

1950 The equilibrium between copper and cupric and cuprous ions and the behaviour of the copper electrode in dilute copper sulfate solutions. *Journal of the American Chemical Society*, 72:3563.

ERICSSON, R., AND T. SYDBERGER

1977 Corrosion products formed on copper exposed to humid SO_2 containing atmospheres. *Werkstoffe und Korrosion* 28(10):755–57.

FENWICK, F.

1926 The equilibrium between cupric ion, cuprous ion and metallic copper. *Journal of the American Chemical Society* 48:860.

FINNEGAN, J. E., R. E. HUMMEL, AND E. D. VERINK, JR.

1981 Optical studies in the dezincification of alpha-brass. *Corrosion* 37(5):256–61.

FRANEY, J.

1987 Degradation of copper and copper alloys by atmospheric sulfur. In *Degradation of Metals in the Atmosphere*, eds. S. W. Dean and T. S. Lee, 306–14. Philadelphia: ASTM.

GARBASSI, F., AND E. MELLO

1984 Surface spectroscopic studies on patinas of ancient metal objects. *Studies in Conservation* 29:174.

GARRELS, R. M.

1960 *Mineral Equilibria*. New York: Harper's Geoscience Series.

GARRELS, R. M., AND C. L. CHRIST

1965 *Solutions, Minerals and Equilibria*. New York: Harper's Geoscience Series.

GILES, F. H., AND J. H. BARTLETT

1961 Anodic behavior of copper in phosphoric acid. *Journal of the Electrochemical Society* 108:266.

GRAEDEL, T. E.

1987a Copper patinas formed in the atmosphere II. *Corrosion Science* 27(7):730–31.

1987b Copper patinas in the atmosphere III. *Corrosion Science* 27(7):759–60

GRAEDEL, T. E., AND N. SCHWARTZ

1977 Air quality reference data for corrosion assessment. *Materials Performance* 16(8):17–25.

GRAEDEL, T. E., ET AL.

1987 *Corrosion Science* 27(7). Special Issue: Copper Patina.

GREGORY, D. P., AND A. C. RIDDIFORD

1960 Dissolution of copper in sulfuric acid. *Journal of the Electrochemical Society* 107(12):950–56.

GUTHROW, C. E., JR., AND G. T. MILLER, JR.

1966 Oxidation of copper single crystals in aqueous solutions of inorganic salts I: Topography of oxide formed in $CuSO_4$ and $Cu(OAc)_2$ solutions. *Journal of the Electrochemical Society* 113:415.

HALPERN, J.

1953 Kinetics of the dissolution of copper in aqueous ammonia. *Journal of the Electrochemical Society* 100:421.

HEMMING, D. C.

1977 The production of artificial patination on copper. In *Corrosion and Metal Artifacts*, 479:98–99, figs. 3–4. NBS Special Publication. Washington, D.C.: National Bureau of Standards.

HILL, G. R.

1953 The kinetics of the initial corrosion of copper in aqueous solutions. *Journal of the Electrochemical Society* 100:345.

HOLM, R., AND E. MATTSSON

1980 Atmospheric corrosion tests of copper and copper alloys in Sweden: 16 year results. *Atmospheric Corrosion of Metals*, 85–104. Technical Publication 767. Philadelphia: A.S.T.M.

HURLEN, T.

1961a Dissolution of copper in acidified ferric sulfate solutions. *Acta Chemica Scandinavica* 15:615.

1961b On the kinetics of the Cu/Cu^{++}_{aq} electrode. *Acta Chemica Scandinavica* 15:630.

1961c Dissolution of copper by oxidizing agents in acid chloride solution. *Acta Chemica Scandinavica* 15:1239, 1246.

IVES, D. J. G., AND A. E. RAWSON

1962a Copper corrosion I: Thermodynamic aspects. *Journal of the Electrochemical Society* 109:447.

1962b Copper corrosion II: Kinetic studies. *Journal of the Electrochemical Society* 109:452.

1962c Copper corrosion III: Electrochemical theory of general corrosion. *Journal of the Electrochemical Society* 109:458.

1962d Copper corrosion IV: The effects of saline additions. *Journal of the Electrochemical Society* 109:462.

JENKINS, L. H., AND U. BERTOCCI

1965 On the equilibrium properties of single crystalline copper electrodes. *Journal of the Electrochemical Society* 112(5):519.

KING, C. V., AND L. WEIDENHAMMER

1936 The rate of solution of copper in dilute aqueous reagents. *Journal of the American Chemical Society* 58:602.

KOSTING, P. R.

1937 Copper patina. *American Society for Testing and Materials Bulletin* 87:18–19.

KRISHNAMOORTHY, P. K., AND S. C. SICAR

1969 Influence of oxygen pressure on the oxidation kinetics of copper in dry air at room temperature. *Journal of the Electrochemical Society* 116(6):734–36.

KRUGER, J.

1959 The oxide films formed on copper single crystal surfaces in pure water I: Nature of the films formed at room temperature. *Journal of the Electrochemical Society* 106:847.

1961 The oxide films formed on copper single crystal surfaces in water II: Rates of growth at room temperature. *Journal of the Electrochemical Society* 108:503.

KRUGER, J., AND J. P. CALVERT

1964 The oxide films formed on copper single crystal surfaces in water III: Effect of light. *Journal of the Electrochemical Society* 111:1038.

LAL, H., AND H. R. THIRSK

1953 The anodic behaviour of copper in neutral and alkaline chloride solutions. *Journal of the Electrochemical Society* 1953:2638.

LAMBERT, R. H., AND D. J. TREVOY

1958 Analysis of films on copper by coulometric reduction. *Journal of the Electrochemical Society* 105:18.

LANGENNEGGER, E. E., AND B. G. CALLAHAN

1972 Use of an empirical potential shift technique for predicting dezincification rates of alphabeta-brasses in chloride media. *Corrosion* 28(7):252–53.

LEIDHEISER, H. JR.

1971 *The corrosion of copper, tin and their alloys*. New York: J. Wiley and Sons.

LEONI, M.

1979 Copper alloys: Their structure and corrosive phenomenon. In *The Horses of San Marco Venice,* 245. G. Perocco, ed. Milan and New York: Olivetti.

LEOTTA, M., AND M. MARABELLI

1984 Le forme di corrosione. In *Marco Aurelio: Mostra di Cantiere. Le Indagini in Corso sul Monumento* 81. Rome: Arti Grafiche Pedanesi.

LINS, A.

1992 The cleaning of weathered bronze monuments. In *Proceedings of "Dialogue 89."* Baltimore: National Association of Corrosion Engineers (NACE).

1990 Report on the bronzes at the Swann Fountain. Submitted to Fairmont Park, City of Philadelphia.

LIVINGSTON, R. E.

1991 Influence of the environment on the patina of the *Statue of Liberty*. *Environmental Science and Technology* 25(8):1400–8.

LU, C. Y., AND W. F. GRAYDON

1954 The rate of dissolution of copper. *Canadian Journal of Chemistry* 32:153.

MARABELLI, M.

1987 Characterization and conservation problems of outdoor metallic monuments. In *Conservation of Metal Statuary and Architectural Decoration in Open-Air Exposure (ICCROM, Symposium 6–8 October 1986, Paris*), 214. Rome:ICCROM.

1992 The environment and the future of outdoor bronze sculpture: Some criteria of evaluation. In *Proceedings of "Dialogue 89."* Baltimore: National Association of Corrosion Engineers (NACE).

MARCHESINI, L., AND L. BADAN

1979 Corrosion phenomena on the Horses of San Marco. In *The Horses of San Marco Venice,* 200–10. G. Perocco, ed. Milan and New York: Olivetti.

MATTSSON, E.

1982 The atmospheric corrosion properties of some common structural metals. *Materials Performance* 21(7):16–19.

MATTSSON, E., AND J. O'M. BOCKRIS

1959 Galvanostatic studies of the kinetics of deposition and dissolution in the copper + copper sulfate system. *Transactions of the Faraday Society* 55:1586.

MILLER, G. T., JR., AND K. R. LAWLESS

1959 An electron microscope study of the formation of oxide on copper single crystals immersed in an aqueous solution of copper sulfate. *Journal of the Electrochemical Society* 106:854, 857, 859.

NÄSÄNEN, R., AND V. TAMMINEN

1949 The equilibria of cupric hydroxysalts in mixed aqueous solutions of cupric and alkali salts at 25 °C. *Journal of the American Chemical Society* 71:1995.

NASSAU, K., P. K. GALLAGHER, A. E. MILLER, AND T. E. GRAEDEL

1987 The characterization of patina components by X-ray diffraction analysis and evolved gas analysis. *Corrosion Science* 27(7):671, 675, 684.

NASSAU, K., A. E. MILLER, AND T. E. GRAEDEL

1987 The reaction of simulated rain with copper, copper patina, and some copper compounds. *Corrosion Science* 27(7):703–19.

NIELSON, N.

[1981] 1984 Examination of copper and copper corrosion product patina from the *Statue of Liberty*. Report from E.I. duPont de Nemours and Co. to the U.S. National Park Service. Reprinted in *Materials Performance* 23(4):78.

PETIT, M. C.

1965 Etude aux faibles tensions, de la valence initiale de dissolution du cuivre. *Electrochimica Acta* 10:291.

PORTERFIELD, W. W., AND G. T. MILLER, JR.

1966 Oxidation of copper single crystals in aqueous solutions of inorganic salts III: Kinetics of oxidation in $Cu(OAc)_2$ solution. *Journal of the Electrochemical Society* 113:528.

POURBAIX, M.

1949 *Thermodynamics of Dilute Aqueous Solutions*. London: Edward Arnold and Co.

1966 *Atlas of Electrochemical Equilibrium*. Oxford: Pergamon Press.

1977 *Corrosion and Metal Artifacts: A dialogue between conservators and archeologists and corrosion scientists*. NBS Special Publication No. 472:13. Washington, D.C.: National Bureau of Standards.

ROBERTSON, W. D., V. F. NOLE, W. H. DAVENPORT, AND F. P. TALBOOM, JR.

1958 An investigation of chemical variables affecting the corrosion of copper. *Journal of the Electrochemical Society* 105:569.

ROENNQUIST, A., AND H. FISCHMEISTER

1960 The oxidation of copper: A review of published data. *Journal of the Institute of Metals* 89:65–75.

ROYER, D. J., J. KLEINBERG, AND A. W. DAVIDSON

1956 The anodic behavior of copper in aqueous solutions. *Journal of Inorganic and Nuclear Chemistry* 4:115.

RUSSEL, R. P., AND A. WHITE

1927 Effect of oxygen concentration on the corrosion of copper by non-oxidizing acid. *Industrial and Engineering Chemistry* 19:116.

SCHMIDT, M.

1970 Observations on natural patination on copper. *Journal of the Institute of Metals* 98:2589.

SCHMITT, H. H.

1962 *Equilibrium Diagrams for Minerals at Low Temperature and Pressure*, 110–26. Cambridge, Mass.: Geological Club of Harvard.

SCOTT, D. B., AND G. T. MILLER, JR.

1966 Oxidations of copper single crystals in aqueous solutions of inorganic salts IV: Effect of pH on oxide topography in $CuSO_4$ and $Cu(OAc)_2$ solutions. *Journal of the Electrochemical Society* 113:883–86.

SILMAN, J. R.

1958 *The stability of some oxidized copper minerals in aqueous solutions at 25 °C and 1 atmosphere total pressure.* Ph.D. diss., Harvard University.

TOPHAM, R. W., AND G. T. MILLER, JR.

1966 Oxidation of copper single crystals in aqueous solutions of inorganic salts II: Topography of oxide formed in $Cu(C1O_4)_2$ and $Cu(CHO_2)_2$ solutions. *Journal of the Electrochemical Society* 113:421.

TOURKY, A. R., AND S. E. S. ELWAKKAD

1948 Studies on some metal electrodes, part I: Oxide-film formation on copper, and the evaluation of the standard electrode potential of the metal. *Journal of the Chemical Society* 70:740.

TUNELL, G., AND E. POSNJAK

1929 On the system $Cu-SO_3-H_2O$. *American Journal of Science* 18:1ff.

VERNON, W. H. J.

1931 A laboratory study of the atmospheric corrosion of metals. *Transactions of the Faraday Society* 27:255, 582.

1932 The open-air corrosion of copper, part III: Artificial production of green patina. *Journal of the Institute of Metals* 49:153–61.

VERNON, W. H. J., AND L. WHITBY

1929 The open-air corrosion of copper: A chemical study of the surface patina. *Journal of the Institute of Metals* 42:181–94.

1930 The open-air corrosion of copper, part II: The mineralogical relationship of corrosion products. *Journal of the Institute of Metals* 44:389–96.

WADSWORTH, M. E., AND D. R. WADIA

1955 Reaction rate study of the dissolution of cuprite in sulphuric acid. *Journal of Metals* (June) 83:755.

WEEKS, J. R., AND G. R. HILL

1956 The kinetic of the corrosion of copper in acid solutions. *Journal of the Electrochemical Society* 103:203.

WOODS, T. L., AND R. M. GARRELS

1986a Phase relations of some cupric hydroxy minerals. *Economic Geology* 81(8):1989–2007.

1986b Use of oxidized copper minerals as environmental indicators. *Applied Geochemistry* 1(2):181–87.

YOON SEUNG RYUL

1971 The corrosion of copper in acid chlorate solution. Ph.D. diss., University of Utah, Department of Mining, Metallurgical and Fuels Engineering.

Biographies

Andrew Lins has held the position of conservator of sculpture and decorative arts at the Philadelphia Museum of Art since 1979.

Tracy Power is a private conservator in San Francisco. Her background includes a master's degree from the University of Delaware.

The Technology of Medieval Jewelry

JACK OGDEN

Despite the extensive research carried out on ancient goldsmithing technology over the last half century, the techniques of medieval goldsmiths have been largely ignored. The following is a preliminary examination of various technical aspects of medieval goldwork, including a comparison of medieval technology with that of Roman and earlier periods. In general, most techniques remained constant. One major exception, however, is the widespread use of drawn gold wire by about the ninth or tenth century C.E. This technique opened up new possibilities for decorative filigree methods which were used to advantage by medieval goldsmiths. Thus an understanding of technological developments is of primary importance for the stylistic study of medieval European gold jewelry.

Early Medieval Practices

Until about 1500 C.E., medieval jewelry techniques in Western Europe, particularly Britain, were related in a multitude of ways to the techniques of their Classical and Byzantine predecessors. Although the stylistic development of jewelry is not the principal concern here, it is important to note that material and technique do influence style to a considerable extent. For example, changes in the style of filigree work—of paramount importance in characterizing jewelry development in the medieval period—cannot be fully considered without understanding the various manufacturing options available to the goldsmith.

From the point of view of the jewelry historian, the distinction between ancient and medieval is based more on terminology than any real discontinuity. A Hellenistic goldsmith could pick up a medieval jewel and be familiar with most of its technology and the sequence of manufacture; the goldsmith might be surprised mainly by the novelty of design and the use of drawn wire. Even the workshops would have been comparable. Wall paintings of jewelers' workshops as seen in the houses of the Vetii at Pompeii, for example, differ little in their essentials from Renaissance or even more recent depictions; the same types of tables, furnaces, anvils, and scales are present.

The medieval period has provided a range of representations of goldsmiths and jewelers at work. Some are religious paintings, such as Taddeo Gaddi's fourteenth-century *St. Elgius as a Goldsmith,* now in the Prado, Madrid. Others illustrate technical treatises, such as the mid-fifteenth-century manuscript of Pliny's *Natural History* preserved in the Victoria and Albert Museum, London (Whalley 1982). This latter manuscript, for example, has a rare depiction of a lapidary drilling precious stones with a pump drill. Representations of jewelers' shops are less common; however an engraving of a jeweler selling rings, plus some rather unorthodox gemmological tests, may be seen in a late-fifteenth-century work by Johannis de Cuba (Kunz [1917:facing 343] 1973). Representations of workshop scenes are far more prevalent from the sixteenth century onward.

Medieval documentary references to goldsmiths are quite common. The prime source is the German monk Theophilus, who compiled a comprehensive account of goldsmithing procedures in the first half of the twelfth century C.E. The third book of his *Treatise on Divers Arts* begins with details on how to construct a workshop and equip it with tables, seats, and forges, and then describes the individual tools and their use. Theophilus is now generally taken to be the pen name of Roger of Helmarshausen, a goldsmith-monk who actually left samples of his work, such as a book cover now in the Domschatz Trier (Ronig 1984:cat. no. 65; Hawthorne and Smith 1979:pl. 8). The identification of Theophilus with Roger of Helmarshausen and the dating of the treatise have been most recently discussed by Dodwell ([1961] 1986) and Hawthorne and Smith (1979).

Near-contemporaries of Theophilus include Jean de Garlande of about 1200, who briefly mentions goldsmiths. He writes that "the craft of these goldsmiths consists of hammering with little hammers on the anvil thin sheets of gold and silver and in setting precious stones in the bezels of rings for use by barons and noblewomen" (Cripps 1880:9). The Englishman Alexander Neckham (1157–1217) describes the necessary tools of a twelfth-century goldsmith (Neckham 1867). These include a forge with top chimney; bellows worked with one hand evenly and with light pressure; an anvil; forceps to hold metal during beating; a hammer for shaping gold; very sharp chisels and another chisel (possibly) for cutting stones, a touchstone to test metal; a whetstone to sharpen iron tools; a hare's foot (*pedem leporarium*) to smooth, polish, and clean gold; a leather bag around the waist to collect grains of metal "lest they vanish away"; a toothed saw (*serram dentatum*); and a file for gold (*limam aurifrigium*). It is interesting that there is no mention of wiredrawing equipment. A hare's foot seems an unlikely implement to use for polishing gold, but perhaps it was really used to clean the bench. Earlier this century Maryon advocated the use of the same implement: "The bench is swept several times a day with a hare's foot, which forms a convenient little brush to which the gold will not adhere" (1971:1).

Theophilus describes how different tools were made and shows that goldsmiths made their own equipment. This was true in antiquity and right up to modern times in many parts of the world. A regulation of the year 1513 states that goldsmiths should not work copper or brass apart from making their tools of the trade (Prideaux

1896–97:39). On this basis, many of the more complex or sophisticated implements, such as elaborate wiredrawing equipment or rolling mills, would be out of the question for small craftshops. The large, fully staffed, and impressively equipped workshops seen in Renaissance and later engravings could hardly have existed earlier outside of royal patronage or the main urban centers.

It was a common practice for goldsmiths to bequeath their tools to their sons, friends, or fellow goldsmiths. Sometimes wills list the tools in some detail. For example, the will of John Colan, a goldsmith of York who died in 1490, contains a comprehensive list of the goldsmith's equipment (Cripps 1914). It enumerates several varieties of hammers, stamps, swages, stakes, files, tongs, wiredrawing implements, a gold balance, and more. But one must not forget a vital truth about any craft: Tools are worthless unless manipulated by someone with skill and experience. As Biringuccio said at the end of his discussion of goldwork, "Having told you the general facts concerning the art of working gold and silver I think I have said enough about it. All the rest that is necessary depends entirely on ingenuity and practice" (Smith and Gnudi 1943).

Much documentary evidence for the medieval jeweler's trade derives from regulations defining their working practices and the limitations on the material they could use. Fraud by means of debased gold, plated base metals, filled hollow-sheet objects, and imitation stones were potential problems for the jewelry-buying public throughout history. Neckham said that the goldsmith should be able to "discern well the pure gold from the latten [brass] and copper lest he buy latten for pure gold" (1867).

Medieval English regulations stipulated that jewelers had to work in certain main streets, not in "lanes or alleys"; only very seldom was work in private houses allowed. Similar restrictions are laid down in the tenth century *Book of the Prefect* from Constantinople (Hendy 1985), and similar rules might well have been widespread. Goldsmiths should be visible, but the actual secrets of the craft should not be divulged (Prideaux 1896–97:doc. no. 44; 150). There are references to goldsmiths being punished because they allowed strangers to "see the privities of the trade" (Prideaux 1896–7:15–16; 22). In the early sixteenth century we are told that no goldsmith could work at night by candlelight or on Sundays.

There was clearly rivalry between goldsmiths. One was punished because he told a customer that a competitor's work was overpriced; another was castigated because he alleged that a rival did not known how to make a cup (Prideaux 1896–97:15–16). Successful goldsmiths could certainly become rich, and this meant they could also amass some gold and silver as stock in trade, rather than just working their patrons' materials.

It has been assumed that a master goldsmith would be skilled in all aspects of the trade (Lightbown 1978:84), but Biringuccio points out that such wide mastery was rare, and that in practice "one has his profession in engraving, another in the hammer, and another in mounting or setting jewels, and in making a ring well" (Smith and Gnudi 1943). There is little evidence for medieval specialization apart from mention of gold beaters in thirteenth-century Paris and enamelers, seal cutters, and burnishers in fourteenth-century London; the burnishers included women (Prideaux 1896–97:8).

In the sixteenth century, however, there is more evidence for specialization. Cellini lists several of his contemporaries and gives their specializations (Ashbee 1967:1–6). Gold wiredrawers are mentioned in 1512 (Prideaux 1896–97:38). In general, the specialized manufacture of components for the jewelry industry is a relatively recent innovation. The 1884 *Fifteenth Annual Report of the Royal Mint* of London states, "A new industry, moreover, is springing up for the supply of rolled or drawn metals to jewelers and dentists."

By the early fourteenth century some distinction is made between wholesale and retail goldsmiths in Paris, and some English regulations appear to distinguish between a goldsmith selling to a merchant and selling to a member of the public. The cost of gold jewelry depended on the materials and labor costs, and these would usually be calculated and listed separately. This is an ancient practice. Plautus, in his play *Menaechmi,* written in about 190 B.C.E., has a dialogue between Menaechmus and a maid. The maid wants a pair of earrings as a gift, and Menaechmus says "sure, you provide the gold, I'll pay for the workmanship." The distinction continued through the medieval period and into the Renaissance. French documentary references to medieval jewelers often mention both the material cost and *la façon.* This is the same as the "fashion" we find in medieval and Renaissance English documents. Shakespeare reworked *Menaechmi* for his play *A Comedy of Errors.* In Shakespeare's version, the plot involved the manufacture of a gold chain, the invoice for which lists everything including the "fashion." Stubbes, writing in the 1580s, gives much of the same information in his description of goldsmiths' tricks: "If you buy a chain of gold, a ring, or any kind of plate, besides it you shall pay almost half more than is worth (for they will persuade you that the workmanship of it comes to so much, the fashion to so much, and I cannot tell what)" (Stubbes 1583). This might seem to imply that workmanship and "fashion" were distinct, but all other indications are that they were one and the same.

Medieval French regulations actually stipulated that workmanship and materials costs should be distinguished on invoices and receipts (Cripps 1880:32–34). Stubbes indicates that the workmanship would be about 50 percent of the cost of the materials. We can obtain some idea of medieval equivalents. For example, a gold ring made for the Duke of Orleans in the mid-fifteenth century cost 14 pounds for the materials and 13 shillings for the manufacture; this means that the goldsmith received less than one-twentieth of the value of the materials he was working. Though low by modern jewelers' standards and even by Stubbes's margin, this is far higher than the goldsmiths' profit margins defined in the maximum-price edict of Diocletian a millennia earlier (Ogden n.d.). On the other hand, other medieval documents show that manufacturing costs could easily be up to Stubbes's level of 50 percent or more of the materials cost. Presumably, fees depended on the nature of the materials, the complexity of work, and the prestige of the individual goldsmith.

Composition

Analyses of Anglo-Saxon and other Dark Age metalwork of the seventh century show that after about 625 C.E. coinage and jewelry alloys were sometimes debased to an extraordinary degree; an alloy containing less than 10% gold is not unusual (Hawkes et al. 1966). There is little published information on the composition of Western European goldwork after this date. Farther east, analysis of Byzantine jewelry shows that a high purity of around 91–92% (perhaps intended to be 22 carats) lasted through the late tenth century, at least. In addition, several gold objects from the tenth to twelfth centuries were found to contain about 80% gold—possibly relating to the gold hyperperon coin of that purity, which was introduced in 1092 and survived almost through the twelfth century (Oddy and la Niece 1986).

In early medieval Western Europe, a wide range of gold purities were used. The goldsmith would refine or debase the gold depending on a variety of factors; at one end of the scale were official purity requirements, at the other the whim or dishonesty of the goldsmith. Goldsmiths would have generally refined their own gold as and when needed. Medieval and early Renaissance sources such as the "Mappae Clavicula" and Theophilus (Smith and Hawthorne 1974), Cellini (Ashbee 1967), Pegolotti (1936), and Biringuccio (Smith and Gnudi 1943) all describe the refining of gold. However, a statute of 1488 says that "finers [refiners] and parters dwell abroad in every part of the realm," which suggests that there were specialist refiners—perhaps mainly for silver—by the end of the medieval period (Cripps 1914).

Some tenth- to twelfth-century Western European goldwork certainly has the color of quite high purity gold, and gold is often referred to in contemporaneous documents as being of high purity. For example, objects are sometimes described as made of *aurum purissimum* (Lehmann-Brockhaus 1938). One may also find Western medieval and even Renaissance use of the Byzantine Greek term *obrysus* and its variants (e.g., *obrizé*) as a description of pure gold. The tenth-century *Alfric's Vocabulary* defines *obrizum aurum* as "real gold."

On the other hand, debasement was acceptable. The "Mappae Clavicula," for example, describes an alloy suitable for making male figures that consisted of 67% gold and 33% silver (Smith and Hawthorne 1974). Female figures were apparently better suited to an alloy of just 20% gold with 80% silver. Even the purer of these two alloys has a high enough silver content to be described as electrum. Other recipes for electrum have survived (Gay [1887] 1928). In general, medieval texts clearly distinguish between gold, electrum, and silver, but one reference to "purest electrum" (*electrum purissimum*) might suggest that not all recordkeepers had a deep understanding of the nature of the alloy (Lehmann-Brockhaus 1938:no. 2551). Analysis of one English gold ring of twelfth century date showed that it was of 60% gold, 30% silver, and 10% copper—a precise ratio that might not be coincidental (John Cherry, personal communication, 1991).

The "Mappae Clavicula" also describes a red alloy with about 57% gold, 29% silver and 14% copper (Smith and Hawthorne 1974:ch. 54). Red gold, or gold alloyed with copper, is quite frequently mentioned in medieval European literature (Kuhn and Reidy 1963) and several extant early medieval ornaments are of distinctly reddish color. In the twelfth century Theophilus describes Arabian gold as having an exceptional red color, and says that workmen imitated its appearance by alloying a fifth-part copper to pale gold (Hawthorne and Smith 1979:ch. 47). Theophilus appears not to have realized that the natural alloy contains copper, and he notes that the presence of copper was a sign of the spurious alloy. On the basis of the admittedly small number of medieval English gold rings studied, the copper content seems to exceed the silver content far more frequently than it does in ancient goldwork. Possibly the red Arabian gold was derived from Crusader gold coins, some of which contain almost 15% copper. On the other hand, the exploitation of Arabian mines is also certain. Al-Hamdani reveals that Arabian mines were exploited in the tenth century (Dunlop 1957), and European corroboration includes the "Song of Roland" which refers to "ten mule loads of gold dug from Arabian mines."

The medieval jeweler generally obtained raw materials from a patron or with money provided by the patron; only rarely did a goldsmith amass enough capital to hold his own stock of precious metal. Gold sources usually included damaged or unfashionable objects returned for remaking. Medieval texts refer to goldsmiths purchasing gold jewelry for resale or reworking. An English charter of 1327 described goldsmiths "who keep shops in obscure streets, and do buy vessels of gold and silver secretly without enquiring whether such vessels were stolen or come lawfully by, and immediately melting it down, make it into plate, and sell it to merchants trading beyond the sea" (Jackson 1905:35). There were less conventional ways in which even the crown could obtain gold. In the fourteenth century "the gold from which the nobles of Edward III were coined was said to have been produced by occult sciences" (Freemantle 1870).

Coinage has long been a major source of gold for goldsmiths. There is documentary proof of this in Byzantine and early medieval times, if not earlier. Even in the nineteenth century, the annual report of the Royal Mint of London noted that "working jewelers are in the habit of making use of large quantities of sovereigns [a British gold coin of 22 carat gold] in the business of their trade, owing to the convenience of being able to obtain by this means gold of a known standard" (Freemantle 1870:24). There appears to have been no British law prohibiting the melting down of gold coinage to make jewelry prior to 1920.

Even if coinage was a main source of gold, the goldsmiths of medieval Europe could still have had to contend with widely differing purities. In the early fourteenth century, for example, Pegolotti of the great Florentine banking house of the Bardi, which had offices across Europe from London to Constantinople, produced a merchants' handbook giving prices and trade details of many countries (Pegolotti 1936). The author lists no fewer than thirty-nine different gold coins that a fourteenth-century merchant might encounter (Pegolotti 1936:287–89). He lists their purity in carats, ranging from debased Tripoli coins of just 11 carats (45.8% gold) up to high purities such as French coins of 23.75 carats. (This is perhaps the earliest recorded European use of the term carats as applied to purity rather than weight.) The Troy ounce was divided, like the Byzantine solidus before it, into 24 carats, which was thus equated with pure gold; 18 carat is 75% gold, and so on. The carat standard of gold makes no distinction between the possible alloying materials. In ancient and medieval times these were typically silver and, to a lesser extent, copper. Today all manner of metals are added, some of which cause problems for goldsmiths.

Other coins listed by Pegolotti include Saracen bezants at 15 carats and *Bisanti vecchi d'oro d' Allessandrie* at 23 carats. Recent analyses have indeed shown that

Islamic and Crusader bezants had a wide range of purities (Gordus and Metcalf 1980). In many ways, little has changed. The 1884 report of the Royal Mint published a list of the various purities of gold coinage issued by contemporary European mints (Royal Mint 1884) revealing variation even within countries—for example, the Austro-Hungarian 8 florin piece was 90% gold, the ducat of the same country was 98.5% gold.

It is hardly surprising that by the thirteenth century some of the authorities in Western Europe had seen the need to control the quality of gold worked and sold by goldsmiths. Such regulations could best be enforced through the goldsmith guilds that had existed in England at least since the twelfth century. In England the "Close Roll 22" of Henry III in 1238 concerning *de auro fabricando civitate Londiniarum* ruled that no one should use gold under a certain purity (defined by price) and that all work was to be carried out in the public street, "not in secret." The same ordinance includes provisions about counterfeit and precious stones plus a ban on the gilding of copper and other base metals. In France, a regulation of 1268 states that "no goldsmith may work in Paris, unless it be of the touch of Paris or better: which touch surpasses all the gold of every part worked on the earth." The "touch of Paris" was actually only 19.2 carats pure, that is, 80% gold. Perhaps the origin of this 80% standard is related to hyperperon coinage of similar purity, as well as goldwork from medieval Constantinople. The French regulations, like the English, include a ban on working at night (except on commissions for Royalty or the Bishop of Paris) and rule against the use of glass-imitation stones and doublets (Boileau 1837:38–39). The "touch of Paris" was also the minimum standard for gold jewelry and other wares allowed in a 1300 English statute of Edward I (Jackson 1905:7–8). According to the same source, goldsmiths were allowed to complete any work in hand made from poorer-quality gold, but if in future they purchased such objects, they had to be melted down and refined, not resold. Later the same century regulations of Edward III in England and John II in France reiterate that no goldsmith could work "gold worse than the touche of Paris" (Prideaux 1896–7:doc. no. 43; Cripps 1914).

A series of analyses of seventeen gold rings from thirteenth- and fourteenth-century Britain shows a wide variation in purity, from about 64 to 96% gold. There are, however, definite clusters at around 80 and 90% gold (both ± 2.5%; John Cherry, personal communication, 1991). Clearly this could be a fertile area for research.

Gold standards were by no means uniform or necessarily followed. One English document of 1339, refers to "parti-gold" (debased gold). Apparently, various debased alloys were also allowed in certain regions of France. For example the Montpellier goldsmiths were evidently allowed to work 14 carat (58.3%) gold (Cripps 1914:22), and in Put-en-Velay, both 14 and 16 carat gold are found (Cripps 1880:16). It appears that itinerant dealers were still selling debased and unmarked goldwork all over France in the late fifteenth century (Cripps 1880:19). A Scottish statute of 1457 gives 20 carat (83.33%) gold as the minimum gold standard for coins (Cripps 1914:161).

The jewelry standard was thus far lower than the contemporary English and French coinage standard of 23 carats, 3.5 grains (99.5%) that survived through the remainder of the medieval period. This is the defined purity of the gold trial piece made as an assay standard in 1477 (during the reign of Edward IV) and described as being of 23 carats 3.5 grains of "perfect fine gold" (Ruding 1812). Chemical analysis of this same trial plate carried out in the nineteenth century showed it to be 99.35% gold, 0.5% silver, and 0.14% copper (Royal Mint 1874).

The minimum gold purity for jewelry was lowered in 1477 when a new regulation stated that "no goldsmith . . . shall work, sell or, set to sale any manner of base gold under the fineness of eighteen carats"; that is, 75% gold (Jackson 1905:12–13). Nevertheless, later in the century an Act of Parliament of Richard II made every goldsmith put his mark on his wares "because gold and silver, which is wrought by goldsmiths in England is often times less fine than it ought to be, because the goldsmiths are their own judges" (Jackson 1905:8–9). Too few analyses have been carried out so far to tell whether the new official 18 carat standard is reflected in the composition of late fifteenth-century gold jewelry from Britain. However, it should be noted that Middleham jewel (Fig. 1) is composed of an alloy with about 75% gold, although there is also [deliberate] gold enrichment up to about 86% in certain areas of the surface (John Cherry, personal communication, 1991). This can be seen in the darker color of the Trinity figures, compared to that of the design on either side. The next change in the gold standard—up to 22 carat (91.7%)—was not introduced in England, Scotland, or in France until the mid-sixteenth century.

The English statute of 1300 had included legislation to ensure compliance with the rules regarding gold purity: "Wardens of the crafts shall go from shop to shop among the goldsmiths to essay if their gold be of [that] touch" (Jackson 1905:7–8). "If gold found in the hands of a goldsmith or dealer is not worth the assay, in buckles, rings or other articles . . . these should be broken, and the man loose his work and rectify his gold" (Prideaux 1896–97:doc. no. 44). The purity of the gold was ascertained by "fire or touch," that is by fire assay or by touchstone. The latter was

FIGURE 1. *The so-called Middleham jewel, a magnificent late fifteenth-century English gold pendant set with a sapphire, found near Middleham Castle in Yorkshire in 1985. The front of the pendant is engraved with a depiction of the Trinity, the back with the Nativity.*

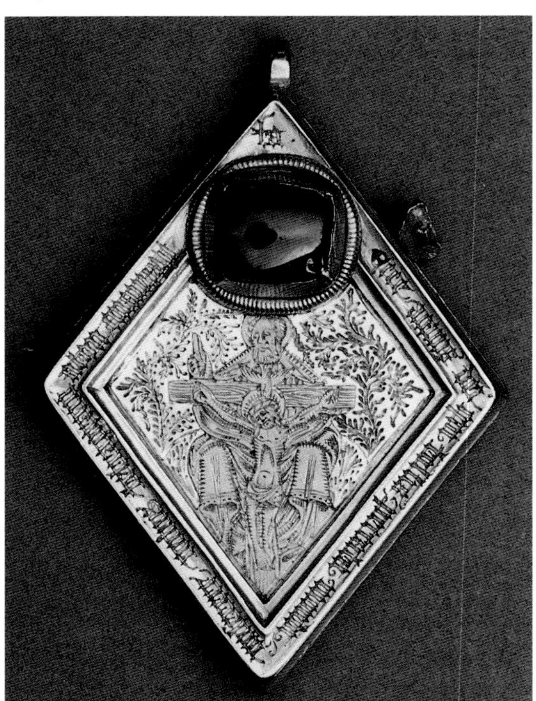

the simplest and "good stones and good needles for to touch be always there ready . . . to make assay of gold of the carat between the merchants and the said master" (Combe 1806). If the gold was not up to standard, it should be "new molten and reforged until it be put to point." Similar rules also applied to stones. Real stones could not be set in base metal, while false (i.e., glass) stones found set in gold were to be broken up and "ground in a mortar" (Prideaux 1896–97:doc no. 44). Notes in the records of the Worshipful Company of Goldsmiths refer to goldsmiths who produced substandard goldwork; this must have been a perennial problem.

We can imagine that similar regulations—in particular restrictions on the manufacture and sale of substandard gold objects—would be enforced as universally as local circumstances allowed. The tenth-century *Book of the Prefect,* dealing with bankers and precious-metal merchants in Constantinople, ruled that those who debased bullion and worked and sold it would have their hand cut off (Hendy 1985:251–53). This set of regulations from Constantinople also contains similar limitations as to places and times of work as found in the later English and French regulations. Possibly the Western regulations derived from the East.

The purity standards were, of course, minimum levels and quite possibly some goldsmiths might have preferred to use gold of higher purity. Certain types of work might have demanded certain purity levels. In the early sixteenth century Cellini notes that gold under 22.5 carats was too hard for fine work and dangerous to solder, while gold over 23 carats was a bit too soft (Ashbee 1967). In recent times jewelers have considered 22 carat gold ideal for fine work. Cellini's ideal was between about 92.5 and 96% pure; this is a fairly precise demand, but within the level of accuracy possible with experienced use of the touchstone. A common purity range for much Hellenistic, Roman, and Byzantine goldwork is 92.5–96%. Some medieval Byzantine goldwork seems to vary in composition depending on function. For example the ninth-century Byzantine gold cross in Tournai is made of sheet gold which is between 90 and 92% gold, whereas the bordering beaded wire, more prone to wear, has a slightly higher silver content (de Cuyper et al. 1987). Biringuccio in his *Pirotechnia,* book 5, chapter 2, noted that silver added to gold will harden it, but in general he seems to have had a cavalier approach to debasement. He notes that if you were a goldsmith you might debase gold with silver if you "begrudge the expense of having it fine, or if you find that you do not have enough, or if you do not care about such fineness" (Smith and Gnudi 1943).

Technology

Sheet Gold

Goldwork was characteristically manufactured from hammered sheet gold and rarely by casting. Hammering has thus always been one of the trademarks of the goldsmith and the most commonly illustrated and described of the goldworker's skills. Changes in the process include the introduction of the hafted hammer for metalwork in the first millennium B.C.E. (Ogden 1982:34). This can probably be linked to the recent research by Nicolini and Bouchet (1990), who used the scanning electron microscope to distinguish between "primitive" or "vertical" hammering of ancient gold and the "progressive" hammering that they contend gradually replaced it.

The greatest potential change in the manufacture of gold sheet came after the period considered here, with the invention of the rolling mill, which rolls the metal between two iron or steel rollers. The invention is accredited to Leonardo da Vinci, who provided a sketch for one (Uccelli 1948). There is no evidence for rolling mills prior to the Renaissance, and these devices were probably rare in ordinary jewelry workshops for several centuries afterward. The earliest apparent documentary reference to rolling sheet gold dates from the seventeenth century (Singer et al. 1957:342). The eighteenth-century *Encyclopédie* of Diderot shows a simple hand-rolling machine for use by jewelers (Diderot and d'Alembert 1751–57:pl.6:25), but his and other contemporaneous workshop scenes are still filled with workers hammering out gold and silver sheet.

The various methods of raising and shaping sheet gold have been described in detail elsewhere; the medieval smith had no major new process at his disposal. Numerous tools, including dies and punches, have survived from the Viking and later worlds. Cellini describes the casting of copper-alloy dies and formers and the use of simple tools of wood and bronze for working sheet gold (Ashbee 1967).

Hollow sheet gold often needs to be filled to provide support and weight. Sulfur was typically used in late Hellenistic, Roman, and early Byzantine gold jewelry but does not appear to have survived into medieval times in Europe, although it is possibly present in some medieval Islamic goldwork. The suitability of the melting temperature and flow properties of sulfur are noted by Biringuccio. Pitch or a pitch-like substance was also used as a filler in early Islamic jewelry (Ogden 1987) but has not been recorded so far in medieval European goldwork. However, pitch or resin might well be expected since pitch is still a common filling material in flimsy modern sheet silver objects. From around the twelfth century onward, filled hollow sheet-gold jewelry was often forbidden in Western European statutes; thus filled hollow work became far rarer than it was in Byzantine or earlier times.

Cutting

Sheet gold was normally cut with chisels, as in antiquity, or with small shears or scissors. The use of scissors or small shears for goldwork in antiquity is generally doubted, and the presence of the tool marks from such implements on purported ancient goldwork is usually taken as a warning sign. However, fine cutting implements of this type came into common use during the course of the early medieval period. Theophilus describes the use of both scissors and shears for cutting metal in his chapters 23, 52, and 76 (Hawthorne and Smith 1979); and a fifteenth-century Greek manuscript, derived from earlier texts, refers to the use of small scissors to cut small pieces of solder (Berthelot 1888:no. 39) Shears are shown being used by a metalworker in an illustration from a mid-fifteenth century manuscript of Pliny's *Natural History* (Whalley 1982).

The fine almost lacelike pierced work of the late Roman and early Byzantine periods does not survive long into the medieval period. Some cruder pierced goldwork may be found, such as that around the side of the Alfred jewel and on several elements of the Egbert Shrine in Trier (Westermann-Angerhausen 1973:figs. 60–64, 66–69, etc.). Theophilus describes a type of pierced work using a chisel and small files which he terms *opus interrasile*. This term, found only in Theophilus, has also been used by modern historians to describe the finer Roman and Byzantine varieties of pierced goldwork. This terminology has its critics, however (Ogden and Schmidt 1990).

Files are mentioned by Theophilus many times, in relation both to their manufacture and use. His near-contemporary Alexander Neckham includes them in his list of essential goldsmith's tools. Files were used in the manufacture of tools and also in some shaping and fitting operations, though perhaps more commonly for silver rather than gold. By their very nature, files remove metal, and thus are to be avoided where possible when working gold. The use of files on goldwork in antiquity is usually doubted, while sure indications of the employment of such a tool is often taken as an indication of a post-ancient origin. In his chapter 26, Theophilus describes the use of a file to smooth the surfaces and edges of a silver chalice (Hawthorne and Smith 1979) and for similar purposes on other silver components, including niello work and openwork. Theophilus's "very small files—square, round, half-round, three-cornered, and flat and thin" do sound like the files found in modern jewelers' workshops (Hawthorne and Smith 1979:94). Cellini describes numerous small files and engravings of jewelers' workshops of the sixteenth century and later can show an assortment of such tools (Ashbee 1967).

The only saw mentioned by Theophilus is an iron saw, fed with sharp sand, used to cut rock crystal. The twelfth-century "Mappae Clavicula" manuscript similarly mentions a copper saw fed with saliva and emery powder for cutting gemstones (Smith and Hawthorne 1974:ch. 191a). In neither of these cases would a toothed saw be implied. Saws with teeth, probably used mainly for cutting wood, were thought by the Romans to have been invented by Daedalus but actually date back to the early Bronze Age in the Near East. For example, copper saws have survived from the First Dynasty of ancient Egypt (Lucas and Harris 1962:449). The use of a toothed saw for cutting gold is highly unlikely in medieval times, as it would result in lost metal and almost certainly would be too coarse for fine work. Nevertheless, Alexander Neckham, writing in the twelfth century, does include a toothed saw (*serram dentatam*) among the equipment to be found in a goldsmith's workshop (Neckham 1867). The jeweler's saw used by craftsmen today would no doubt have been appreciated by medieval goldsmiths, but it is hard to imagine that fine, handmade iron-piercing saws would have been worth the time and effort prior to their mass production and resulting affordability. They were difficult to make by hand and, like their modern mass-produced counterparts, all too vulnerable to fracture. Until good evidence exists to the contrary, the presence of characteristic parallel-sided, fine saw marks on gold or silver work purported to date from the Renaissance or earlier times should be viewed with great caution.

Wire

The early medieval period saw the introduction of wiredrawing—an innovation that was to have a widespread effect on both the construction and design of jewelry. Prior to this period, gold wire was generally formed by hammering—a process better suited to relatively large-diameter wires such as earring hooks—or by twisting narrow gold strips and then rolling the twisted wires to compact them. Wires of this general type have characteristic spiral seam lines, so to speak, along the wire, the number and spacing of which largely depend on the ratio of width to thickness of the original strip. The resulting wires have been classified under various names, including "strip twisted" and "block twisted," but I prefer to use the generic term *strip twisted* for all such wires. The various ancient technologies for making such twisted wires and their chronology have been dealt with at length by several authorities and will not be discussed here (Oddy 1977; Ogden 1991). Suffice it to say that the majority of wires in gold jewelry made prior to early medieval times show the characteristic spiral seam lines resulting from their twisted construction.

In wiredrawing, a rod or metal strip is pulled through consecutively smaller holes in an iron of steel drawplate; thus, it is literally drawn out in longer and thinner wires. Simple hand drawing, with strong forceps or tongs and a drawplate mounted on a sturdy piece of wood, is seldom practical for wires over about 2 mm in diameter. More complex drawing machines—termed *draw benches*—provided greater leverage and strength and allowed long lengths of wire to be made in a great variety of diameters.

The use of drawing for the production of Roman and pre-Roman gold wires has not been demonstrated despite the microscopic examination of thousands of ancient ornaments. On the basis of current knowledge, it is possible to conclude that gold wiredrawing probably first appeared in the Western world in about the seventh or eighth century C.E. Pre-seventh-century examples of drawn gold wire from Europe or the Mediterranean world have not yet been identified with certainty (Ogden 1991). The use of certain drawn wire on Korean gold earrings of the fifth to seventh centuries C.E. might suggest that the origins of wiredrawing lie in the East (Ogden 1991)

Duczko's study of Viking filigree work has demonstrated the general adoption of wiredrawing by the ninth century C.E. (1985). More recently Whitfield (1990) studied the various surviving drawplates and examined many examples of northern European filigree work. She concludes that wiredrawing was used in Northern Europe by around 800 C.E. According to Craddock (Youngs 1989), Celtic gold wires of the sixth to ninth centuries were made by twisting. Duczko's work on Viking jewelry has not found drawn wire dating from before the ninth century. On the other hand, it would be surprising to find that the braided silver wires on the eighth-century Derrynaflan paten were not drawn. The braided Trichinopoly work (named after the town in India where this type of braided chain was a traditional craft) of the ninth-century Trewhiddle silver scourge is certainly drawn (Fig. 2).

A well-known lunate earring type that straddles the Byzantine and early Islamic world, and which can be dated fairly accurately to the tenth century C.E., sometimes

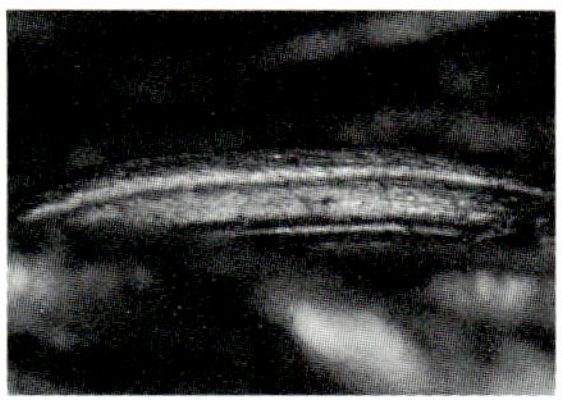

FIGURE 2. *Detail of drawn silver wire from the scourge in the Trewhiddle hoard. Anglo-Saxon, 870 C.E. British Museum, MLA 80.4-10.4.*

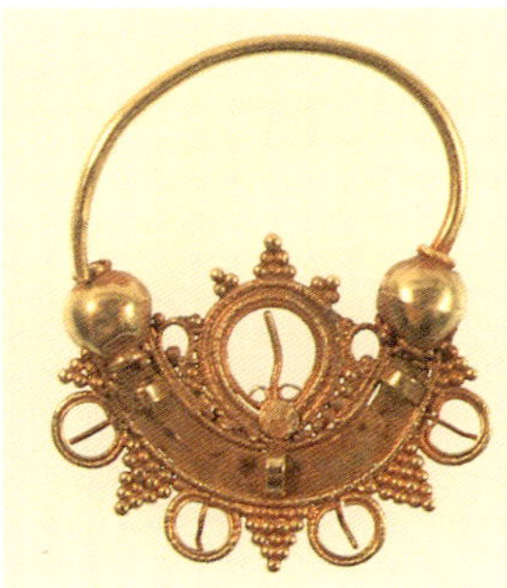

FIGURE 3. *A gold lunate earring exemplifying a well-known Byzanto-Islamic earring type dating from the tenth century C.E. This earring incorporates drawn round wire and spiral-beaded wire. The loops originally bore pearls.*

includes drawn wire (Fig. 3). Drawn and strip-twisted gold wires have both been noted in Fatamid Islamic work of the eleventh century (Ogden 1987). Current research in progress by the author suggests that drawn wire was the most common type used for Fatamid work and that examples of this work presumed to be strip-twisted should perhaps be reexamined.

The earliest description of wiredrawing might lie in a Greek compilation of goldsmiths' recipes. The surviving manuscript dates from 1478 but is almost certainly, for the most part, a copy of far earlier sources (Berthelot [1888] suggests possibly the eleventh century and earlier). This provides a good, but sadly ambiguous early account of drawing silver wire: "Take fine silver, hammer it and cut it into strips and twist [?] it with a rounding [?] iron [?]. Then put it in the drawplate and draw once" (Berthelot 1888:doc. no 33). The part about the "rounding iron" (perhaps roller) is unclear; the original Greek is και γυσε το εις σιδερον στρογγυλον. The Greek word used for drawplate is συρτεν.

The earliest datable description of wiredrawing is by Theophilus in the twelfth century. His book 3, chapter 8 describes the manufacture of the iron plates through which wires are drawn. Once into the thirteenth century, many documentary references to drawn wire may be found; for example, *filum auri tractum* (Lehmann-Brockhaus 1960:nos 2741, 2743, 2745). Thirteenth- and fourteenth-century English literary texts often refer to gold wire. Although these are almost invariably poetic descriptions of a heroine's hair, the allusion would better suit fine-drawn gold wire than shorter lengths of strip-twist wires. From the second half of the thirteenth century, details of iron and brass wiredrawers in Paris (Boileau 1837:61–63) and a reference to what may be a corporation of gold wiremakers may be found in a document dealing with "Des Batteurs d'or et d'argent à filer." It has been assumed that these were gold wire drawers, and certainly they are distinct from the beaters of gold foil, which are described separately (Boileau 1837:74–78). Unfortunately, the word "filer" could apply to gold threads as well as wires; and the mention of striking with a hammer in the text perhaps makes gold threadmakers more likely than wire-drawers.

Wiredrawing and some quite complex wiredrawing machinery are described at some length by Biringuccio in the early sixteenth century. He notes that wire "can be made in any thinness and length that the craftsman wishes, especially that which is made of gold and silver" (Smith and Gnudi 1943).

The earliest surviving representation of wiredrawing is probably in a manuscript of 1389 C.E. called the "Mendel Brothers Hausbuch," now in Nüremberg. This depiction, illustrated by Oddy (1977), shows a monk drawing wire. The drawplate is held on a heavy wood block or tree stump. The monk, holding the wire with a large pair of tongs, is seated on a swing to allow full leverage by pushing on the ground or wood with his feet. From the sixteenth century onward, representations of jewelry workshops typically show the use of draw benches. Theophilus states that round wires of relatively large diameters for conversion into beaded wires were made by hammering. This might mean that during the twelfth century only thin gold wires were being drawn. If this is so, it would place the invention of the draw bench and other more powerful drawing equipment into the thirteenth to fifteenth centuries—after Theophilus but before Biringuccio and his contemporaries. On the other hand, some Fatamid Islamic goldwork dating from the eleventh century includes drawn wires of more than 2 mm in diameter.

Old processes can be continued for generations, particularly in small family workshops where tools and skills are handed down from generation to generation. Wiredrawing also suggests the preparation of some stock of raw materials. The fourteenth century monk on the swing, mentioned above, is shown with several coils of wire he had made. A goldsmith would hardly draw just a couple of centimeters of gold wire needed to decorate a single ring bezel.

Biringuccio makes a very perceptive remark when he notes that "goldsmiths also draw gold wire to make the ornamentation of their work easy and more attractive" (Smith and Gnudi 1943). The introduction of wiredrawing clearly facilitated certain decorative and functional ideas. Several types of filigree work only came into their own, in fact, when wiredrawing permitted the easy manufacture of long, thin lengths of gold wire, while such components as suspension loops and hinges made from springlike coils of wire tended to replace the sheet-gold equivalents.

Decorative Wire

A major change in gold filigree decoration during the course of the medieval period was the change from beaded wires to twisted wire ropes. A rope, in jeweler's terminology, is a tight twist of two thin wires. With wiredrawing, it was easier to draw thin wires and twist these into ropes than to hammer or draw a thicker wire and convert this to a beaded wire. Ropes became more and more common during the course of the medieval period and, in the Renaissance, almost totally superseded beaded and other decorative wires in gold jewelry.

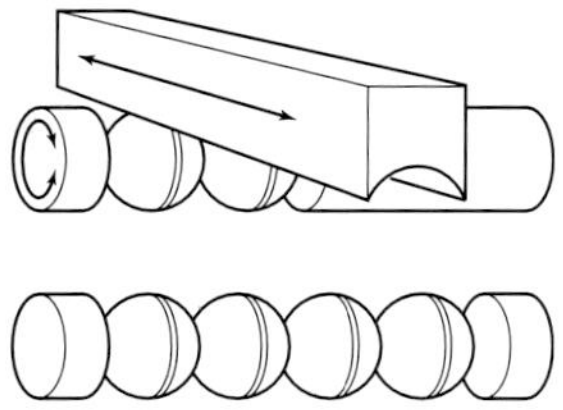

FIGURE 4. *Diagram of the two-edged implement for making beaded wire as described by Theophilus in the twelfth century and used in Roman times, if not earlier.*

Beaded wires had been popular in jewelry from about the seventh century B.C.E. onward. They were originally made by rolling a round section wire under a single or double edge, which would leave one or two grooves around the wire. By rolling the wire and moving the tool sequentially along its length, a series of grooves could be made that would produce a beaded effect. Theophilus describes the use of a double-edged tool for making beaded wires and this ancient process is illustrated schematically in Figure 4. By the late Roman period, some beaded wires were also being produced by swaging; that is, the wire would be stamped between two metal plates, each bearing a line of hemispherical hollows. If the wire were rotated and moved through the swage blocks, a very regular beaded effect could be produced. Swaged, beaded wire typically has very spherical beads, because hemispheres are the easiest depressions to cut in the swage blocks. The heavy beaded wires on the fourth-century Huntress plaque in the British Museum has lateral tool marks that would point to the use of a swage (Ogden and Schmidt 1990). Swaged, beaded wires certainly became usual in early Byzantine and Dark Age times. Theophilus describes the swage for making beaded wires, a tool he terms an *organarium,* literally "little organ," a term found in no other medieval writings.

Beaded wire is very common in European jewelry well into the medieval period and is found in Northern European goldwork up to the twelfth century, but it is far rarer after that and practically absent in Renaissance goldwork. Duczko points out that beaded wire almost totally disappears in Slavonic jewelry after the early tenth century (1985). There is also an almost total lack of beaded wires in early medieval Islamic jewelry (Ogden 1987).

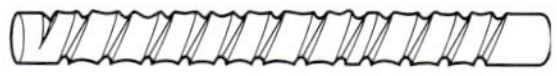

FIGURE 5. *Diagram of spiral-beaded wire. A plain round wire is rolled under a single edge at an oblique angle to produce a groove similar to that of a screw thread.*

An alternative type of decorative beaded wire has a groove running around the wire rather like a screw thread (Fig. 5). This type of wire was produced by rolling a round wire under a single edge held at a slightly oblique angle. The process was ideal for thin-gauge wires where true beading was more difficult to produce. This type of wire, which may be called *spiral-beaded* wire, is found from around the middle of the first millennium B.C.E. and, apart from an apparent break during the Roman period, continues well into the Middle Ages. Spiral-beaded wire is characteristic of Viking jewelry (Fig. 6). It is neither mentioned by Theophilus, nor is it found on his book cover in Trier, though it is found on the work of some of his contemporaries, and the technique may have survived into the thirteenth century. Further east we see spiral-beaded wire on some of the Byzanto-Islamic lunate earrings of the tenth century but not in Fatamid or other medieval Islamic goldwork from the mid-eleventh century onward. Just how widespread the technique was remains to be studied, but we can note the presence of some spiral-beaded wires (made from strip-twisted round wires) in Javanese gold jewelry of about tenth to twelfth centuries (Fig. 7).

In Northern European medieval jewelry of the tenth to twelfth centuries one often finds a combination of beaded and spiral-beaded wires, the spiral-beaded wires

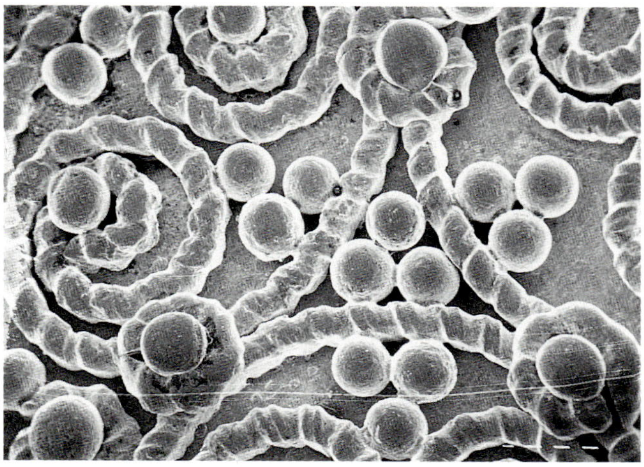

FIGURE 6. *An example of spiral-beaded wire and granulation on a Viking silver pendant from Birka, Sweden, ninth century C.E. (Duszko 1985, no. Bj 943).*

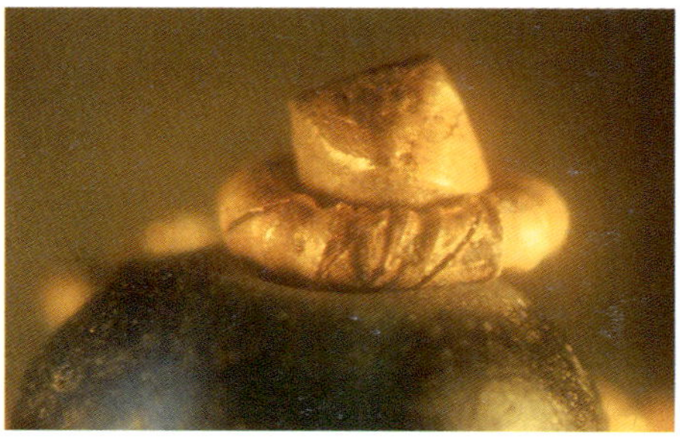

FIGURE 7. *Detail of the wire collar on an ear decoration from Central Java. The wire was made by a strip-twist technique and then rolled under a single edge to form a spiral-beaded wire, tenth century C.E.*

being of finer gauge. We can see this on the Dowgate Hill brooch found in London and dated to the early eleventh century (Figs. 8, 9). This brooch was traditionally seen as a German product, but the flat, nonarchitectural style of the object has led to a recent reassessment of the object as Anglo-Saxon (Westermann-Angerhausen 1983–84). The Townley brooch (Figs. 10–13), of about the same date and said to have been found in Scotland, also incorporates both beaded and spiral-beaded wires, but its more architectural structure points to an Ottonian German origin. It is of interest that neither of these brooches, among all their copious filigree, incorporate any plain, round-section wire.

The use of hollow-beaded wires—in effect, beaded tubes—in Byzantine goldwork has been examined (Ogden and Schmidt 1990). This is another decorative technique that found occasional use well into medieval times. Examples include the famous Konrad crown in Vienna (along the lower edge of the panels and also for the hinges) and a Viking silver ornament from Enggarda (Stenberger 1947:no. 435).

Another type of decorative wire-work that is visually very similar to beaded wire consists of a compact, springlike coil (Fig. 14). This type of coil work, another product of the wiredrawing revolution, is rare prior to medieval times (there are some Etruscan precedents) and seems not to have been widespread. One example is the

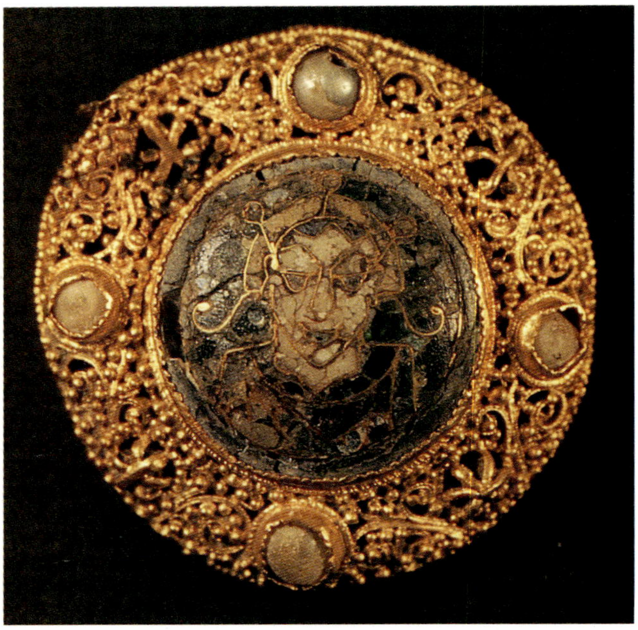

FIGURE 8. *A cloisonné enamel and gold filigree brooch found in London at Dowgate Hill, late tenth to eleventh centuries* C.E. *British Museum, MLA 56.7-1.1461.*

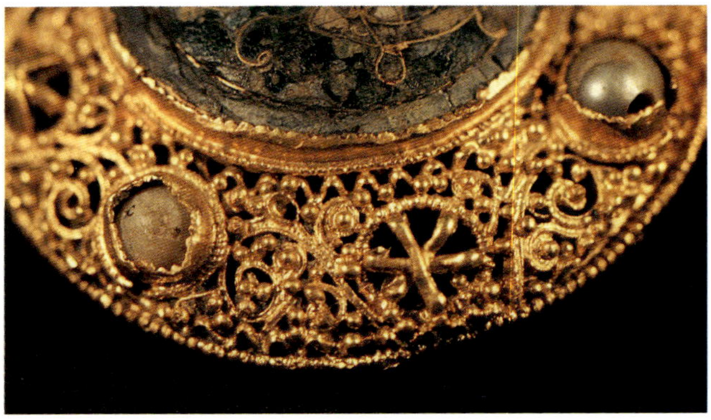

FIGURE 9. *Detail of the Dowgate Hill brooch in Figure 8, showing the filigree.*

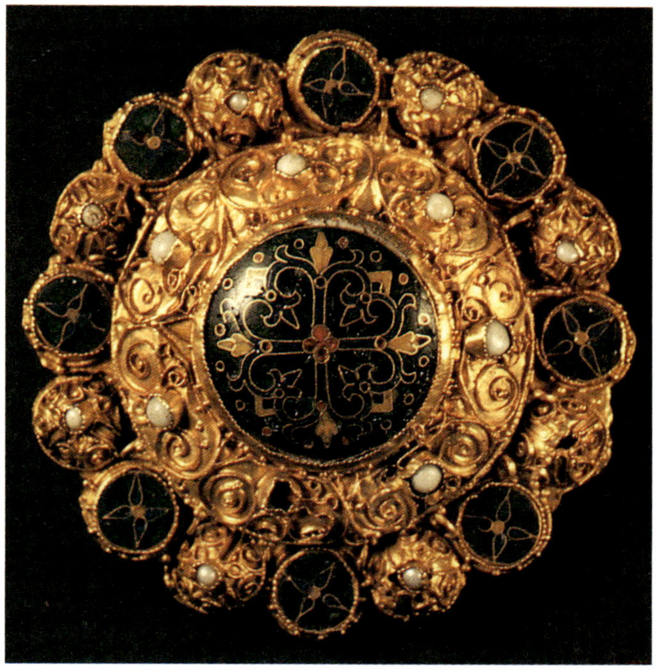

FIGURE 10. *Ornament of cloisonné enamel and applied filigree, the so-called Townley brooch, probably Ottonian German, early eleventh century C.E. British Museum, MLA Townley collection.*

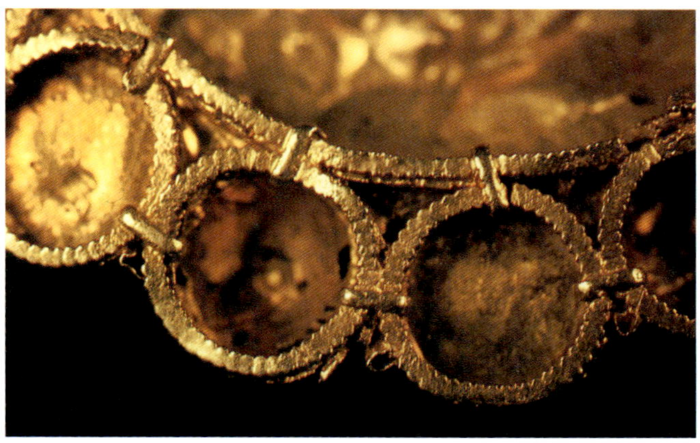

FIGURE 11. *Detail of the back of the Townley brooch in Figure 10, showing the flattened beaded wire forming the framework.*

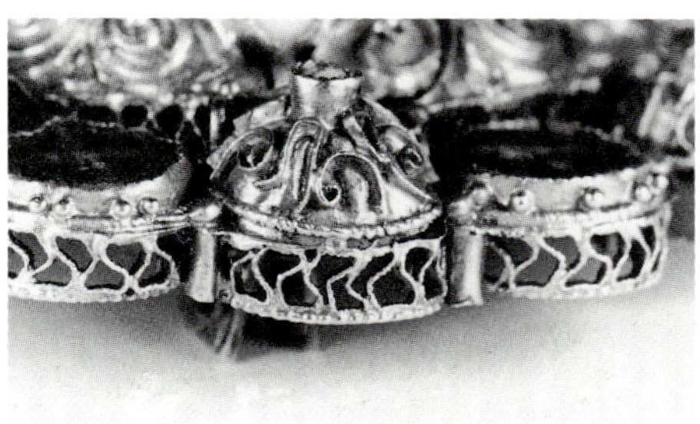

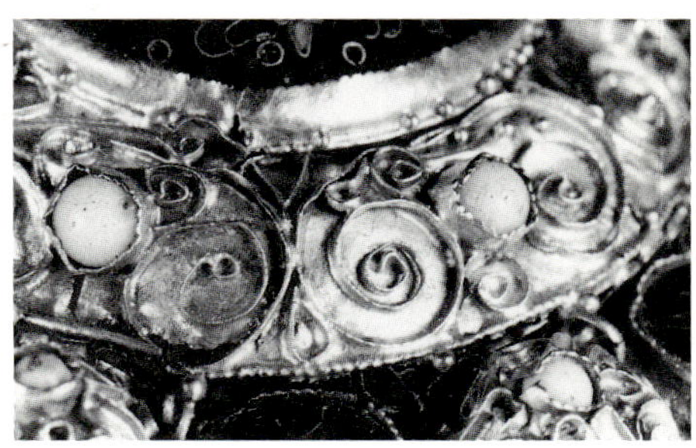

FIGURE 12. *Detail of the side of the Townley brooch, near right, showing open scrollwork of flattened spiral-beaded wire.*

FIGURE 13. *Detail of the upper surface of the Townley brooch, far right, showing applied filigree scrollwork of flattened spiral-beaded wire.*

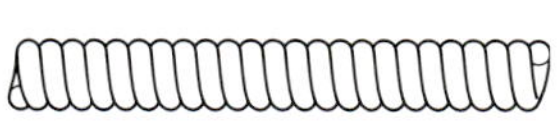

FIGURE 14. *Diagram of helicoidal spiral wire.*

border of the stone setting on the fifteenth-century Middleham jewel (Fig. 1). This technique might be seen as a late medieval and Renaissance characteristic. A more open springlike coil is also found in medieval goldwork; and a squashed version, providing a series of flat loops, forms a border to stone settings, rims, and other features from about the tenth century onward.

Both beaded and spiral-beaded wires (and occasionally ropes) were flattened for use in filigree work (Fig. 15). In the twelfth century Theophilus describes the process in his book 3, chapter 52, titled *The Art of the Metalworker*, as follows: "Pick up the fine [beaded] wires and hammer them lightly on an anvil so that they become

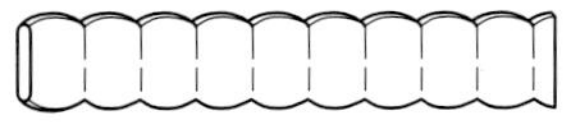

FIGURE 15. *Diagram of flattened beaded wire.*

somewhat flat and thin, but without the top and bottom of the beads losing their shape" (Dodwell 1986). The wires were shaped into the required forms and soldered to their background. By flattening the filigree wires in this way and then attaching them on edge to the background, the goldsmith could produce precise, fine detail with good depth and, thus, definition. The book cover by Roger of Helmishausen (Theophilus) in the Domschatz at Trier shows just this type of flattened beaded-wire-filigree scrollwork with the occasional granule highlight. These flattened wires were often termed *serrated* in older works when their mode of manufacture from wire was not understood.

Flattened beaded wire and, in particular, flattened spiral-beaded wire, is often found in ancient Iberian and late Hellenistic work. After the end of the Hellenistic period the technique apparently almost disappears until the Dark Age and Viking periods. An early post-Classical example of flattened beaded-wire filigree is on the Hunterstone brooch of about 700 C.E. (Stevenson 1974:pls. 12b, 13b, etc.). There is also flattened beaded wire on the eighth-century Derrynaflan chalice and on the silver trefoil brooch from Kirkoswald, which dates to the mid-ninth century or slightly earlier. Flattened beaded and spiral-beaded wires are common in Viking jewelry of the ninth and tenth centuries (Duczko 1985:210), and one occasionally also finds flattened ropes from this same period, such as on a silver ornament from Eastern Europe (Benda 1967:pl 74). Both flattened beaded and spiral-beaded wires are common on European decorative precious metalwork of the tenth to twelfth centuries, but not much after this. Both the Dowgate Hill and Townley brooches include flattened filigree. The Dowgate Hill brooch has an unusual openwork filigree composed of doubled, flattened spiral-beaded wires highlighted with some granulation. This is reminiscent of the Fatamid openwork with its double wires surmounted with granules (Fig. 9). The entire back framework of the Townley brooch is made up of flattened beaded wire (Fig. 11), while the serpentine openwork scrolls on the side and the scroll filigree on the upper surfaces are made from flattened spiral-beaded wire (Figs. 12, 13).

The gold brooch set with an earlier cameo, shown in Figure 16, is usually seen as an Ottonian product of about the same date as the Townley brooch, but the whole approach to the filigree is different.The cameo brooch employs no beaded or spiral-beaded wire, or flattened wires. Instead, the decoration includes fine twisted-wire ropes, plain (drawn) round-section wires, and perfect lines of regular granules (Fig. 17). The major technical distinctions between the filigree work on the Townley and cameo brooches also serve to separate other supposed Ottonian gold objects. Further research is needed to confirm the origins and relative chronologies of these pieces.

The choice of wires for filigree depended on both practical reasons and personal whim. For example, a gem-set gold cover for a cup in the Fitzwilliam Museum, Cambridge, probably from the late thirteenth to fourteenth century (Fig. 18), employs true beaded wire for the main decoration around the central setting but spiral-beaded wire for the finer gauge decorative wire around the individual stone settings. Here, as in comparable ancient and medieval objects, the choice might well have been practical: The thinner wires were easier to make this way. In other cases, personal choice must have played a part, and perhaps further study will allow more

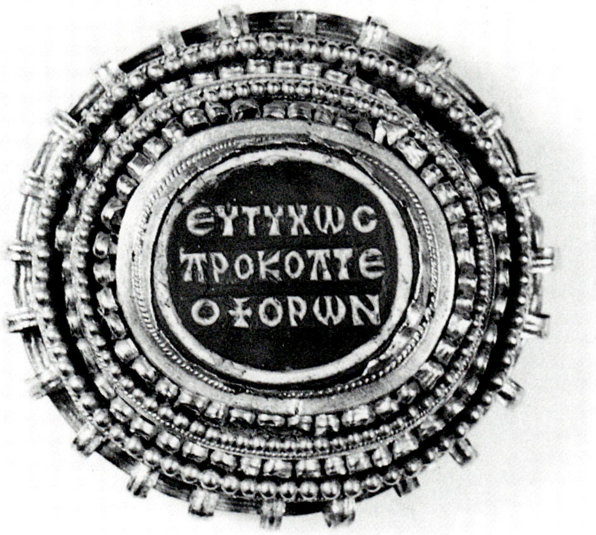

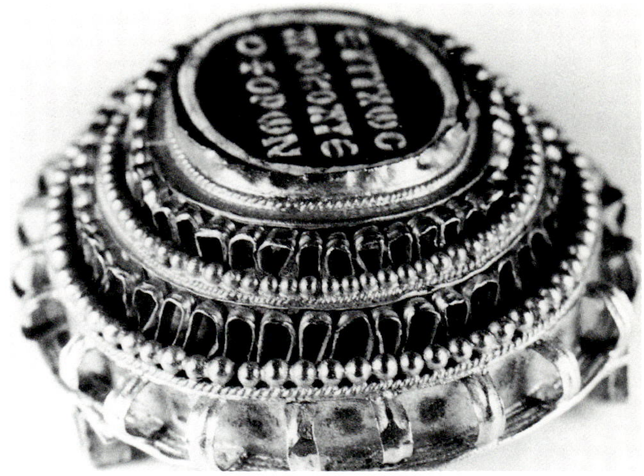

FIGURE 16. *Granulated and filigree brooch, above, set with an earlier cameo. The goldwork is Ottonian German, tenth century C.E.; the cameo is of late Roman date. British Museum, MLA AF352.*

FIGURE 17. *Detail of the side of the brooch shown in Figure 16, revealing granulation and twisted-wire ropes as opposed to beaded and spiral-beaded wires.*

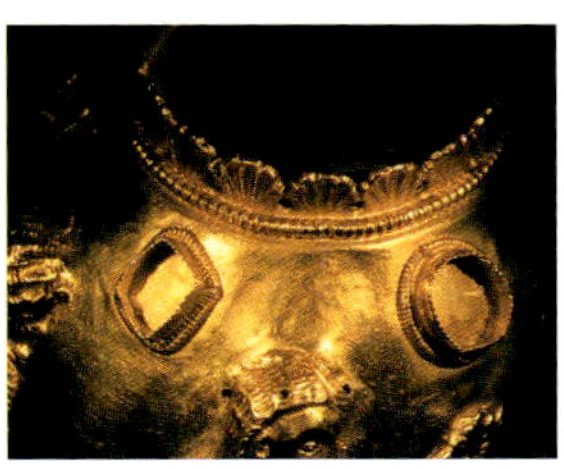

FIGURE 18. *Detail showing the use of beaded and spiral-beaded wire on a gold cover. Probably German, twelfth century C.E. Fitzwilliam Museum, inv. M/P.1.1921*

characterization of the workshops in which they were produced. For example, Theophilus's book cover in Trier uses flattened beaded wire. A contemporaneous book cover in the same collection (Ronig 1984:no. 68) uses flattened spiral-beaded wire. The well-known tenth-century portable altar, known as the Egbert shrine, also in Trier (Ronig 1984:no. 23) has flattened beaded-wire filigree but no granulation. The nail reliquary of the same date and supposedly from the same workshop (Ronig 1984:no. 24) has filigree of ordinary beaded wire, not flattened, but including some granulation.

One type of decorative wire work that disappears totally in medieval times is wavy ribbon, which is last seen in Dark Age work of the seventh century C.E. Wavy ribbon was made by coiling a narrow gold strip into a tube, flattening it and then unfolding it. The strip has a series of regular diagonal creases, creating an attractive, wavy appearance. It might not be coincidental that a type of decorative strip made by twisting died out just at the time that round wire made from strip twisting was superseded by wiredrawing. A type of undulating strip that might be seen as a successor to wavy ribbon, again with precedents in Etruscan work, became popular in medieval jewelry right across Europe during the medieval period. An early post-Roman example is seen on a series of gold pendants from Cyprus that probably dates from the seventh century and that also includes what is possibly the earliest drawn-gold wire that has been identified thus far from the ancient Old World (Whitfield 1990). This undulating strip is common in medieval jewelry. For example, it is seen on the Townley, Dowgate Hill, and cameo brooches discussed earlier, and on examples of the lunate type of earring from the eastern fringes of Europe. The gold strip used to form this undulating decoration is often of shallow D-section, which in at least some cases appears to have been produced with a one-sided swage. The serpentine-scroll type of openwork, as seen on the side of the Townley brooch (Fig. 12), is also common in about the eleventh century throughout Europe and in the Islamic world. Sometimes the serpentine scroll is composed of narrow strips of gold sheet. In other cases, as on the Townley brooch, it is formed from flattened spiral-beaded wire.

Cellini notes that skilled exponents of filigree could work without a preliminary drawing, but the best work resulted from careful designing first (Ashbee 1967). Biringuccio also noted that a goldsmith should be a good designer. Guidelines for filigree were sometimes lightly scored on the gold or silver surface. The author has seen lightly incised lines laying out the position of filigree on Anglo-Saxon goldwork. The positions for the settings on the Cambridge gold cover were also marked out by scoring fine lines—perhaps by tracing around the stone, since the settings were clearly designed to receive existing cut gems. Fuller jewelry designs sketched out on other materials, such as paper or vellum, are rare prior to Renaissance times; early examples include flat stones from Ireland with the scratched designs for brooches (Youngs 1989:nos. 154, 155)

Chains

In the ancient world, precious-metal chains were characteristically of the loop-in-loop type. Simple or plain chains—the form most common today—were seldom used, and when they were they were often reserved for particular decorative purposes. During the Byzantine period the ratio of loop-in-loop chains to simple chains changed, but loop-in-loop chain was still typical of much Dark Age and Viking jewelry. After about the tenth-century loop-in-loop chains were rare in goldwork, and the early Renaissance saw their almost total demise. One late example is the chain in the mid- to late-fifteenth-century Fishpool hoard from London (Fig. 19). Yvonne Hackenbroch's corpus of almost one thousand jewels dating between about 1500 and 1600 only includes a single example with any loop-in-loop chain (Hackenbroch 1979:no. 633), and even here the chain might not be original to the piece. Loop-in-loop chain is also surprisingly rare in medieval Islamic jewelry (Ogden 1987).

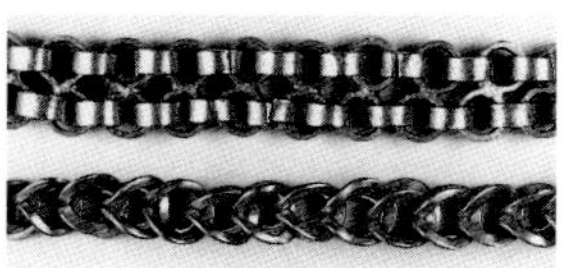

FIGURE 19. *Detail of two gold chains from the Fishpool hoard, found in London. One chain is of loop-in-loop type, the other of conjoined figure of eight links (the links joining the figures of eight are left unsoldered to avoid possible fusing during the final assembly of the chain). British Museum, MLA 67.12-8.1-9.*

The demise of gold loop-in-loop chains was probably a result of changing taste. The Renaissance chains on the flamboyant figural jewels are attempts at looking mechanical and structural; they bind a monster or constrain a vessel. Loop-in-loop chains are a fairly extravagant use of gold wire, but their massive and elaborate Renaissance successors hardly represented a savings in metal. Loop-in-loop chains continued to be used for silver and base-metal jewelry, particularly the so-called European peasant jewelry. Gold loop-in-loop chains reappeared with late eighteenth-century Neoclassical goldwork and then became popular with the archaeological jewelry styles of the nineteenth century.

The multiple loop-in-loop chain, which has a tightly linked cylindrical form with an almost plaited appearance (often erroneously termed *plaited*), might have inspired the wire-mesh tubes made from a continuous length of wire, like the so-called French or spool knitting with wool produced by generations of children. This type of braided chain, sometimes called Trichinopoly, first appears in Europe in about the eighth century. It is seen on the Tara brooch and a long length forms the principal part of the Trewhiddle scourge. The curious herringbone-braided wire decorations on the Derrynaflan paten and the Ardagh chalice are closely related, if not of identical construction. It has been suggested that Theophilus describes the manufacture of Trichinopoly work in his book 3, chapter 62, when he writes: "First draw thin or thicker wires of copper or silver and intertwine them, using an awl, in three braids, or in four, five or six, depending on the thickness you want." This braided wire was then drawn through a series of circular holes of diminishing size in a wood block to produce an even, round length. Trichinopoly, made possible by wiredrawing, does not appear to have been produced much after the tenth century in Europe. But similar work in drawn-gold wire can be seen in jewelry from medieval Southeast Asia, and the technique has been used there ever since.

Solder

Several medieval solder recipes have survived ranging from those in the "Mappae Clavicula," and by Eraclius and Theophilus in the West to al-Hamdani and Kashani in the Islamic world. Typically, the solders were binary or ternary alloys of gold, silver, and copper. The choice of proportions depended on the color and melting temperature of the alloy being worked and the inherent or perceived benefits of particular alloy combinations. Solders would generally have been made up as required, usually by alloying a little silver and/or copper to the gold being worked—a procedure recommended in jewelry text books up to recent times. Kashani, writing in the fourteenth century, gives a solder alloy specifically for red gold of the popular Islamic and European type. This solder was composed of two-thirds gold alloyed with one-sixth part each of silver and copper. This would assume that Islamic red gold contained less than 17% copper—which would suit the coinage alloys with up to 15% copper mentioned earlier. Silver solders were typically silver-copper alloys; Biringuccio recommends 75% silver, 25% copper, while a seventeenth-century English silversmith says that silver solder was 67% silver and 33% copper (Jackson 1905).

Solders could be produced as alloys or as copper-salt mixtures. The former could be used as fillings or as cut platelets, the latter called paillons today. A medieval Greek recipe for "royal solder," which is 75% gold and 25% silver, explains that filings would be used for thin gold, paillons for thicker gold (Berthelot 1888:doc. no. 9.) The subsequent recipe says the same about silver solder. Biringuccio also writes of solder cut into small pieces.

Copper-salt solders are assumed to have been used for at least some of the fine granulation work in antiquity and must relate to Pliny's discussion of *chrysocolla* (literally "gold solder" and certainly some type of copper mineral or compound) for soldering gold. The "Mappae Clavicula" and Theophilus describe solders based on a copper salt, probably copper oxide. In one case this was mixed with olive-oil soap, in another, alum. Cellini used a verdigris mixture, which he noted was really not soldering at all because the parts really just fuse together without any solder alloy (Ashbee 1967). A similar technique of soldering without solder is described by Biringuccio. We can note that the granulation and filigree work on medieval goldwork tends to be more flooded than its earlier counterparts.

The "Mappae Clavicula" also contains recipes for gold-mercury solders. The use of mercury amalgam solders for early goldwork is certainly a possibility, but examples have not been detected on extant goldwork. Taruashvili of the Georgian National Museum of Fine Arts suggests that the gold cell walls of the finest cloisonné

FIGURE 20. *Detail of the interior of the brooch in Figure 16, showing two types of join in the sheet metal (indicated by arrows).*

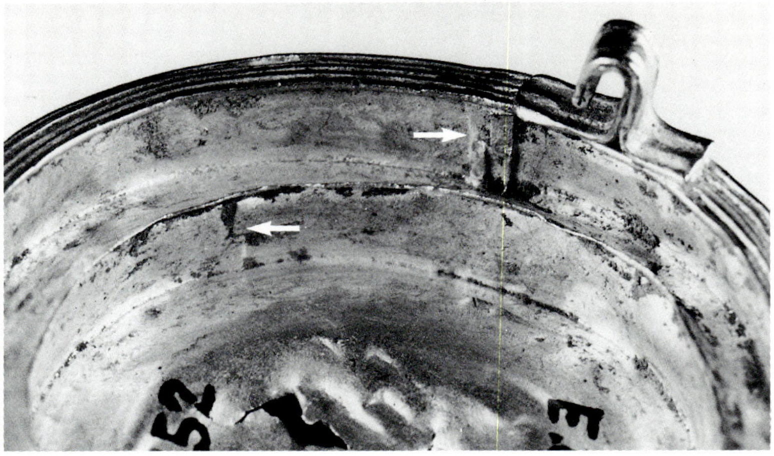

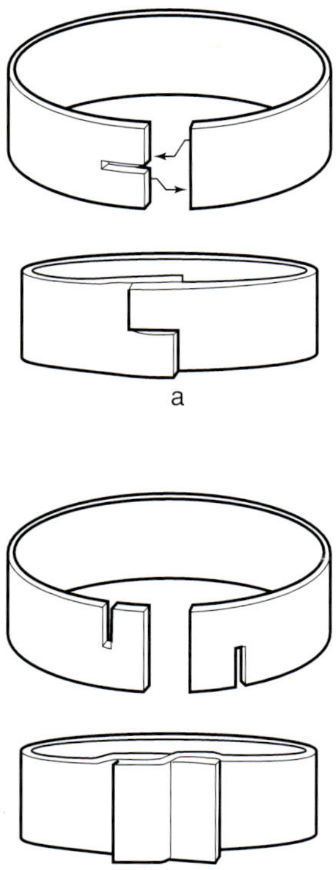

FIGURE 21a, b. *Diagram of the joins in Figure 20.*

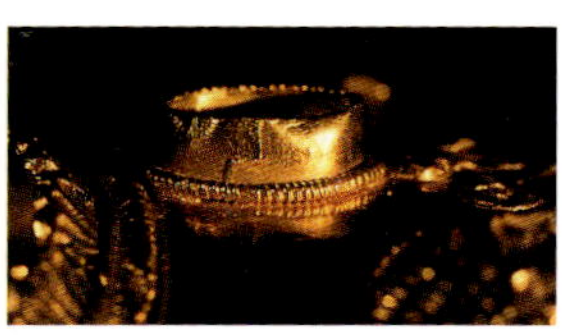

FIGURE 22. *Detail of the side of a stone setting on the gold cover in Figure 18, showing sheet metal join.*

enamel work was produced with an amalgam solder. Taruashvili produced successful replicas in this way, but seemingly no analysis work on the ancient enamels was undertaken (Amiranashvili n.d.:18). However, a silver-mercury solder has been identified on a silver spoon from the late Roman Mildenhall Treasure (Lang and Hughes 1984) and examples in gold might well be forthcoming.

By the early Renaissance, borax as a flux for solder was in general use. Borax was noted by the Islamic writer Ibn Hawqal in the late tenth century, and Theophilus in the twelfth century described its use as a flux for niello. According to Biringuccio, the components of filigree work were attached to the substrate with quince-seed paste or gum arabic and then soldered with borax and a gold or silver alloy solder in a little furnace. Bran or dry elder twigs provided the heat. In earlier times chaff or straw were usual sources of heat for goldsmiths.

A variety of medieval work, ranging from some sections of the Konrad crown to the little lunate earring shown in Figure 3, made use of small clippings of wire to fill gaps between components during soldering. The small section of beaded wire so used on the Konrad crown has been fairly well placed to minimize disfigurement. This is not so true of the earrings.

One type of soldered join seen on some medieval goldwork is worth noting. The sheet gold setting borders on the Ottonian cameo brooch in Figure 16 are formed from strips of sheet gold that are slit and overlapped in a manner more frequently seen in sheet-copper and brass work. The cameo brooch shows two type of join (Fig. 20). One is a simple opposed-slit type, the other a more complex interslotting. Both of these assemblies are shown in Figure 21a, b. It is hard to see how the additional strength of such a join would be needed in goldwork, but the technique is not unique; the simple opposed-slit join is also seen on the settings of the Cambridge gold cover (Fig. 22), also supposedly of German origin, and many other examples presumably await discovery. The technique was not universal; it is not seen on the Townley brooch for example, which has the more usual overlap join.

Granulation

Granulation work is just one subsection of soldering. The persistent belief that granulation died out during the Dark Ages in Europe and was only rediscovered by the diligence of nineteenth- and twentieth-century goldsmiths is a product of European chauvinism. Granulation is found to fine effect well into the medieval period, and some ninth- and tenth-century work is of remarkable quality. In England the Alfred jewel is a fine example, and there are some superb Viking and Eastern European examples from the ninth and tenth centuries. Some eleventh-century Fatamid Islamic granulation work, including the lines of granules soldered on double-wire frameworks, is of exceptional quality.

Later in the medieval period, granulation work became less prominent, perhaps largely displaced by drawn wire filigree; but it was not forgotten, and it often forms highlights on filigree work. We see such granulation on Theophilus's book cover in Trier. Theophilus does not explain the technique of granulation, but he does describe small tweezers for manipulating the individual grains (Dodwell [1961]1986: book 3, ch. 7). A mid-thirteenth-century description of the treasures in St. Paul's Cathedral, London, describes goldwork as "cum margatiti et grani auri," which might be a reference to granulation; small gold beads seem less likely (Lehmann-Brockhaus 1960:no. 2748). The Dowgate Hill, Townley, and cameo brooches (Figs. 8–9, 10–13, 16–17) all have granulation work. In Renaissance goldwork, granulation is relegated to the odd grain and is seldom part of the main decorative scheme.

Casting

As in antiquity, casting was very seldom employed by goldsmiths apart from the initial production of the ingot or blank. Solid and often quite massive gold ornaments that manufacturers would find natural to cast today were generally fabricated from wrought gold. Casting was potentially wasteful of metal and provided little advantage in days when workmanship was relatively inexpensive. Neckham (1867) does tell us that the goldsmith should be skilled "in the work of casting," but no medieval gold jewelry is unequivocally known to have been made by casting. Possibly some of the solid rings were cast, but many undoubtedly were not.

Theophilus describes the manufacture of ingots and blanks by casting—the usual starting point for vessels or other objects—but the only mention of cast gold is for decorative handles and mounts such, as "a cast lion's head very elegantly carved" that could grace the end of the handle of a "gold or silver strainer" (Hawthorne and Smith 1979:ch. 57). Cellini talks about casting silver seals and decorative mounts for silver vessels and also mentions lost-wax casting for gold and silver, but he does not give examples of the type of objects cast (Ashbee 1967).

Cellini also describes the use of cast lead ornaments that would then be worked on by the master and then used as models for the casting of precious metal (Ashbee 1967). Base-metal ornaments were often cast and numerous molds have survived, including those for the ubiquitous lead Pilgrim badges.

Engraving

The medieval use of true engraving, in which a sliver of metal is gouged from the metal surface, contrasts with the methods of earlier periods. The depth and sharpness of engraving on medieval signet rings is noteworthy; the precision possible can

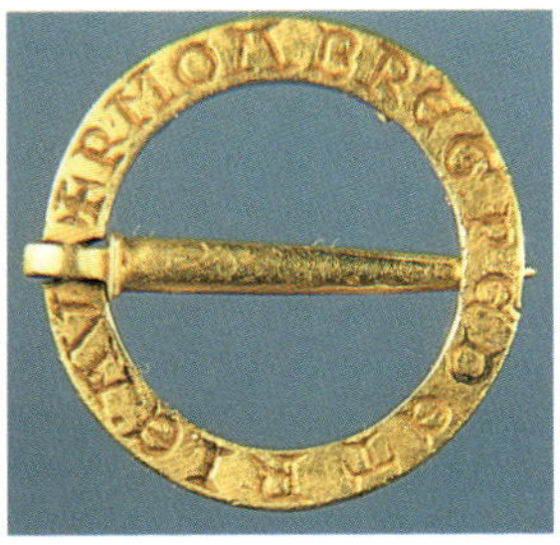

FIGURE 23. *Medieval gold ring brooch with engraved inscription. (The other side says "I am here in place of a friend.")*

be seen in the fine Middleham pendant in Figure 1. The use of engraving in antiquity was probably more common than generally assumed (examples include Mycenaean, Phoenician, and Hellenistic gold rings), but it was not widespread.

Theophilus describes the manufacture of engraving tools with sharp, oblique cutting edges that are nearly indistinguishable from their modern counterparts. An English statute of 1300 underlines the potentially wasteful nature of engraving in its ruling that "gravers, cutters of stone and of seals shall give each [i.e., return to the patron] their weight of gold and silver as near they can" (Jackson 1905:7).

Engraving could be used for functions other than just decorative designs or inscriptions. For example, on the Cambridge gold cover of around 1300, the goldsmith scored a fine line to mark the position of the settings and then used an engraving tool to push up small curls of gold to act as guides to hold the settings in place during soldering.

The texts engraved on medieval jewelry are not always readily readable. In some cases this is due to current lack of understanding of ancient conventions or convolutions. For example, the side of the ring brooch in Figure 23 is not immediately readable and at one stage was assumed to be of magical significance. In fact, the letters are probably the alternating letters of the names Robert and Marger [?]. In some cases, the literacy of the engravers must legitimately be doubted. From 1487 we hear of problems when an apprentice engraver noted for his expertise in engraving "all manner of letter" was found to be unable to read or write (Prideaux 1896–97:28).

Another form of engraving, the *pointillé* (dot) engraving that became common toward the end of this period, had a long history and can be traced back to dotted inscriptions on Classical vessels.

Enamel and Niello

In medieval work, three main types of enamel may be distinguished. These are champlevé, in which the enamel is applied in hollows in the metal; cloisonné, in which the enamel is held in cells bounded by thin walls of metal strip; and enamel *en ronde boss,* in which there is an overall coating of a three-dimensional figure, usually with white enamel. Cloisonné enamel, employing thin, soldered strips of metal, was best suited to gold, while champlevé enamel, using engraved depressions, required thicker metal and was thus most common with copper alloys. These manufacturing differences can often obscure the stylistic similarities of the final work. The Dowgate Hill and Townley brooches (Figs. 8–9, 10–13) both have fine cloisonné enamel centers.

Medieval European texts often describe enameled precious-metal objects in the most glowing terms, describing it as of the finest work and of robust beauty (Lehmann-Brockhaus 1938:2838, 2842). References are made to Anglo-Saxon enameled regalia and to enameled objects from Constantinople (Lehmann-Brockhaus 1938:2972, 2857). We can assume that most of this enamel was of the cloisonné type which can be seen on Dark Age goldwork and which was fairly widespread through medieval times from Britain to the Byzantine and Islamic worlds. Early medieval English examples include the late ninth-century Alfred jewel. Fine cloisonné enamel in late ninth- to early tenth-century contexts in Eastern Europe includes recently excavated gold lunate earrings with circular cloisonné enamel motifs (Vazarova 1980:pl. 7).

Westermann-Angerhausen (1983–84) argues for a native Anglo-Saxon origin and an early eleventh century date for the Dowgate Hill brooch. The even more recent study and publication of a group of English cloisonné enamel copper alloy brooches of late tenth to early eleventh-centuries leads Buckton to state (1986:16):

> This would establish the existence of a previously unsuspected centre of *cloisonné* enamelling in England at a time when *cloisonné* enamel in Europe, particularly in the German and Lotharingian centres of the Rhineland, Regensburg and Trier, was at its peak.

Champlevé enamel was an old European tradition dating back into the Iron Age. In medieval times champlevé work on copper alloys was associated with the town of Limoges, and documentary references to *opere Levovitico, labore Limogiae, opere Lemovicensi,* and so on, occur from the mid-twelfth century onward (Gay [1887]1928:s.vv. *email* and *Limoges*). Champlevé enamel was used on medieval goldwork and possibly can be seen as largely taking over niello on goldwork. Examples of champlevé work on gold include the Middleham jewel (Fig. 1), although only traces of the blue enamel filling the engraved lines now remain. Champlevé enamel is also found in silver and copper-alloy rings of medieval Islamic origin (Ogden 1987). A regulation of Edward II dated 1370 notes that "cutters of seals and enamelers shall return to each his weight of gold and silver as near as they can save it" (Prideaux 1896–97:doc. no 440). This certainly implies champlevé enamel, where the precious metal would need to be cut out with engraving tools.

The mention of enameled silver is interesting since it is almost entirely absent in ancient work. We can assume that this was due to problems relating to adhesion and relative coefficients of expansion. Something of the sort is suggested by Biringuccio who states, "each kind of gold, silver or copper requires enamel harmonious to its nature else it does not respond." In Paris in the early fourteenth century, enamelers agreed to enamel only on silver and gold of high standard and not to use lead glass, which was suitable only for silver of fraudulent alloy (Lightbown 1978:83). The continued use of niello on silver, compared to black champlevé enamels on gold, must relate to the difficulties in enameling silver.

The final coating of a three-dimensional gold object with a layer of enamel (enamel *en ronde boss*) was known in Hellenistic Italy but became popular in the second half of the fourteenth century. This type of enamel is particularly ubiquitous in Renaissance jewelry, where the general delicate and piecemeal construction of the underlying goldwork can usually be contrasted with the cast and often coarse work of much of its nineteenth-century imitators.

Filigree enamel, in which filigree wires form the boundaries of the enamel, was typical of Hellenistic work, but is very rare in medieval jewelry and only reappears in the Renaissance when it was supposedly introduced into Italy from the Balkans by the Venetians (Hackenbroch 1979). A text of about 1400 seems to refer to a type of mastic inlay used to imitate enamel (Gay [1887]1928:s.v. mastic); this might relate to the use of bitumen as an imitation of niello in some medieval Islamic jewelry.

Niello is a black sulphide of silver or a mixed sulphide including silver and lead. This material was fused into designs engraved on a gold or silver surface and polished. In some case a form of niello was used to paint designs straight onto a metal surface. This attractive technique is described in the "Mappae Clavicula" and can be seen quite frequently on medieval Islamic gold jewelry. Niello appears in late Roman gold jewelry and continued to be popular through the Dark Ages and into the medieval period. A fine mid-ninth-century English example is a ring with the name of King Ethelwulf (British Museum, MLA 29.11-14.1).This miter-shaped ring is made from heavy hammered gold sheet with an overlap join at the back. The design is deeply engraved to hold the niello.

The addition of lead to the niello makes it far more fluid and improves the bond with the surrounding metal. As far as analyses show, the use of leaded niello in Europe was a development of the medieval period and perhaps does not predate the tenth century (Oddy et al. 1983). Both the "Mappae Clavicula" and Eraclius provide niello recipes that include lead (Smith and Hawthorne 1974:chs. 56, 58, 195, 196; Merrifield 1849:ch. 48).

Both silver and gold with niello decoration are mentioned in late thirteenth- and fourteenth-century documents; one account of 1396 tells how a French goldsmith had to reniello a goblet when he repaired it (Lightbown 1978:75). Cellini notes in the introduction to his book that the art of niello had fallen into disuse by the early sixteenth century, although the process was still discussed by a few old men (Ashbee 1967).

The Dawn of the Renaissance

In many ways the goldsmiths of the early Renaissance continued the same traditions of their forebears, which in turn harked back to the Hellenistic and earlier times. Goldwork was usually assembled from sheet gold and decorated with filigree. However, there were considerable stylistic changes and changing fashions in decorative processes.

The Renaissance signaled the rebirth of an affection for the ancient Classical world; Renaissance jewelers looked back on classical goldwork with awe and admiration. The highest praise, quoted by the ever-vain Cellini, was Michelangelo's statement that one of Cellini's jewels was "so beautiful that I do not think any goldsmith of the ancient world ever fashioned anything to match it" (Ashbee 1967). In view of this, one must ask why those perennials of Hellenistic and Etruscan work—loop-in-loop chain, beaded wire, and granulation—were almost totally banished from the Renaissance jeweler's repertoire.

Acknowledgments

The author would like to thank John Cherry and Christopher Entwhistle of the Department of Medieval and Later Antiquities at the British Museum for their help in allowing the author to examine and photograph objects in their care. John Cherry kindly also provided unpublished analyses of some medieval British jewelry and discussed various aspects of the Middleham Jewel.

References

AMIRANASHVILI, S.

n.d. *Medieval Georgian Enamels of Russia*. New York: Abrams.

ASHBEE, C. R.

1967 *The Treatises of Benvenuto Cellini on Goldsmithing and Sculpture*. New York: Dover.

BENDA, K.

1967 *Ornament and Jewellery*. Prague: Artia.

BERTHELOT, M.

1888 *Collection des anciens alchimistes grecs*. Paris: G. Steinheil.

BOILEAU, E.

1837 *Réglemens sur les Arts et Métiers de Paris rédigés au xiii siècle*. G. B. Depping, ed. Paris: Impr. de Crapelet.

BUCKTON, DAVID

1986 Late 10th- and 11th-century cloisonné enamel brooches. *Medieval Archaeology* 30:8–18.

COMBE, T.

1806 Copy of a Indenture, made in 1469, between King Edward IV and William Lord Hastings, Master of the Mint. *Archaeologia* 15:164–78.

CRIPPS, W. J.

1880 *Old French Plate*. London: J. Murray.

1914 *Old English Plate*. 10th ed. London: J. Murray.

DE CUYPER, F., G. DEMORTIER, C. J. DUMOULIN, AND J. PYKE

1987 *La croix byzantine dy Trésor de la Cathédrale de Tournai* (*Aurifex* 7). Louvain: Université Catholique de Louvain.

DIDEROT, D. AND J. D'ALEMBERT

1751–57 *Encyclopédie ou dictionaire Raisonne des sciences, des arts et des metiers. Recueil de planches sur les sciences, les arts libéraux et les arts méchaniques, avec lur Explication: Orfèvrerie joaillerie*. Paris: Chez Briasson.

DODWELL, C. R.

[1961] 1986 *Theophilus: De Diversis Artibus*. Reprint. Oxford: Oxford University Press.

DUCZKO, W.

1985 *Birka V: The Filigree and Granulation Work of the Viking Period*. Stockholm: Kungl. Vitterhets Historie och Antikvitets Akademien.

DUNLOP, D. M.

1957 Sources of Gold and Silver in Islam according to al-Hamdani. *Studia Islamica* 8: 29–49.

FREEMANTLE, C. W.

1870 *First Annual Report of the Deputy Master of the Mint*. London: Royal Mint.

GAY, V.

[1887] 1928 *Glossaire archéologique du moyen âge et de la Renaisance*. 2 vols. Reprint. Paris: Libraire de la Société Bibliographique.

GORDUS, A. A., AND D. M. METCALF

1980 Neutron activation analysis of the gold coinages of the Crusader states. In *Metallurgy in Numismatics* 1:119–150. D. M. Metcalf and W. A. Oddy, eds. London: Royal Munismatic Society.

HACKENBROCH, Y.

1979 *Renaissance Jewellery*. London: Sotheby, Parke Bernet.

HAWKES, S. C., J. M. MERRICK, AND D. M. METCALF

1966 X-ray fluorescence analysis of some Dark Age jewellery. *Archaeometry* 9:98–138.

HAWTHORNE, J. G., AND C. STANLEY SMITH

1979 *Theophilus: On Divers Arts*. New York: Dover.

HENDY. M. F.

1985 *Studies in Byzantine Monetary Economy c. 300 –1450*. Cambridge: Cambridge University Press.

JACKSON, C. H.

1905 *English Goldsmiths and their Marks*. London: Macmillan.

KUHN, S. M., AND J. REIDY

1963 *Middle English Dictionary*. Ann Arbor: University of Michigan Press.

KUNZ, G. F.

[1917] 1973 *Rings for the Finger*. Reprint. New York: Dover.

LANG, J., AND M. J. HUGHES

1984 Soldering Roman silver plate. *Oxford Journal of Archaeology* 3(3):77–107.

LEHMANN-BROCKHAUS, O.

1938 *Schriftquellen zur kunstgeschichte des 11 und 12 Jahrhunderts für Deutschland, Lothringen und Italien*. Berlin: Deutscher Verein fuer Kunstwissenschaft.

1960 *Lateinische Schriftquellen zur kunst in England, Wales und Schottland vom jahre 901 bis zum Jahre 1307*. Munich: Veröff. d. Zentralinst. f. Kunstgesch.

LIGHTBOWN, R. W.

1978 *Secular Goldsmith's Work in Medieval France*. London: The Society of Antiquaries.

LUCAS, A., AND HARRIS, J. R.

1962 *Ancient Egyptian Materials and Industries*, 4th ed. London: Arnold.

MARYON, H.

1971 *Metalwork and Enamelling,* 5th ed. New York: Dover.

MERRIFIELD, M. P.

1849 *Original Treatises Dating From the XIIth to XVIIth Centuries on the Arts of Painting,* vol 1. London: J. Murray.

NECKHAM, A.

1867 *De Nominibus Utensilium.* A. Scheler, ed. In *Lexicographie latine du XIIe et du XIIIe siecle.* 114–116. Leipzig. (Translated in Lightbown 1978:4–5.)

NICOLINI, G., AND B. BOUCHET

1990 Les feuilles d'or de placage des casques celtiques d'Amfreville et d'Agris. *Gallia* 47:67–77.

ODDY, W. A.

1977 The production of gold wire in Antiquity. *Gold Bulletin* 10(3):79–87.

ODDY, W. A., M. BIMSON, AND S. LA NIECE

1983 The composition of niello decoration on gold, silver and bronze in the antique and medieval periods. *Studies in Conservation* 28(1)29–35.

ODDY, W. A., AND S. LA NIECE

1986 Byzantine gold coins and jewellery. *Gold Bulletin* 19(1):19–27.

OGDEN, J. M.

1982 Jewellery of the ancient world. London: Trefoil.

1987 Islamic goldsmithing techniques in the early medieval period. In *Islamic Rings and Gems,* 408–27. D. J. Content, ed. London: Philip Wilson.

1991 Classical gold wire: some aspects of its manufacture and use. *Jewellery Studies* 5:95–105.

n.d. *Gold Jewellery in Ptolemaic, Roman and Byzantine Egypt.* Leuven: Peeters. In press.

OGDEN, J., AND S. SCHMIDT.

1990 Late antique jewellery: pierced work and hollow beaded wire. *Jewellery Studies* 4:5–12.

PEGOLOTTI, F. BALDUCCI

1936 *La Pretica della Merctura.* A. Evans, ed. Cambridge, Mass.: The Mediaeval Academy of America.

PRIDEAUX, W. S.

1896–97 *Memorials of the Goldsmith's Company: Being gleaned from their records between the years 1335 and 1815.* London: Privately printed.

RONIG, F. J.

1984 *Schatzkunst Trier.* Trier: Spee.

ROYAL MINT

1874 *Fourth Annual Report of the Royal Mint.* London: Royal Mint.

1884 *Fifteenth Annual Report of the Royal Mint.* London: Royal Mint.

RUDING, R.

1812 Some accounts of the trial of the Pix. *Archaeologia* 16:164–170.

SINGER, C. J., E. J. HOLMYARD, AND A. R. HALL

1957 *A History of Technology,* vol. 3. Oxford: Oxford University Press.

SMITH, C. STANLEY, AND M. T. GNUDI

1943 *The Pirotechnia of Vannoccio Biringuccio.* New York: The American Institute of Mining and Metallurgical Engineers.

SMITH, C. STANLEY, AND J. G. HAWTHORNE

1974 *Mappae Clavicula: A Little Key to the World of Medieval Technology.* Translated by the American Philosophical Society. n.s. 64(4).

STENBERGER, M.

1947 *Die Schatzfunde Gotlands der Wikingerzeit: II Fundbeschreibung und Tafeln.* Lund: Kungl. Vitterhets Historie och Antikvitets Akademien.

STEVENSON, R. B. K

1974 The Hunterston Brooch and its Significance. *Medieval Archaeology* 18:16–42.

STUBBES, P.

1583 Tricks of goldsmiths and Vintners. In *Anatomy of the Abuses in England in Shakspere's Youth,* part 2.1, series 6, no. 16. New Shakespeare Society, London: Trübner.

UCCELLI, A.

1948 *Storia della Tecnia dal Medio evo ai nostri giorni.* Milan: U. Hoepli.

VAZAROVA, Z

1980 Parures d'or du tombeau 27 à Pliska. *Archeologiya* 1:52–56.

WESTERMANN-ANGERHAUSEN, H.

1973 *Die Goldschmiedearbeiten der Trierer Egbertwerkstatt.* Trier: Spee.

1983–84 Ottonischer Fibelschmuck, neue Funde und Überlegungen. *Jewellery Studies* 1:20–36.

WHALLEY, J. I.

1982 *Pliny the Elder: Historia Naturalis.* London: Sidgwick and Jackson.

WHITFIELD, N.

1990 Round wire in the early Middle Ages. *Jewellery Studies* 4:13–28.

YOUNGS, S.

1989 *The Work of Angels: Masterpieces of Celtic Metalwork, 6th–9th centuries* A.D. London: British Museum Press.

Biography

Jack Ogden is a specialist in the materials and technology of early goldwork and has written and lectured widely on the subject. He is currently director of the Cambridge Centre for Precious Metal Research and editor of *Jewellery Studies,* the journal of the Society of Jewellery Historians of which he was a founder in 1977.

Gold Foil, Strip, and Wire in the Iron Age of Southern Africa

ANDREW ODDY

Gold has been an important metal in the southeastern part of Africa for at least the last millennium, and possibly longer, and many sites show signs of primitive (although usually undated) mining activity (Summers 1969). There are three main sites in Southeast Africa where gold artifacts from the later Iron Age have been found: Mapungubwe in the northern Transvaal, which flourished in the eleventh and twelfth centuries; Great Zimbabwe, which flourished in the thirteenth and fourteenth centuries; and some rich burials at Ingombe Ilede in Zambia, which date from the late fourteenth and/or fifteenth centuries.

Gold was one of the commodities that brought early explorers and traders south on the Red Sea. Although in the time of Solomon the ships had probably reached only as far south as the Gulf of Aden, by the first century C.E. at the latest, Greeks from Egypt had rounded Cape Guardafui and sailed down the East African coast, probably as far as Rhapta opposite the island of Zanzibar (Casson 1989). The map (Fig. 1) shows this ancient route. It is likely, however, that Arabs had penetrated into the Indian Ocean well before this time; Arab traders subsequently reached as far south as Sofala, establishing trading posts along the coast, at least from the ninth century C.E. These posts were outlets for the products of the interior, of which gold was probably one of the most important.[1]

As a result of the settlements by Arabs and their intermarriages with the indigenous peoples, the Swahili culture developed along the coast, and this culture had maritime trading connections with the Red Sea, the coast of Arabia, and ultimately with India. Even shards of Chinese ceramics dating from the Sung Dynasty (C.E. 960–1279) onward have been found at Great Zimbabwe (Summers 1963:46, fig. 14).

The tenuous contacts with the cultures of Arabia and India, which are proven by imports from these regions, raise the question of when external influences first penetrated the African interior. The cultural origins of the great stone ruins in modern Zimbabwe, the best example of which is Great Zimbabwe, were much debated in the earlier part of this century. It has long been recognized, however, that the ruins were

FIGURE 1. *Map of Africa showing principal sites mentioned in the text.*

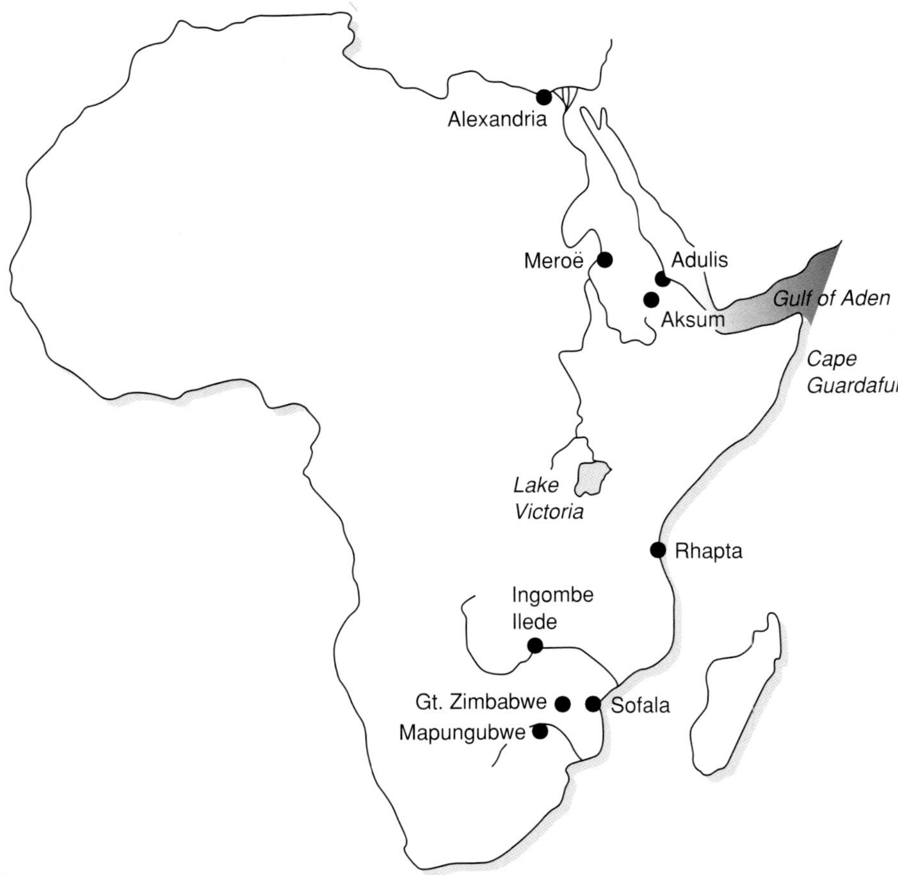

built by a wholly indigenous (Bantu) civilization. This recognition stands, despite ongoing arguments for the penetration of the gold-producing areas by Indians and even Indonesians at least two thousand years ago (Hromník 1981).

In fact, the history of gold use is intimately bound up with that of other metals. It is also well known that the exploitation and use of metals by the peoples of the African continent varies greatly from region to region. The great natural barriers created by geography and climate meant that the use of metal did not show the same pattern of development in West and East Africa, nor in the Mediterranean coastal regions north of the Sahara. During the early first millennium C.E., iron use occurred at sites in eastern sub-Saharan Africa as far south as the Transvaal, but its method of introduction has been much debated and is still unclear. The beginnings of food production and the first working of iron appear to be roughly contemporary, however. At about the same time, a movement of peoples took place, accompanied by the spread of the Bantu family of languages. Whether these concurrent cultural changes all resulted from the same event or were purely coincidental is still being argued. In the past it has even been suggested that iron smelting was an independent discovery in sub-Saharan Africa, but the increasing availability of radiocarbon dates for Iron Age sites now supports the view that knowledge of iron was introduced from the north and spread southward (Phillipson 1985:149).

One of the links in the chain was, presumably, the kingdom of Meroë, which was established in the sixth century B.C.E. in what is now Sudan, where vast deposits of slag from iron smelting have been found. Meroë is separated from Egypt by desert. For this and for political reasons, communications were established in other directions, particularly eastward toward the Red Sea and southeastward toward the emerging Aksumite civilization in the highlands of Ethiopia. The heyday of iron production at Meroë was apparently during the last centuries B.C.E. when iron was exported in return for luxuries from the north and east, which were imported via the Red Sea route. Iron use appears to have been established in the area around Lake Victoria during the last few centuries B.C.E., and reached as far south as Natal by about C.E. 300. It is noteworthy that over an enormous area of eastern and southern Africa, the earliest iron-using communities show a remarkable degree of cultural homogeneity (Phillipson 1985:171ff).

This all began to change in about the tenth century C.E. when the Iron Age population began to expand. Domestic animals became more common, and actual mining for metal ore supplanted—or at least augmented—collection from the surface. At the same time, regional differences in the material culture became marked in the archaeological record. Powerful kingdoms began to emerge in which trade and the resultant wealth were controlled by an elite stratum of society. In the area of present-day Zimbabwe, many sites preserve remains of stone-built monuments dating back to the early centuries of the present millennium (MacIver 1906). Few of these sites have been excavated systematically, and many were deliberately ransacked in the nineteenth century by treasure hunters searching for gold; but some archaeological investigations have been carried out that have preserved evidence for the use of gold by the indigenous Africans. Gold artifacts are known to have been found—either by excavation, treasure hunting, or accident—on about twenty sites altogether, all of which are within or close to the main gold fields in present-day Zimbabwe.

The earliest site to have produced gold in any quantity is the acropolis at Mapungubwe, a site in the northern Transvaal, about two kilometers south of the Limpopo River. The acropolis consists of a long, narrow, steep-sided sandstone hill, which has only two paths to the top and was, therefore, very easy to defend. At the top are traces of stone walls and rock-cut cisterns. It is thought that the thin covering of soil must have been transported to the top in baskets and that the hilltop was the living area as well as the burial place of the nobility.

Serious excavations at Mapungubwe began in 1933. Several of the site's rich graves, thought to date from the twelfth century, were found to contain iron-wire bangles, ivory bangles, bone tools, pottery, and gold. Excavations on the site have continued intermittently since then (Fouché 1937; Gardner 1963). Four different types of gold objects have been identified: coiled-strip anklets, wire circlets, objects made of gold foil, and beads.[2]

Gold Sheet (Foil)

The best known of the finds from Mapungubwe is a small rhinoceros (Fig. 2) made of gold sheet. Although rather crude in appearance, the body and legs are all made from one piece by hammering, indicating a passing familiarity with the principles of *repoussé* work. The front hips are nicely delineated, but on the whole, the execution of the rhinoceros does not exhibit much skill, as the shaping has been partly achieved by folding and creasing the gold.

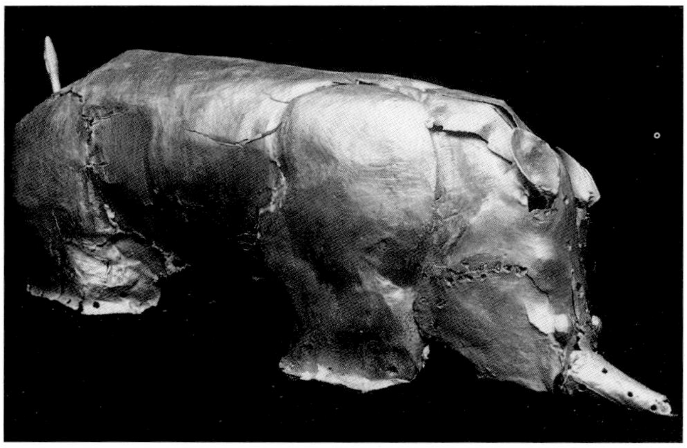

FIGURE 2. *Gold-foil rhinoceros from Mapungubwe. Pretoria University. Inv. no. DAPU 33.170.*

One feature of the rhinoceros that has raised questions about the manufacture of the goldwork is the animal's single horn. It thus appears to be an Indian rhinoceros rather than the African species, which has two horns. This discrepancy clearly presents a problem of interpretation, since the figurine could also represent a young animal that has not yet started to develop a second horn. However, the more likely explanation may simply be that the artisan lacked the skill necessary to fit the head with more than one horn. The head is, in fact, made from a separate piece of gold to which the ears and horn have been added. The tail is also separate. No archaeological evidence is available to indicate the methods of manufacture of the gold sheet from which the rhinoceros was made, but since the technology was already available for smelting iron, it was certainly available for melting gold. One might postulate, then, that gold was collected from rivers and streams and possibly obtained by mining, then melted and cast into flat sheets that were subsequently hammered on an anvil to form the thin foil out of which the sheet-gold objects were formed.

The inside of the rhinoceros is not visible, but when the other sheet-gold finds are examined, a clear difference may be seen between the surface textures on the two sides of the foil. The surviving finds from Mapungubwe include gold-sheet fragments of several other incomplete animal figurines. Microscopic examination shows that the outer surfaces were polished with a coarse abrasive, which left deep scratches on the surface (Fig. 3). The inner surfaces, however, have a rough, granular-looking finish which shows no sign of polishing (Fig. 4). The most plausible explanation is that the inner surfaces represent the underside of the gold sheet as it was hammered out, and the granular texture reflects the texture of the surface of the anvil, which was most likely made of stone. To ascertain whether a stone anvil was used, it is necessary to carry out some replication experiments; in this case, by beating out thin gold sheet using a large boulder from Mapungubwe to support the gold.

A stone anvil was in fact excavated "behind Mapungubwe Hill" in 1937 in an area that was interpreted as an iron-smelting site (Gardner 1963:12–13). Even though gold beads were found there "under a small red rubble floor," the site was postulated to be Venda. This presumably means that the excavator thought that the anvil should be dated considerably later than the rich burials containing most of the gold objects. This conclusion may well need to be revised.

The hammers used for beating out the gold sheet were most likely made of metal, probably iron, although no hammers have been identified among the exten-

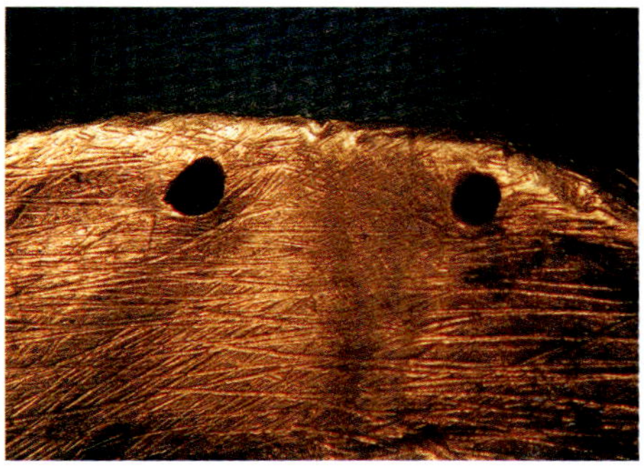

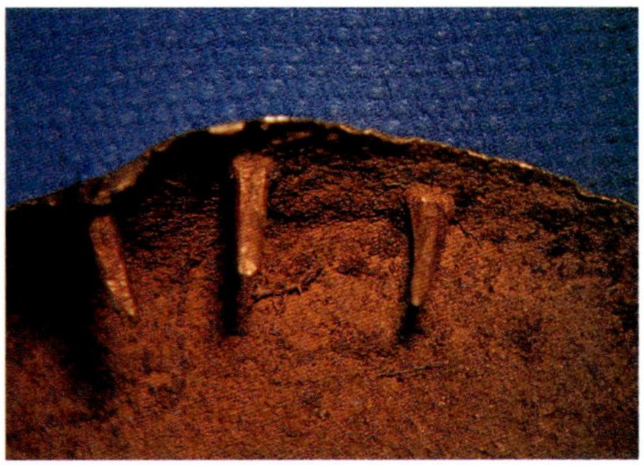

FIGURE 3. *Outer surface of a piece of gold foil, near right, showing deep polishing scratches.*

FIGURE 4. *Inner surface of gold foil, far right, showing granular texture. Note the presence of small gold nails originally used to attach the gold to a wood former.*

sive range of metal tools, weapons, and articles of personal adornment excavated at Mapungubwe. Iron chisels, awls, spikes, and lumps of iron have been found, however (Gardner 1963:93–168), so it is reasonable to assume that iron hammers were also known. Hammer stones from the site have been identified (Gardner 1963:93–168).

Once the gold sheet was made and formed into the required shapes, the artisan needed to join the various components of the object. How this was accomplished is uncertain. No evidence exists for the use of soldered or fused joins on any of the gold objects from Mapungubwe or its related sites. The gold sheet was supported by being attached to an organic former with small gold nails. Since gold nails are soft, the former must have been made of wood, which was certainly worked at this period (Voigt 1983:77–79, 111–19). Nail holes can be seen on many of the gold-sheet fragments, in some cases with the nails still in situ (Fig. 4). This proves that the nail holes were made by the process of hammering the nail through the gold sheet into the wood. Although it is clear that the nail holes were not preformed, the gold sheet may have been pricked with an awl to facilitate the first penetration of the point. Nails have been found in considerable numbers, and an examination of empty holes and of loose nails sometimes shows that the nail head has been deformed by the hammering process. Nails were tapered and either round or rectangular in cross section. Typical dimensions (in millimeters) are as follows:

length	*diameter just below head*	*diameter near tip*
4.9	1.1	0.6
4.6	1.0	0.5
3.8	0.9	0.7
4.9	1.2	0.9
3.5	0.9	0.6
3.7	1.1	0.5
4.5	1.3	0.8
4.9	1.0	0.6

A few tapering nails similar to those described above have also been found at the later (seventeenth- to eighteenth-century) site of Dhlo Dhlo and are now preserved in the Queen Victoria Museum in Harare. They are among a find of metalworking debris consisting of globules of gold, nuggets, and fragments of wire and foil, much of which is made of debased gold that clearly contains a considerable proportion of copper. The British Museum contains similar metalworking waste from an unknown site in Zimbabwe, which includes about forty similar nails (Fig. 5); some of these have round shanks and some have square ones.[3]

When two pieces of an object needed to be joined, the gold was overlapped and the pieces were nailed together. Patterns of nail holes in the center of sheet fragments attest to this practice. Evidence for an overlapping join is visible on the scepter (Fig. 6), one of the other more complete gold-sheet objects. The scepter is mounted on a modern wooden former, which is not quite the right size, resulting in the failure of the corresponding nail holes along the overlapping edges to align correctly (Fig. 7). The edges of the gold sheet are very rough and do not show evidence of having been cut with a metal (iron) blade. The original staff of the scepter was carved with a barley-sugar twist, to which the gold was fitted with some difficulty, resulting in creasing to improve the fit (Fig. 8). Despite the lack of skill shown in the three-dimensional shaping of the gold, the scepter is decorated with lines of indented "dots" (Fig. 9), made by pressing a blunt-pointed tool, perhaps made of bone, into the surface. Pointed-bone tools have been preserved from the site (Voigt 1983:77; 101ff). The head and shaft of the scepter are covered with separate pieces of gold sheet.

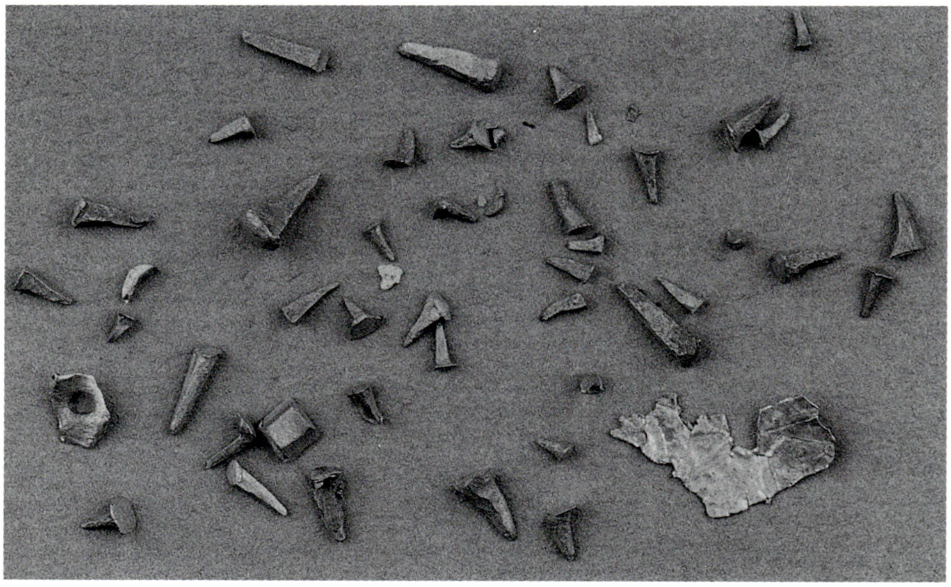

FIGURE 5. *Small gold nails from an unidentified site in Zimbabwe. British Museum, Ethno 1923:12-11(9).*

FIGURE 6. *Gold foil-covered scepter from Mapungubwe, below left. Inv. no. DAPU 33.174A.*

FIGURE 7. *Detail of Figure 6, below right, showing overlapping join of the gold foil.*

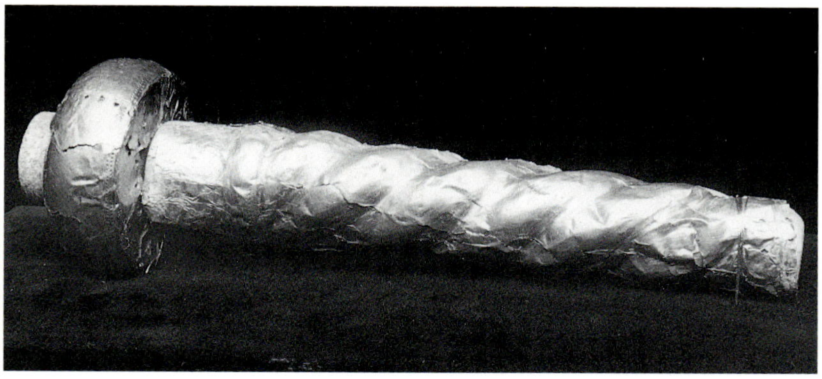

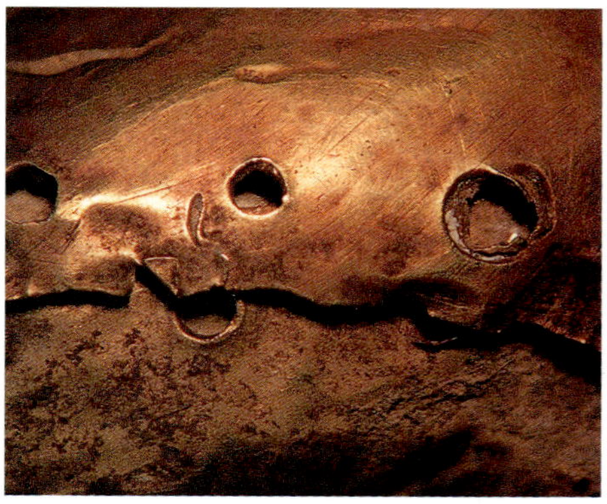

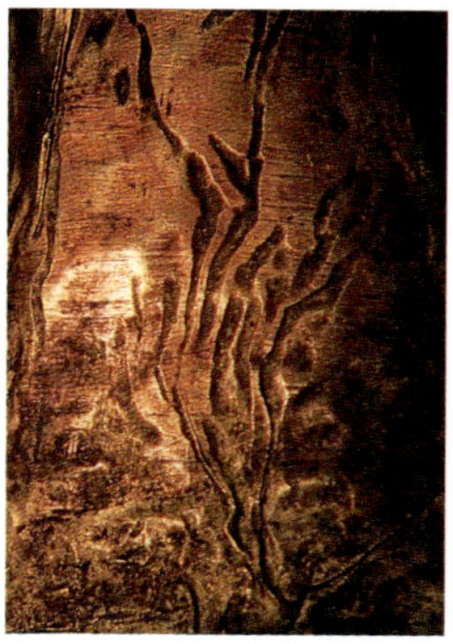

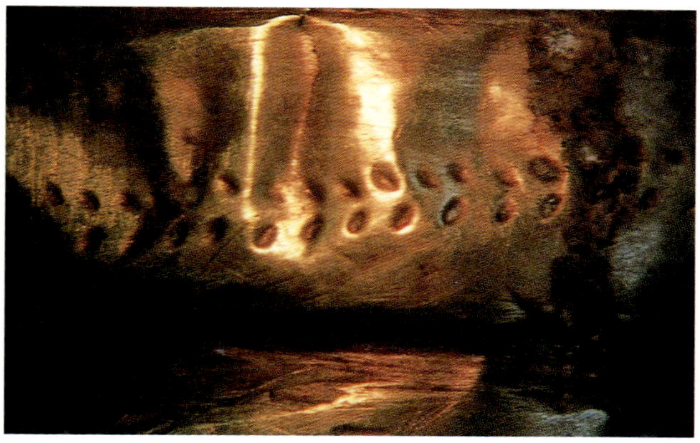

FIGURE 8. *Detail of Figure 6, showing creasing of the gold foil.*

FIGURE 9. *Detail of Figure 6, showing indented dot decoration of the gold foil.*

One other essentially complete object has survived: a gold bowl which is now, and presumably was originally, mounted on a wooden former. It was partly shaped by *repoussé* work, but it is also pleated near the rim.

Other fragments of gold sheet are sometimes recognizable as parts of animals, but some appear merely to be decorative plaques, perhaps originally attached to wooden household objects or used as jewelry. The most intriguing are three parts of what seem to be the same animal, from which one pair of legs has been recovered (Fig. 10), together with a head (Fig. 11). This animal is interesting for three reasons: It has been much more skillfully shaped by the *repoussé* technique than was the rhinoceros; its patterns of indented decoration are more advanced in design, and its species is uncertain. The animal has sometimes been interpreted as a horse (Voigt 1983:130), indicating either contact with horse-riding traders from the coast as early as the twelfth century, or a later period for these particular fragments. This interpretation relies on the presence of ears and the decoration along the head, which is assumed to represent a harness. But the elongated snout is also characteristic of a crocodile, in which case the decoration along the snout could represent teeth. Yet the ears are difficult to reconcile with those of a crocodile, and, furthermore, the teethlike decoration on the snout reoccurs on the torso. Perhaps the animal is supposed to be a hippopotamus after all, a creature with short legs, ears, and plenty of teeth!

Microscopic examination of the chevron pattern on the torso and legs reveals markings made by scribing the outside of the sheet with a blunt-pointed tool while the gold was resting on a pliable surface. The same conclusion is reached by

FIGURE 10. *Gold-foil legs of an unidentified animal from Mapungubwe. Inv. no. DAPU 33.173D.*

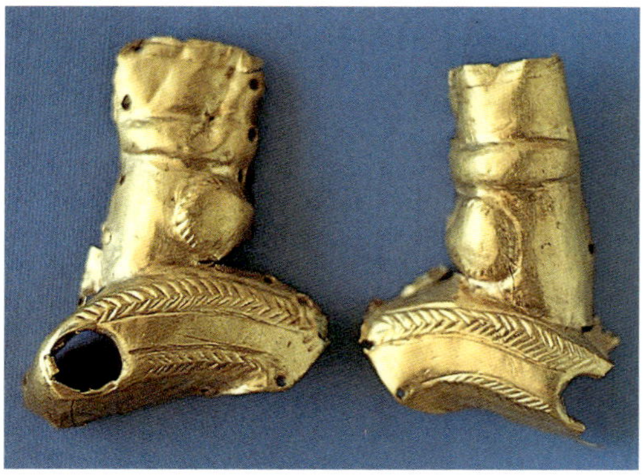

FIGURE 11. *Gold-foil head of the unidentified animal in Figure 10. Inv. no. DAPU 33.173E.*

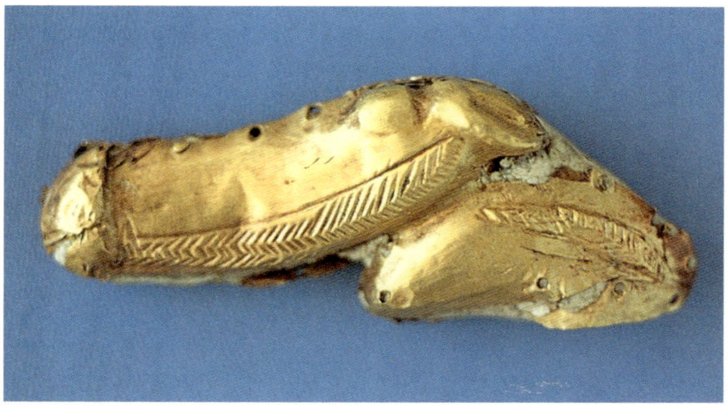

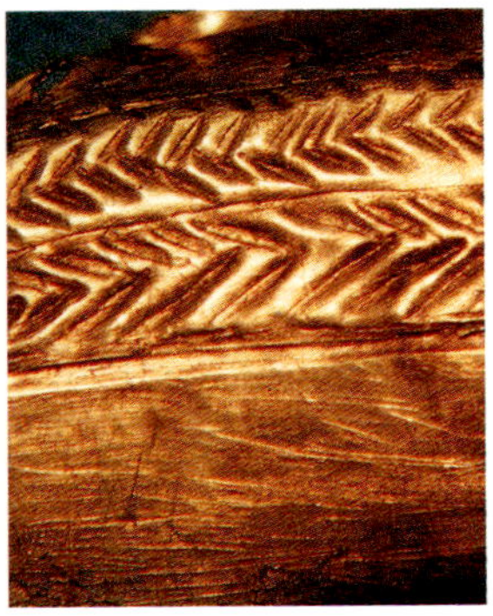
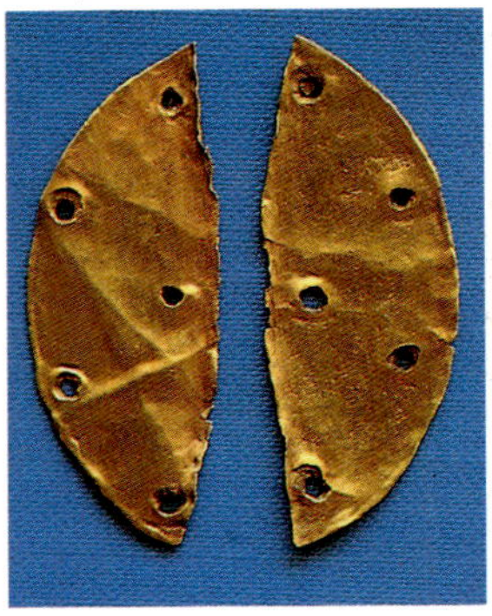
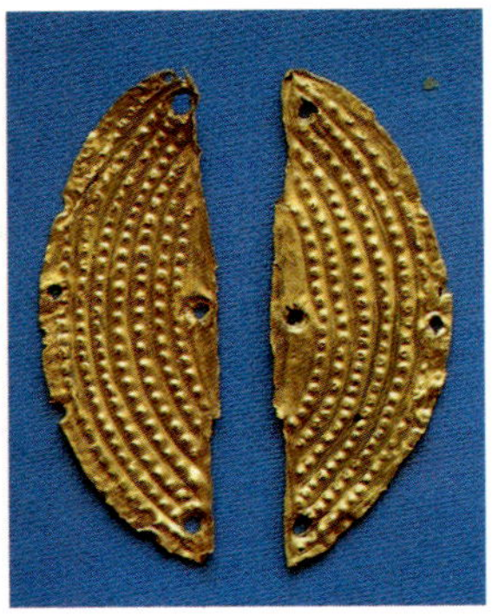
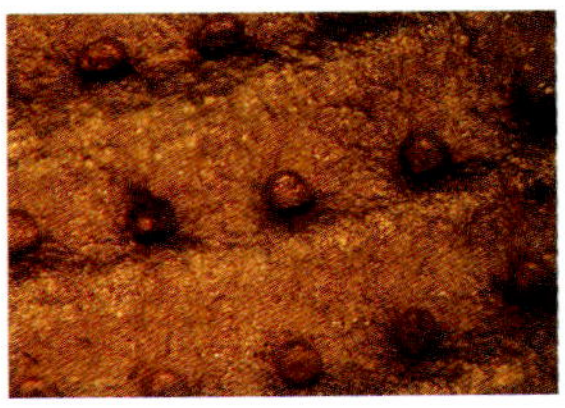

FIGURE 12. *Detail of Figure 11, showing indented chevron pattern on the animal's head.*

FIGURE 13. *Half-moon pieces of gold foil from Mapungubwe, second from left. One piece appears to be included in inv. no. DAPU 33.170. The other is part of inv. no. DAPU 33.173.*

FIGURE 14. *Decorated half-moon pieces of gold foil from Mapungubwe, third from left. Inv. no. DAPU 33.173.*

FIGURE 15. *Detail of the back of a decorated half-moon piece of gold foil, top right, showing the sharpness of the circular punch marks made from the same side of the sheet and the diffuseness of the lines made from the other side.*

FIGURE 16. *Fragments of gold foil, near right, from Great Zimbabwe. Inv. no. QVMH 7297(6)(a)(2).*

examining the pattern on the head (Fig. 12). However, at a high magnification, the pattern is seen to include scratches that suggest the tool used to indent the gold was metal rather than bone. In fact, some bone tools fitted with iron points have survived at Mapungubwe (Voigt 1983:104–5).

The gold-sheet fragments from Mapungubwe also include two half-moon-shaped pieces, pierced with five holes (Fig. 13), and some similarly shaped pieces which are also decorated with impressed dots and lines (Fig. 14). The dots were impressed from the back of the gold sheet using a blunt punch with a diameter of 0.3 mm (Fig. 15). The lines have been rubbed onto the front surface with a pointed tool. Again, the presence of scratches suggests the use of a metal tool.

Gold foil is rarely seen among the finds from the other sites, although the material from Great Zimbabwe in the Queen Victoria Museum in Harare includes a group of crumpled fragments (Fig. 16) that are also pierced with holes and have one rough and one smooth side. Also found at this site were a fragment from the Ridge ruins at Great Zimbabwe[4] and pieces from at least three other unspecified locations at this site.[5] One of these pieces is folded; another has a polished front, but its back is covered in slight, longitudinal ridges, suggesting that it might have been originally burnished on a piece of wood.

One very small tubular gold foil bead (approximately 9 mm long and about 2 mm in diameter) was excavated from the Chiwona ruins by Caton-Thompson in 1929 and is thought to be contemporary with Great Zimbabwe (Caton-Thompson 1931:127). A similar but much larger tubular bead is made of rolled-up gold sheet (Fig. 17). This bead (7.3 mm long and 4.25 mm in outside diameter) was found in one of the rich graves at Ingombe Ilede in Zambia (Fagan 1969:64–78), which are now dated as late-fourteenth and fifteenth centuries (Phillipson and Fagan 1969). These graves are most important for their gold beads (Oddy, in preparation).

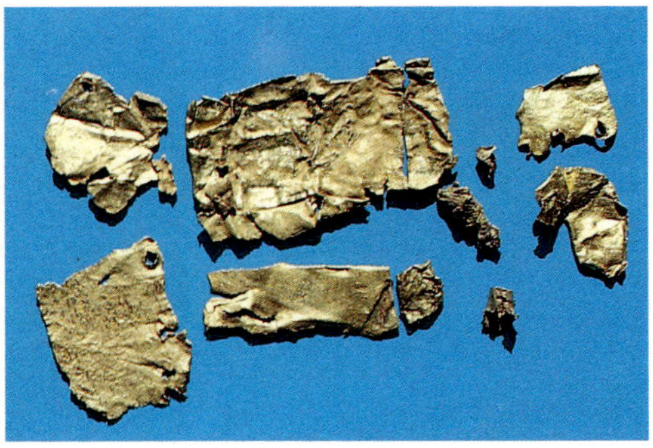

FIGURE 17. *Detail of beads excavated from burials at Ingombe Ilede in Zambia. The tubular bead is made of rolled-up gold foil. LMZ, in a glass bottle labeled II/1.*

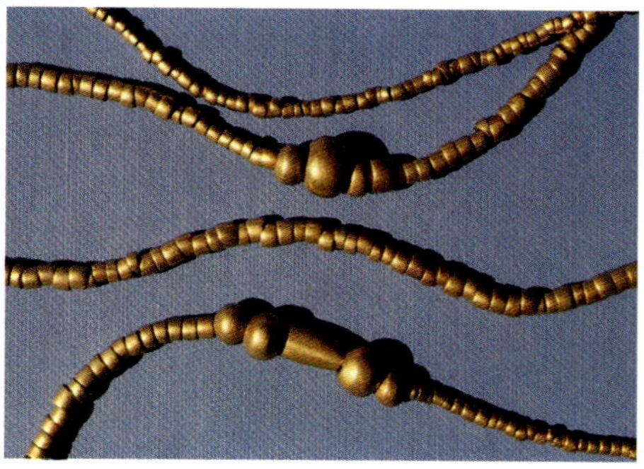

FIGURE 18. *Goldsmiths' waste from an unknown site in Zimbabwe. Note the presence of a very small, tubular, gold-foil bead. British Museum, Ethno 1929:12-1(11).*

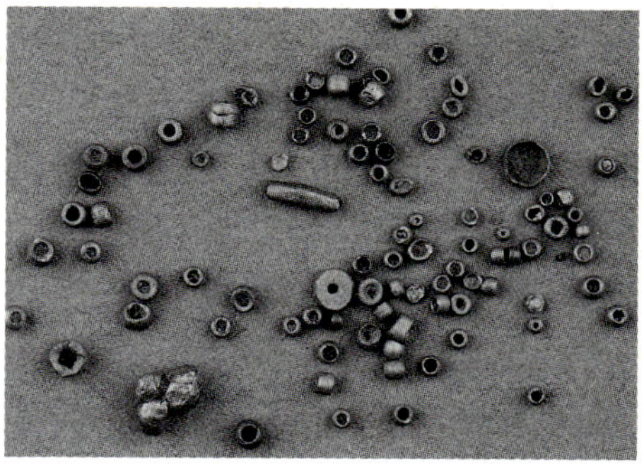

Both the Queen Victoria Museum in Harare and the British Museum contain quantities of gold-casting waste, half-made beads, and small cast spheres. The British Museum material[6] is known to originate from somewhere in modern Zimbabwe, while the Harare material[7] is known to have been found at Dhlo Dhlo, a site that flourished in the seventeenth and eighteenth centuries. Both lots contain fragments of foil, and there is yet another tubular foil bead (5 mm long and 1.5 mm in outside diameter) at the British Museum (Fig. 18).

In addition, two gold-wire anklets from the Belingwe Hills (southeast of Bulawayo) have added rectangles of gold sheet wrapped around the twisted and coiled wire (Figs. 19, 20). Finally, a string of gold beads from an unspecified site in Zimbabwe includes one large iron bead covered with gold foil (Fig. 21).

FIGURE 19. *A pair of gold-wire anklets, near right, from the Belingwe Hills in Zimbabwe. The anklets have pieces of gold foil wrapped around them. British Museum, Ethno 1929:11-14(1+2).*

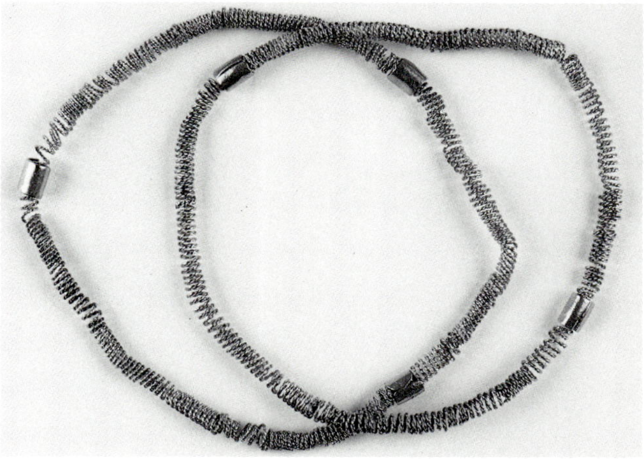

FIGURE 20. *Detail of gold-wire anklets, far right, shown in Figure 19.*

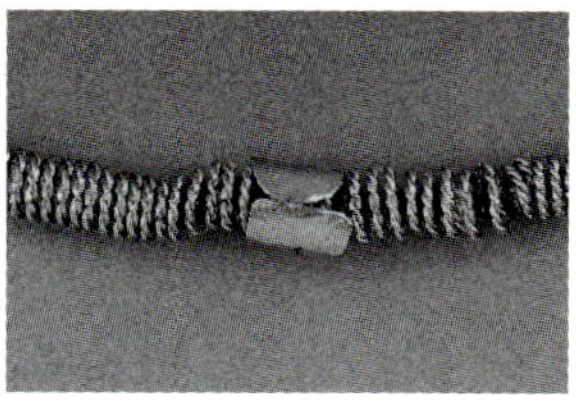

FIGURE 21. *Strings of beads from an unknown site in Zimbabwe. One of the beads is made of iron and covered with gold foil. British Museum, Ethno 1929:12-11(1).*

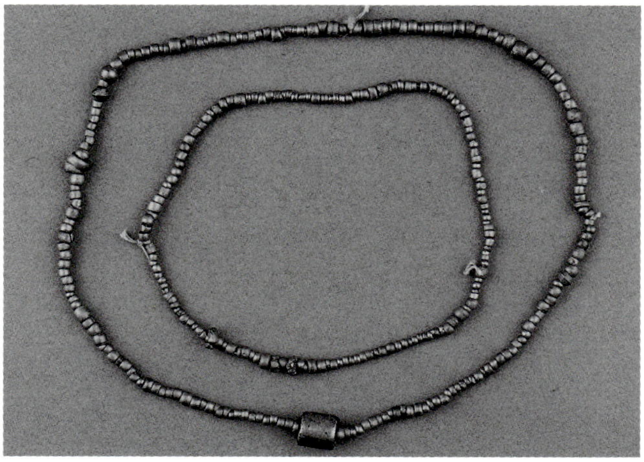

Gold Strip and Wire

Apart from the gold sheet, the finds from Mapungubwe also contain long lengths of spirally wound gold strip (Figs. 22, 23), the spirals having an outside diameter of 12–15 mm. The spirals are very neatly made (Fig. 24), but where they have been deformed it is possible to see how the strip was cut and how it was made into spirals. Examination of the backs of the strips (inside the spirals) shows that they were cut from a sheet of gold foil by repeatedly scoring or cutting with a metal blade (Fig. 25). In some cases this process was not carried out to completion, and the final separation of the strip appears to have been achieved by tearing along the partially scored line. Definite evidence for the scoring, presumably using a straightedge and a metal blade, can be seen in some places where repeated cuts have not followed exactly the same line. The cutting was always carried out from the rough side of the gold sheet, as the smooth side of each strip is always wider than the rough side. The strips were wound into a spiral around a core made of some sort of fibrous material, traces of which survive. Lengths of strip were not joined together physically, but simply intercoiled for a few turns when a new length was needed.

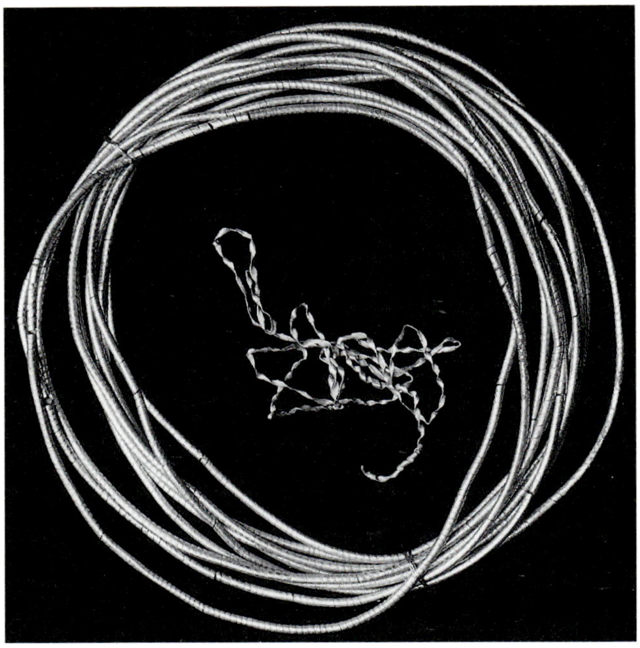

FIGURE 22. *Spirally wound gold strip from Mapungubwe. Inv. no. DAPU 33.168.*

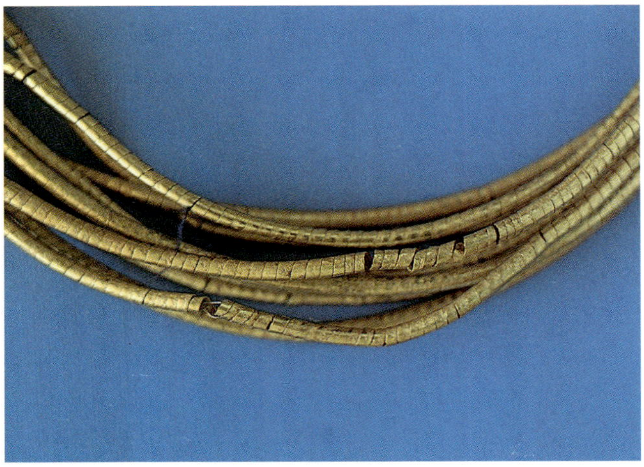

FIGURE 23. *Detail of spirally wound gold strip shown in Figure 21.*

FIGURE 24. *Magnified detail of spirally wound gold strip shown in Figure 21, near right.*

FIGURE 25. *Detail of the back of the gold strip from Figure 21 showing scored lines made when the strip was cut from the edge of a sheet of gold foil, far right.*

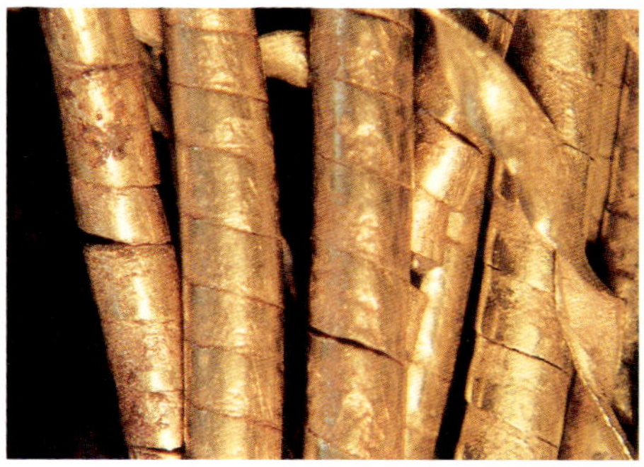

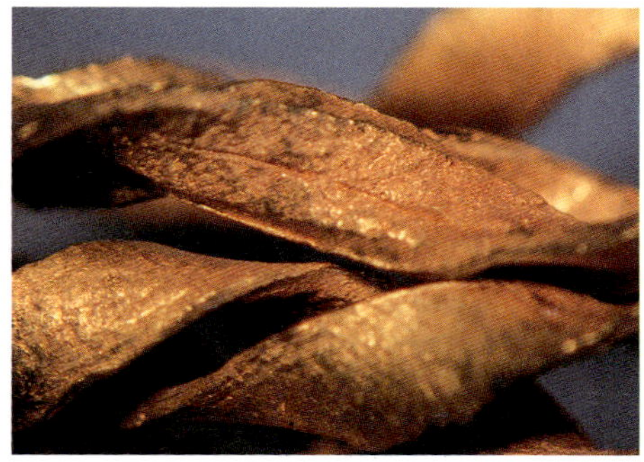

Apart from the long lengths of coiled spiral strip from Mapungubwe, one similar fragment from Zimbabwe, now in Harare, is stored in a modern glass vial with numerous pieces of wire.[8] Fragments of spiral made of copper strip that were cut from sheet in exactly the same way as the gold spiral was made have also been found at Mapungubwe.[9]

Finally, there are the artifacts that are made out of wire. In Europe, the modern method of making wire by drawing is thought to have been invented in the early Middle Ages (Oddy 1983). Before this date, wire was made by hand using one of several different techniques (Oddy 1977, 1987); therefore, it is of interest to identify the techniques used in the African Iron Age and, in particular, to note when wiredrawing was introduced.

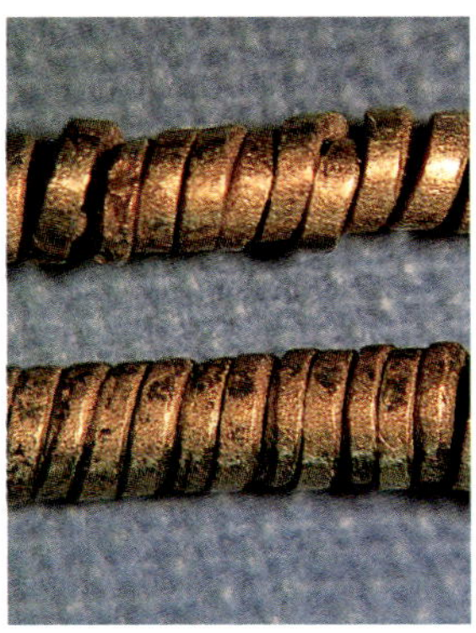

FIGURE 26. *Detail of coiled gold-wire bracelets from Mapungubwe. Inv. no. DAPU 33.169 and part of the contents of a cigar box labeled "B Mapungubwe sites S_1, $P.S_2$."*

Four bracelets from Mapungubwe are made of gold wire that has been coiled into a spiral and then bent into a loop (Fig. 26). The outer surface is flat, but whether this is the result of manufacture or of wear over a long period has not been determined. The surface within the spiral is rough, with longitudinal striations. This cannot be taken to indicate wiredrawing, as the wire varies in diameter on the four bracelets as follows: 0.40–0.55 mm, 0.60–0.65 mm, 0.63–0.70 mm, and 0.50–0.60 mm.

Two short lengths of gold wire were found to have a D-shaped cross section (measurements across the flat part of the circumference are 0.45–0.55 mm). It is most likely that this wire was made by hammering, and that the striations are the result of a smoothing process. The D-shaped wire may have been swaged (hammered into a semicircular groove cut into the surface of an anvil).

Finds of gold wire are more common from Great Zimbabwe.[10] Some of these are obviously hammered, with facets in some cases, and are uneven in diameter. Typical diameters of twenty-three different pieces of wire are (in millimeters):

0.40	0.55	0.40	0.40	0.40
0.35	0.40	0.25	0.40	0.45
0.40	0.20	0.30	0.30	0.45
0.40	1.40	0.40	0.40	
0.55	1.40	0.70	0.30	

Some of the wires have been twisted together as pairs, and some have been coiled into a spiral. However, the most interesting aspect of the gold wire from Great Zimbabwe is that some of it appears to be drawn.[11] The diameters of two of these drawn pieces are 0.55 mm each and a third is 0.50 mm. Some of the wire has a

D-shaped cross section and appears to have been made by drawing and swaging.[12] Four of the seven pieces are 0.30 mm in diameter, two are 0.35 mm, and one is 0.20 mm.

Of the other sites in Zimbabwe, the gold-manufacturing debris from Dhlo Dhlo (seventeenth to eighteenth centuries) included small pieces of wire. Some were hammered and some drawn. The objects, presumed to be anklets, and found in the Belingwe Hills southeast of Bulawayo, are made of hammered wire (Fig. 20).

One other important site for gold is a series of burials dating from the late fourteenth and/or fifteenth centuries that was found at Ingombe Ilede in Zambia. Most of the gold consisted of beads, but wire was found in one burial and consisted of several lengths of paired wires that were apparently twisted together and then wound into spirals (Fig. 27). Found at the left elbow of the skeleton, these appear to be parts of armlets. The outer surfaces show evidence of wear but the inner surfaces, where visible, are rough and uneven with facets and other marks typical of hammering. The wire is nevertheless remarkably even in diameter, varying between 0.33 mm and 0.45 mm. As with all the wire objects, there is no soldering, and lengths are mechanically joined by interlocking separate pieces together for a few turns of the spiral.

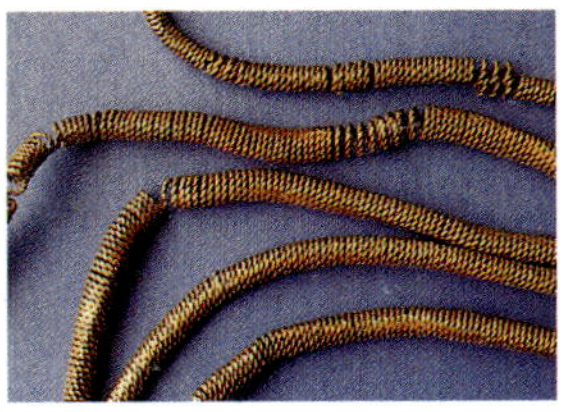

FIGURE 27. *Detail of wire spirals from Ingombe Ilede. These consist of pairs of wires twisted together and then wound into a spiral.*

In addition, the Ingombe Ilede burials contain spools of drawn copper wire and wiredrawing dies made of iron (Fagan 1969). These tools are of a type that has been recorded as having been used by the Bantu in relatively recent times. It must be assumed that wiredrawing had reached the interior by about the fourteenth or fifteenth centuries at the latest. Unfortunately, it is impossible to date the drawn-wire fragments from Great Zimbabwe, which could well be earlier than this.

Conclusion

In its earliest phase, as represented at Mapungubwe, gold working by the indigenous Bantu was at a rather primitive stage. Gold wire and thin sheet, or foil, were made by simple smithing techniques, and the latter was cut with a metal knife to form gold strip. The wire and strip were wound into spirals, and pairs of wires were often twisted together. Some knowledge of *repoussé* techniques and impressed decoration are evident on the gold sheet, but neither process shows any degree of sophistication.

Even after wiredrawing was introduced, no knowledge of soldering is evident. Wiredrawing may, therefore, have been an independent invention in this part of Africa, but since it occurred at a period that has yielded plenty of evidence of trading contacts with the coast, this innovation more likely reflects those increasing influences.

Acknowledgments

The author wishes to express sincere appreciation to Dr. Elizabeth Voigt (formerly of the Transvaal Museum and now Director of the McGregor Museum in Kimberley) for arranging his visits to the museums in Africa and to the International Gold Corporation for its support. The objects were kindly made available by Professor J. F. Eloff and Dr. Andrie Meyer at Pretoria

University; N. M. Katanekwa of the National Monuments Commission, Livingstone; M. Mukela at the Livingstone Museum in Zambia; and Carolyn Thorp at the Queen Victoria Museum in Harare. To all these colleagues he is very grateful. Copies of the detailed report, and of the 126 color slides which illustrate it, have been deposited at all three museums and at the Department of Ethnography in the British Museum, London, whose successive keepers, Malcolm McLeod and Dr. John Mack, kindly made their relevant material available for examination.

Notes

1. Evidence for the chronology of the gold trade along the coast of East Africa has recently been summarized by Horton (1986).
2. All the Mapungubwe gold that was examined is in the collection of the Department of Archaeology, Pretoria University (DAPU). The gold from Ingombe Ilede is in the Livingstone Museum, Zambia (LMZ); and that from sites in Zimbabwe is in the Queen Victoria Museum, Harare (QVMH). An outline of the gold technology has recently been published (Oddy 1984), and a fuller description of the beads is in preparation (Oddy in preparation). It is the purpose of this chapter to describe the gold sheet (foil), strip, and wire from these and other sites in greater detail.
3. The British Museum's register records that this material was presented by Franklin White, Esq., and that it is "from ancient ruins, Rhodesia, Zimbabwe."
4. Inv. no. QVMH 7297 (6)(c).
5. Inv. nos. QVMH 7297 (6)(K); 7297 (1); 7297 (6)(M).
6. British Museum, Ethno 1923:12-11(10-11).
7. QVMH, not numbered.
8. Inv. no. QVMH 7297 (6)(a)(4).
9. In a cigar box labeled "B Mapungubwe sites S_1, P.S_2."
10. Inv. nos. QVMH 7297 (6)(a)(1); 7297 (6)(a)(4); 7297 (6)(c) "Ridge ruins"; 7297 (6)(M).
11. Inv. nos. QVMH 7297 (1); 7297 (6)(M).
12. Inv. no. QVMH 7297 (6)(a)(4).

Bibliography

CASSON, L.

1989 *The Periplus Maris Erythraei*. Princeton: Princeton University Press.

CATON-THOMPSON, G.

1931 *The Zimbabwe Culture*. Oxford: Clarendon Press.

FAGAN, B. M.

1969 Ingombe Ilede: The burials. In *Iron Age Cultures in Zambia*, 64–78. B. M. Fagan, D. W. Phillipson, and S. G. H. Daniels, eds. London: Chatto & Windus.

FOUCHÉ, L.

1937 *Mapungubwe: Ancient Bantu Civilisation on the Limpopo*. Cambridge: Cambridge University Press.

GARDNER, G. A.

1963 *Mapungubwe Volume II: Report on excavations at Mapungubwe and Bambandyanalo in Northern Transvaal from 1935 to l940*. Pretoria: Van Schaik.

HORTON, M.

1986 The Swahili corridor and the southern African Iron Age. Typescript of a seminar delivered at the School of Oriental and African Studies, University of London.

HROMNIK, C. A.

1981 *Indo-Africa: Toward a new understanding of the history of sub-Saharan Africa*. Cape Town: Juta.

MACIVER, D. R.

1906 *Mediaeval Rhodesia*. London: Macmillan.

ODDY, W. A.

1977 The production of gold wire in Antiquity. *Gold Bulletin* 10(3):79–87.

1983 Gold wire on the Thetford jewellery and the technology of wire production in Roman times. In *The Thetford Treasure*. C. Johns and T. Potter, eds. London: British Museum Publications.

1984 Gold in the southern African Iron Age: A technological investigation of the Mapungubwe and other finds. *Gold Bulletin* 17(2)70–78.

1987 Does "strip-drawn" wire exist from Antiquity? *MASCA Journal* 4 (4):175–77.

In press Gold beads of the southern African Iron Age. *Jewellery Studies*.

PHILLIPSON, D. W.

1985 *African Archaeology*. Cambridge: Cambridge University Press.

PHILLIPSON, D. W., AND B. M. FAGAN

1969 The date of the Ingombe Ilede burials. *Journal of African History* 10:199–204.

SUMMERS, R.

1963 *Zimbabwe, a Rhodesian Mystery*. Johannesburg: Nelson.

1969 *Ancient Mining in Rhodesia and Adjacent Areas*. Salisbury: Museum Memoir No. 3 of the National Museum of Rhodesia.

VOIGT, E. A.

1983 *Mapungubwe, an Archaeozoological Interpretation of an Iron Age Community*. Pretoria: Transvaal Museum Monograph No. 1.

Biography

Andrew Oddy is Keeper of Conservation at the British Museum, London, where he has worked for twenty-seven years. His research interests are the conservation of metals and stone and the technological investigation of metal antiquities. He is particularly interested in gilding and wire making from the Bronze Age to the medieval period. He is the author of more than 150 papers and has recently edited a book, *The Art of the Conservator* (British Museum Press 1992), on the conservation of eleven major works of art.

Conservation of Architectural Metalwork: Historical Approaches to the Surface Treatment of Iron

FRANK G. MATERO

The history of the development and use of structural and ornamental metalwork in the building arts represents a long and complex interrelationship between function, aesthetics, and design played against the backdrop of architectural theory and scientific invention. In this regard, the great technological developments and economic growth experienced during the latter half of the nineteenth century, especially in North America, brought about an unprecedented increase in the use of metals—particularly iron and steel—for all aspects of building. An article titled "Art Iron at the Fair" in the July 1893 issue of *Ornamental Iron* describes this trend (1893:65):

> Iron in its various uses, constructional and ornamental, occupies a greater prominence in architecture than in Europe. Iron skeleton construction for high buildings is essentially an American conception. Also iron stairways, store fronts and elevator inclosures are here used to an extent without a parallel anywhere else in the world. This has chiefly resulted from the necessities of modern American office building.

To meet the demand for new construction, architects and engineers sought immediate, inexpensive building methods that would at the same time satisfy current design trends. Cast and rolled iron and sheet metal satisfied these requirements because they could be quickly and inexpensively fabricated and assembled into almost any architectural form desired. Historians of architecture and building technology have long been fascinated by these materials—structural iron and steel, in particular—for their contribution to the development of prefabrication and the evolution of the modern steel-frame skyscraper.[1]

While the promise of iron and steel as new structural materials for building became a major issue both in North America and in Europe during the nineteenth century, of equal interest was the subject of their surface treatment and finishing for protection against corrosion and for decoration. In a relatively short period of time, the technological developments reported in the numerous scientific and metal trade journals became commercially available and were embraced by the building profession, as the following excerpt indicates (Swank 1892):

> Mention has been made of the artistic finish of some of our iron and steel work, but the subject seems worthy of further notice. It is not only in stove-founding, in the graceful designs of bridges and elevated railways, and in the delicate combination of iron and steel with other materials in the construction and ornamentation of buildings that American ironworkers have displayed both taste and skill. . . . Some of the castings have been colored to represent bronze and others to represent steel, while others again preserve the natural color of iron.

The surface treatment of architectural metals has been largely neglected, especially in the preservation of many significant nineteenth- and twentieth-century structures, resulting in a poor understanding of the original appearance and function of these metallic surface finishes. The current restoration of Burnham and Root's Rookery in Chicago, Illinois, has allowed an in-depth study of the finishing techniques of what was considered in its day to be the largest, most exuberant display of ornamental ironwork set within a building (Fig. 1), according to a period account in *Ornamental Iron* (1893:[1]94–98).

Although iron founding and casting are known to have been practiced in the ancient world, cast iron arrived on the modern scene in the early eighteenth century as one of the outcomes of the Industrial Revolution (Smith 1981:127–73). Previously limited in production due to the requirement for sufficient quantities of iron ore and charcoal in close proximity, cast iron became more readily available after 1709 with Alexander Darby's use of coke, a distillate of pit coal, as a suitable and available charcoal substitute in the smelting of iron. Within a short period of time, domestic arti-

FIGURE 1. *The Rookery, light court, view looking northwest, 1888.*

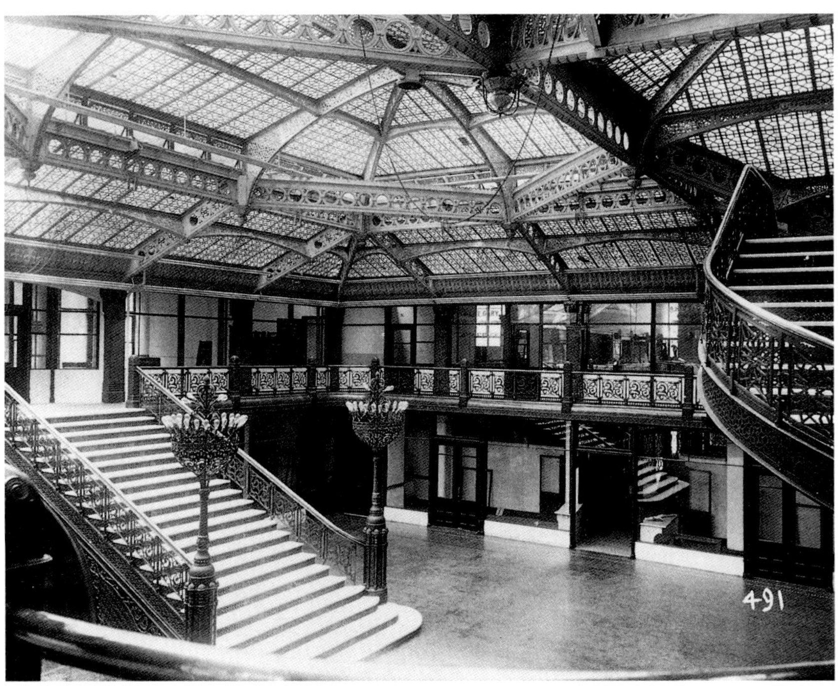

cles and industrial components such as steam engine cylinders, rails, and tram rail plates became common manufactured items. In 1755 Sir John Smeaton employed the first large pieces of cast iron for mill and engine work (Lee 1983:100–1). However, it was not until 1779, when the first major cast-iron structure, the Iron Bridge at Coalbrookdale, was constructed across the Severn River in Shropshire, England, that the tremendous potential for the new material was demonstrated for the first time (Condit 1964:5).

The exploitation of cast iron as a structural and decorative building material during the first half of the nineteenth century was largely based on the knowledge gained through its early experimental manufacture and use in production (casting and cooling), component design (shape), and assembly (joining techniques).[2] Unlike traditional wood and masonry construction, cast and rolled iron possessed a combination of advantages, including excellent compressive and tensile strengths, respectively; limited fire resistance; the ability to be molded or rolled in a wide variety of shapes and designs; and reproducibility. These qualities made it an attractive building-material alternative, especially for utilitarian structures such as bridges, fire towers, markets, and mills (Condit 1964:6). But perhaps one of the most lavish applications of its architectural potential was demonstrated to the public with the 1851 erection of Joseph Paxton's Crystal Palace in Hyde Park, London, and Carstensen and Gildemeister's 1853 Crystal Palace in New York City (Condit 1964:6–7).

In the early decades of the nineteenth century, American iron foundries produced only a limited amount of cast wares and building elements, despite the availability of iron ore. The rapid rise of the industry began with the development of the hot-blast furnace in 1829, coupled with the discovery that the abundant supplies of anthracite coal in Pennsylvania and New Jersey could be utilized for the production of pig iron (Porter 1915:78–79). These technological developments were largely responsible for the significant increase in the tonnage of pig iron produced in America after 1850[3]; however, other factors contributed to the rapid and widespread use of the material after 1840.

Improvements in furnaces, foundry machinery, and workshop appliances allowed for larger production of molten metal. In addition, increased knowledge of iron metallurgy and fuel, together with advances in pattern making and molding produced finer and larger castings (Spretson 1878:v). A growing national transportation network of railroads not only provided a more efficient means of transport for the raw materials of iron ore, flux, and fuel required for production, but it also allowed manufactured goods to reach even the most remote cities and towns. Manufacturers were quick to recognize the potential of iron's availability and quickly developed the trade catalogue as a persuasive marketing device for promoting and selling building elements and even entire architectural assemblies.[4] These factors, in conjunction with relevant building issues such as the growing concern for fireproof construction and the enactment of building codes for the country's overcrowded and expanding cities, encouraged the use of both rolled and cast iron; this use reached immense proportions by the end of the century. Foundries specializing in the production of architectural ironwork rapidly developed to satisfy the nation's building needs. As one architectural foundryman remarked (Fryer 1876:1):

> This branch of iron manufacture has increased enormously within the past fifteen years, and the possibilities are that the future will develop a still greater proportional growth. It is a commonplace saying that as a nation we have but just begun to use iron.

Architectural Use

The second half of the nineteenth century witnessed rapid technological developments in the manufacture and assembly of architectural iron and steel. The early experimental structural systems of cast-iron lintels on piers and columns were soon replaced by midcentury with larger rolled-iron beams, which in turn were replaced at the close of the century by steel frames of immense dimension, span, and strength. According to one period source, improved manufacturing techniques were also responsible for creating large architectural castings of great artistic finesse which would not have been possible earlier (*The Inland Architect and News Record* 1889:18):

> The demand for [ornamental ironwork] has improved the processes of manufacture, until we have . . . castings which a few years ago would have been impossible. Large castings for centuries have been made, but the castings of pieces large in area and so thin that they become only a veneer, at the same time keeping them from warping, and so free from blow holes that the lines of a delicate tracery ornament are not broken, is another matter, and requires the highest skill and the greatest facilities for handling.

In addition, foundry mechanization through the invention of the first hand-operated molding machine in 1879 by Frank Reinholds of New Haven, Connecticut, allowed for increased production of castings and eventually led to fully mechanized molding around 1900 (Worrilow 1949:478).

As a result of these material improvements and innovative production methods, cast iron became an accepted and even preferred material not only for structural systems but for interior and exterior cladding, stairways, railings, balustrades, and ornamental details of all types. In 1884 James Swank, in his *History of the Manufacture of Iron in All Ages* (1892:527), commented:

> We [Americans] make liberal use of both cast and wrought iron and steel in the construction of public and private buildings. Our use of iron and steel for these purposes has rapidly increased in late years, and in much of our work of this character we have given truly artistic effects to these metals. We probably excel all nations in the use of iron and steel for ornamental purposes in connection with masonry, brick-work, and wood-work. Fine illustrations of the artistic combination of iron with other materials ten years ago may be seen in the interior of the new State Depart-

ment building [Old Executive Office Building, 1879] at Washington and in the interior of the passenger station of the Pennsylvania Railroad at Philadelphia, while upon every hand are seen later illustrations of our artistic uses of iron and steel.

As noted earlier, a lavish use of structural and ornamental ironwork was especially evident in the interiors of large public and commercial buildings under construction in industrial cities such as New York, Washington, and Chicago during the last quarter of the nineteenth century. Because of their multistory height and volume and the inadequate illumination levels delivered by gas and electric lighting, many such buildings required open courts with stairs and galleries designed to bring the natural light into the core of the building. In these situations, cast iron had immediate applicability as a low-fire-resistant material easily cast into the requisite repetitive designs, and possessing the necessary strength for use on interior stairs, railings, and skylights. In addition, the incorporation of the newly invented elevator in high-rise structures required the installation of elaborately ornamented metal enclosures and grilles.[5]

Traditional Protective and Decorative Surface Treatments

Despite the technological advances in smelting and founding of iron and steel, and improved casting techniques and increased production, the corrosion of ferrous metals remained a major problem and concern for their use as both structural and decorative architectural elements.

The noble metals most commonly employed in architecture—copper, bronze, and brass—generally did not require any additional surface protection or enhancement since their natural or artificially induced patina was considered an aesthetically acceptable and even desirable finish. Where unique conditions such as highly polished or colored surfaces were desired, clear or tinted coatings were sometimes applied for protection, enhancement, and ease in maintenance of the finish.[6]

The less noble metals—tin, zinc, and iron—were almost always protected against atmospheric corrosion, usually by painting. However, painting was only one of three possible systems available at this time for surface protection and decoration of iron and steel. The other two systems—protection by metallic deposition such as hot dipping (tinning) or electroplating, and protection by oxide surface conversion such as Barffing (discussed later)—were direct results of the application of scientific discoveries and did not come into widespread commercial use until the latter part of the nineteenth century.

Except for strictly utilitarian situations, most architectural surface treatments were selected in accordance with current aesthetic trends. During the second half of the nineteenth century, architectural critics formed strong opinions concerning what they believed to be the proper finish treatments for architectural cast iron and sheet metal. These authors identified with one of three groups of opinion: those encouraging the use of imitative finishes such as stone and bronze, the imitation restricted only by good taste; those who rejected imitative finishes as being deceptive but did not denounce metal finishes that referred to another material obliquely in order to achieve an intended effect; and those who sought a treatment for metals that would be expressive of the material in its own right, as well as of the architecture that it served.[7]

Paints and Coatings

Protective coatings of linseed-oil-based paints pigmented with red lead and iron oxide were the most commonly recommended primers for iron and steel during the nineteenth century. Of the two pigments, red lead was considered by most technical writers to be superior to iron oxide.[8] Although less effective, iron oxides were used more extensively as primers due to their low cost and light stability. Other pigments such as zinc oxide were not seriously considered or available until their production in America around 1850; carbon graphite, introduced later in the century, was received with cautious optimism.[9]

Other brush-applied surface coatings such as black or colored asphalt, linseed oil, vulcanized rubber, hot coal tar, fossil-resin (coachmaker's) varnish, spirit varnish, white wax or paraffin, and lime water were also recommended.[10] With the exception of lime water, all these treatments relied on the hydrophobic nature of the film to isolate the reactive surface of the metal from moisture. Depending on the pigments employed, many treatments were also believed to reduce or prevent corrosion by galvanic action of the metallic oxides and carbonates.

A more utilitarian treatment for ferrous metalwork was the black japanned finish commonly used for wrought-iron and hardware grilles and fences. Lacquer for wrought iron could consist of either shellac or an oil varnish, while japan was commonly made with lamp-black pigment in a copal or similar oil varnish.[11] The glossy nature of the japan tended to highlight the metal's tool marks or imperfections, whereas a flat-black finish, achieved by thinning with turpentine, could be used to obscure these markings.

Surface protection and decoration could also be achieved by gilding and bronzing. These applied surface treatments were frequently employed on interior architectural ironwork, although generally only as accents or highlighting. Gilding could be performed at the foundry as mercury or fire gilding.[12] It could also be applied after installation in situ as oil gilding with an oil or oleoresinous size and a wide variety of metal leafs, including gold and gold-and-copper alloys, as well as silver, aluminum, and nickel.

Bronzing involved the use of brass powders of various composition and color, brushed or pressed onto drying oil or oleoresinous mordants applied to the iron surface and then varnished for luster and protection.[13] By varying the color of the mordants, the powders, and the varnishes or glazes, diverse variations in the imitation of bronze, copper, and other metals could be achieved.[14] Bronzing powders could be applied over the entire element or, more typically, used as highlights on painted surfaces.[15] Despite criticisms of their instability, bronzing powders were widely advocated in the literature and used (Gardner 1888:114).

Electroplating

Of all the surface finishing techniques utilized for architectural ironwork, perhaps none was more versatile in its applications than electroplating. Developed in the late 1830s and 1840s,[16] the process, along with electrotyping, or so-called galvanoplastic work, was in commercial use for architectural ironwork by the late 1870s in America.[17] Promoted for its decorative as well as protective benefits, electroplating enjoyed increased use in finishing an infinite variety of small and medium-sized elements, the latter easily assembled into large architectural features such as staircases and enframements.

In the commercial advertisements and descriptions of projects published in the trade journals and catalogues of the day, manufacturers listed a wide variety of plated and oxide-surface finishes for ironwork.[18] Of the electroplated finishes in vogue at the end of the nineteenth century, copper and the cuprous alloys—brass and bronze—were generally considered the most versatile for their ability to be patinated in any number of imitations of cast bronze.[19]

Where cost allowed, bronze was often the preferred metal for ornamental use on interiors and exteriors, however less expensive imitations abounded, including plated and painted ironwork and even glazed terra-cotta.[20] The process of electroplating was expensive relative to other surface treatments and required a high degree of skill; to be effective under adverse conditions, a heavy deposit of the finishing material was required. "Duplex and triplex electro-bronzed" ironwork, as it was termed by its manufacturers, was widely promoted and employed in numerous commercial buildings after its introduction sometime around 1890 (Wilson 1915:68–69).[21]

Ornamental Iron was a trade periodical published by Winslow Brothers Company to promote the use of architectural metalwork in general and the company's own production in particular. The magazine published a series of articles discussing the development and introduction of electroplated ironwork for exterior use, describing the technical evolution as follows (*A Front of Bronze* 1893:22)[22]:

> Electroplating on cast iron has been attempted heretofore, both in this country and in Europe, but never with success, as the wrecks of a number of otherwise fine structural objects in this city [Chicago] and abroad attest . . . where the plating has so entirely failed of its ornamental purpose as to require painting in order to hide its ugliness and retain its preservative qualities. As a result of a long series of experiments these obstacles have been overcome and a method of treatment has been perfected whereby the surface of the iron receives the deposit directly, and an adhesion secured which is impervious to all attacks and gives an enduring quality to the work never before secured. We speak advisedly in asserting that this process is peculiar to the duplex bronze on iron products by the Winslow Brothers Company, and that, by no other process yet devised or at present in use can or has a satisfactory result been secured. The metal itself is a form of bronze, practically indestructible . . . except that a slow oxidation gradually brings about a change in the surface color,

> which enhances rather than diminishes its beauty of tone, giving to it an added artistic charm.

Although bronze was listed as the most desirable of the electroplated finishes, true bronze-plating, the deposition of bronze—an alloy of copper and tin—was not largely practiced due to difficulties in achieving even plating. According to one period source, "What is erroneously misnamed 'bronze-plating' is really a light-colored brass deposit formed by taking a good copper-cyanide solution and adding to it just sufficient zinc solution to give the required shade" (Field and Weill 1951:471).

Copper alone was also used as the electrodeposited base for chemical patinations simulating a wide variety of so-called bronze finishes "from red to bright yellow and dark to light green."[23] However, copperplating on iron by means of direct, acidic copper-sulfate baths was limited due to the spongy nature and poor adhesion of the copper deposited. Therefore most copperplating on iron and steel required an initial thin plating, or "strike," of copper deposited in an alkaline copper-cyanide bath prior to continued plating in the acid-sulfate bath (Burns and Bradley 1967:258).

In addition to chemical patination, other modulations of color and luster could be easily achieved by light abrasion and polishing of the patinated surfaces. A variety of surface qualities could also be achieved by the application of bronzing powders, glazes, and gilding. These effects may be observed in buildings with preserved metalwork such as the Rookery, which was described in the original manufacturers' literature (Poulsen and Eger n.d. [ca. 1890]):

> On our stair work, and other work of a similar nature, we electroplate in Bronze, Brass and Antique Brass, the railings, newels and ornamental parts, and as much as possible design our work so that all high relief lines and mouldings are polished to afford artistic contrast. The electro-plate is thoroughly substantial and durable.

Oxide Conversion Coatings and the Bower-Barff Rustless Process

In addition to the techniques already described for the coloring and protection of iron and steel, thermal oxidation of ferrous metals received much attention during the late nineteenth century for its value in producing corrosion-resistant or so-called rustless coatings. These stable oxidation films, formed directly on the surface as integral components of the parent metal, could range in color from brown to black to blue-black. Depending on their formation process, these films produced a hard, wear-resistant skin of varying thickness (Burns and Bradley 1967:577). Several distinct methods of forming oxide films on ferrous metals by heating were in use during the late nineteenth century, each producing finishes of varying color and thickness. For general coloring, low-temperature heating (200–500 °C), which produced interference tints in combination with oil or other organic preservatives, could produce thin colored surface films of limited protective value. Surface conversion could also be induced by chemical means in conjunction with heat.[24]

However, by far the most significant protective oxidation coating to be developed and employed on architectural iron and steel was the so-called Bower-Barff Rustless Process invented by Professor Frederick S. Barff of Kilburn, England, who published his method in 1877.[25] The process was later modified by George Bower. Despite variations in the processes later developed by Bower, Wells, Gesner, and others, the treatment essentially consisted of heating the cleaned iron or steel to dull or bright red in contact with air in special ovens or furnaces, and then admitting superheated steam to the furnace. This process produced an oxide layer of magnetite (Fe_3O_4) or magnetite and wustite (FeO). Later improvements in the process allowed complete conversion to magnetite by reduction through continued heating in a producer-gas atmosphere of hydrogen and carbon monoxide until the desired thickness of magnetic oxide coating was obtained.[26]

Barff was not the first to observe the formation and stability of magnetite on ferrous surfaces.[27] Antoine Lavoisier (1743–94) first noted the formation of magnetitic mill scale and commented on its stability and impervious character. Later Michael Faraday (1791–1867), the British chemist and physicist, similarly described the protective action of the oxide formed in the tubes of a steam superheater (Wilson 1915:47). According to Barff, his interest in magnetite as a protective measure for iron and steel began around 1875 during his observations of a pipe carrying superheated steam (Barff 1877:255). His process, which was first made known in 1876, consisted of heating the ironwork to 1000 °C and passing steam superheated to 538 °C over it.

The process developed by Barff and later perfected by Bower was immediately publicized by the architectural, engineering, and metal trade journals as a solution to the problem of iron corrosion.[28] Barff clearly recognized the value of the process to architects and engineers, stating in his first lecture to the Society of the Arts (Barff 1877:256):

> I cannot over-estimate the advantages which the employment of this process must confer on architects, who will be by it enabled to employ iron, whether wrought, or cast, much more largely, not only in the decoration, but in the construction of their buildings.

The following year, on Wednesday June 5, 1878, Barff presented his discovery, as well as treated samples, at the General Conference of the Royal Institute of British Architects. The process was received with great enthusiasm and considered "of incalculable value . . . elevating the metal [iron] almost to the level of bronze and the costlier metals," according to the conference proceedings (1878).[29]

Early reports on Barff's process described problems with film adhesion and a lack of film uniformity; however, with improvements in the furnace, large-scale commercial application was reported in London by 1880.[30] George and Anthony Bower, of St. Neats, England, attempted to improve Barff's process by using air instead of steam, but they were unable to obtain the desired results until they employed producer gas of hydrogen and carbon monoxide to reduce the red oxide formed by heating the parts in air. The air treatment lasted 40 minutes and the gas treatment 20 minutes, these being repeated alternately four to eight times. Finally they purchased the Barff patent, and the Bower-Barff process was patented in 1881. That same year, George Bower delivered a paper before the Iron and Steel Institute of the United Kingdom describing his process and the large-scale works already in operation in France, Germany, and the United States.[31] The right to use the process in the United States was acquired by George W. Maynard of New York, and the first furnace was erected at the Hecla Architectural Iron Works in Brooklyn, New York (Wilson 1915:47–48), the same company that provided all the ornamental ironwork for the Rookery in Chicago. The process was described several years later after its acceptance and widespread use, as follows (Brannt 1896:297–98):

> Barffing is so called from the name of its discoverer; and its purpose is to render the surface of metallic articles treated, resistant to acids and impervious to humid oxidation. Iron and steel that have been properly Barffed will not rust. The process belongs in no sense to the chemical laboratory, and requires no particular scientific knowledge or technical skill for its success. Highly polished work retains its finish after the process, although the treatment changes the shining polish to a dark slate color. But working parts, such as sliding and rubbing surfaces, retain their smoothness, and work as freely after as before the process; plugs, valves, and stems for cocks, gates, and faucets, for instance, working just as smoothly as though they had not been Barffed.

Although the Bower-Barff process was considered expensive and its use restricted to elements that did not require precision fitting due to the slight volumetric increase (machinery parts), the process offered excellent corrosion resistance for architectural ironwork, which could be painted or coated with paraffin or oil for additional protection. Where necessary, the magnetite coating could be built up for maximum protection by repeated treatments[32] (Fig. 2).

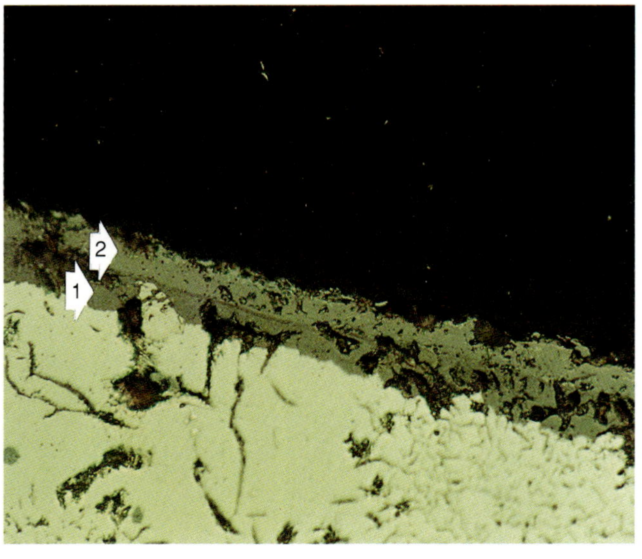

FIGURE 2. *Photomicrograph of cross section of a Bower-Barff finished window handle (Appendix: Sample 1) under visible light taken at ×400 (Zeiss metallographic microscope). Note the 25–30 micron double layer of the artificially formed oxide film (1, 2).*

Subsequent modifications in the basic process resulted in a number of patented treatments, including the Wells Process, the Gesner Process, the Bradley Process, the Bontempi Process, and Coslettizing. These commercial processes produced magnetite coatings of better color, uniformity, and thickness for less cost (Wilson 1915:48–49).

The Bower-Barff Rustless Process afforded a durable, efficient finish for protection against corrosion and for the decoration of wrought and cast iron and steel. Used alone or in combination with other finishes such as electroplating and bronzing or gilding, the deep black finish quickly became a popular treatment for interior ironwork and, in particular, hardware and ornamental wrought iron. By avoiding the disadvantages of paint-film obfuscation and repeated maintenance, and providing a nonimitative black appearance, the Bower-Barff process was considered "the best and only proper finish for art ironwork."[33]

By the last decade of the nineteenth century, a full array of protective and decorative finishes for architectural ironwork was available and in widespread use. Although painting continued as the most common means of protection and decoration, especially for exterior work, electroplating and the Bower-Barff process rapidly became the most popular finishing techniques for common and ornamental interior and exterior ironwork as indicated by the many significant commercial buildings employing these finishes.[34] The situation was well summarized in the introduction to the *Illustrated Catalogue [of the] Hecla Iron Works* of about 1890 (Poulsen and Eger):

> We desire to call attention to our several methods and styles of finish of our Ornamental and Architectural Cast and Wrought-Iron Work as illustrated herewith, where such finishes are desired beyond mere painting.
>
> Our processes are comparatively new in their application to this class of work and go far toward enhancing its artistic value; and at the same time compare favorably in expense with the usual methods of bronze-painting and decorating. In the past few years we have been successful in introducing these finishes in combination with stair and ornamental ironwork furnished by us in most of the prominent buildings recently erected in New York, Chicago, and most of the principal cities in the West.
>
> Our standard of work is high and by intelligent application of these various finishes, combined with good workmanship, it is possible to most nearly approach the artistic in architectural and ornamental ironwork.
>
> Electroplating—On our stair work and other work of a similar nature, we electroplate in Bronze, Brass and Antique Brass, the railings, newels and ornamental parts, and as much as possible design our work so that all high relief lines and mouldings are polished to afford artistic contrast. The electro-plate is thoroughly substantial and durable.
>
> Bower-Barff Rustless Process—This process imparts to the surface of the iron a blue-black color on rougher castings, while on polished work the iron retains its lustrous appearance with a beautiful steel-blue shade that harmonizes well with electroplated work where used in combination. . . .

It not only prevents rust but ornaments the surface of the iron as well, and has a wide application for interior and exterior ornamental cast and wrought iron work.

Case Study: The Rookery

The investigation and analysis of the architectural finishes of the principal interior spaces—light court, lobbies, and library—of Burnham and Root's Rookery (1885–88) were undertaken from 1989 to 1990 in preparation for the long-awaited restoration of the building. It was the focus of these investigations to examine and analyze the exterior and interior architectural materials to ascertain the type and extent of surviving original finishes, significant alterations made over time, and their existing conditions. Assessment was based on in-situ investigation and laboratory analysis supplemented by written and photographic documentary sources.

Based on the scope of the project and the time allotted for site work, paint sampling and surface-finish investigation were confined primarily to selected areas within the locations noted above. Wherever possible, known elements dating to the building campaigns of Burnham and Root (1885–88), Frank Lloyd Wright (1905–7), and William Drummond (1927–32) were examined and sampled for comparison. More than 150 samples of existing paints and coatings and representative metalwork were removed for laboratory examination and analysis.

Once these materials were characterized, identified, and documented, a program of field testing was conducted to determine the feasibility of retaining those original or subsequent finishes identified for conservation. Overpainted or lost finishes were studied to replicate their established historical appearance. The original surface finishes of the metalwork—the electroplated and Bower-Barff finishes on the ornamental ironwork—were studied further, and methods of overpaint removal, cleaning, and presentation were tested and assessed. In each test situation, a broad range of materials and conditions was considered, and treatments were selected which were found to be simple but effective in their application, involved minimum health risks, and conserved as much of the original surface as possible.

Burnham and Root's Rookery of 1885–88 (Figs. 1, 3–9) is widely considered to be one of the major masterpieces of commercial architecture built in America during the last decades of the nineteenth century. From the beginning, it was considered "a design of great fertility: bold, original, and inspiring with its glass-covered light court" and flying staircase (Condit 1964:64). In context, the design and materials of the Rookery relate to the progression of commercial buildings built by Burnham and Root during the 1880s and 1890s. Their preference was for intensely colored red brick, terra-cotta, and natural building stone for the exterior, and a contrasting use of light-colored marble and ornamental metalwork for the large open vestibules and light courts of the interior. These contrasting elements became hallmarks of their commercial architecture and were quickly emulated elsewhere.

As dark, warm, and massive as the Rookery's exterior was, its internal light court was the complete antithesis—light, cool, and reflective (Fig. 5). In addition to providing much-needed light to the inner tiers of offices, Root's light-court designs, first

FIGURE 3. *The Rookery, light court, view looking southwest, 1888.*

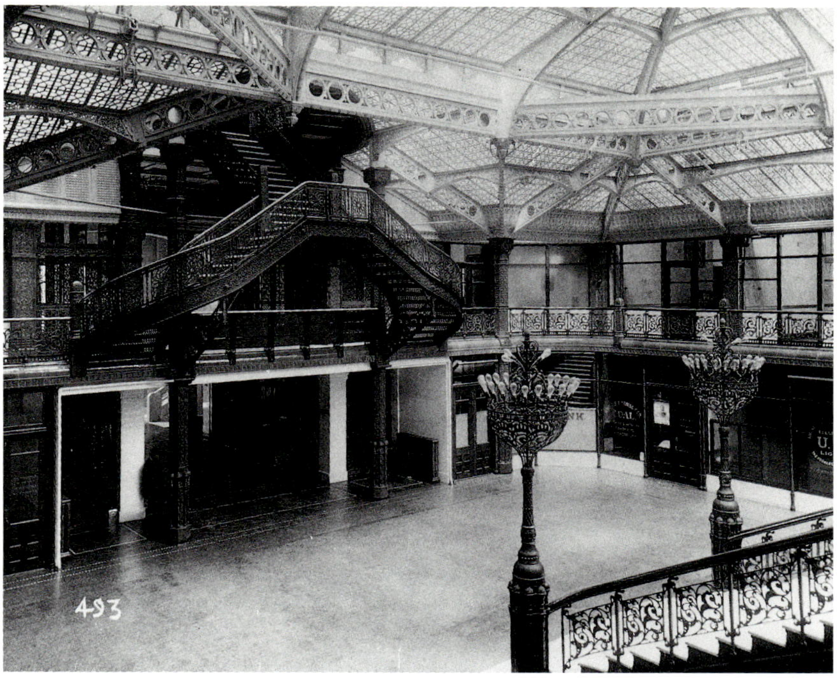

seen in the Burlington and Quincy General Office Building of 1882–83 (destroyed) and later developed in the Insurance Exchange Building of 1884–85 (destroyed), the Phenix Building of 1885–87 (destroyed), and the Rookery, "made light thematic . . . and an arena in which to explore the aesthetics of motion through light and space" (Hoffman 1973:32).

In the case of the Rookery, this effect was largely achieved through the construction of an independent skeleton wall of cast-iron columns joined by wrought-iron spandrel beams, thus maximizing the amount of light entering the inner offices. The architects also made use of light-colored reflective materials—white-glazed bricks and gold-glazed terra-cotta lintels, sills, and copings; gilded cast-iron mullions; and a glossy white-painted skylight and exterior oriel stair. This orchestration of light with highly reflective white, black, and "gold and dark bronze" metallic-colored surfaces must have created a dazzling effect and one which contrasted greatly with the exterior.

FIGURE 4. *The Rookery, light court, below left, view looking southwest, 1985, before restoration.*

FIGURE 5. *The Rookery, light court, below right, view looking southwest, 1991, during restoration. Note exposure of the copper electroplated and Bower-Barffed finishes.*

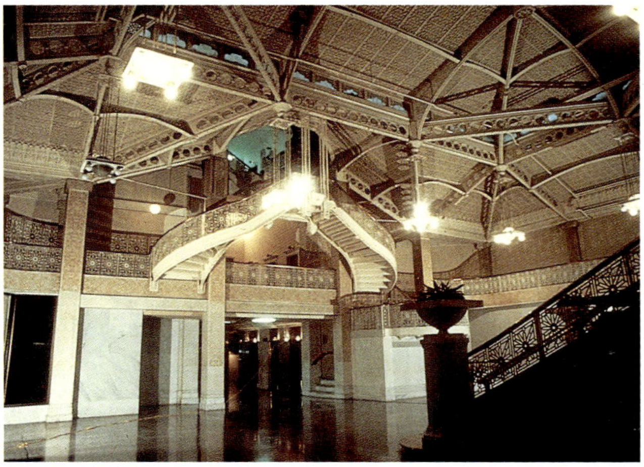

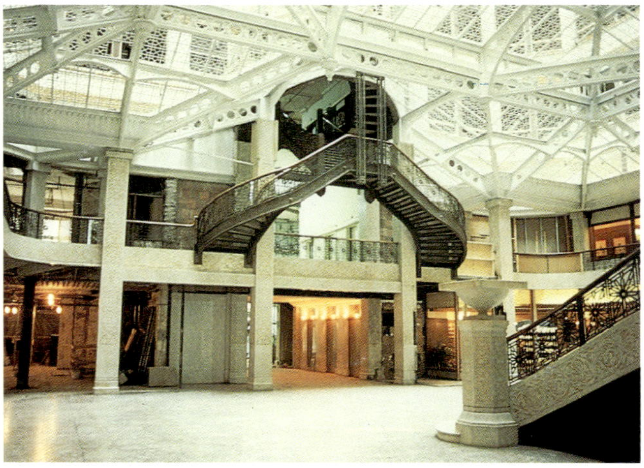

Root's achievement of dynamic balance between the exterior and interior of the building is dramatized at each entryway. Entrance from either Adams or LaSalle streets opened onto large, bright, double-height vestibules with elaborate marble stairways leading to sweeping balconies and the elevators beyond. Below these stairways, the vestibules became compressed past the elevators, suddenly opening up to the light and space of the light court beyond. The light court was a masterpiece of circulation planning. A grand staircase, centered on the east side, allowed access to the glazed office suites lining the mezzanine balconies of the light court's perimeter. On the west wall a double flight of stairs was flamboyantly cantilevered into the court, connecting the mezzanine via a single-run return staircase to the third floor (Fig. 5). This rose up through the building as a unique hemihelix oriel staircase, heavily ornamented with cast ironwork and glass (Fig. 8). The entire light court was enclosed with a complex skylight composed of ornamental cast-iron trusses and braces and ribbed glazing to provide directed diffuse light.

As in Burnham and Root's other commercial projects, the finest materials were used in the decoration and finishing of the Rookery's interior public spaces, including polished Italian marble walls, polychromatic floor mosaics, and elaborate cast- and wrought-iron metalwork. However, by far the most lavish ornamental work in the interior was the decorative ironwork—including the light court's main and double stairways, skylight framing, elevator cars and grilles, newel posts, electroliers, and oriel staircase—all furnished by the Hecla Iron Works Company of New York.[35]

It is difficult to comprehend the novelty and impact this exalted use of ironwork must have had at the time, given the almost immediate popularity and widespread use it enjoyed in other commercial buildings in the ensuing decades. The significance of the design, quantity, and finish of the Rookery's ironwork can be best understood through a feature article on the building's ironwork published in *Ornamental Iron* in November 1893, five years after the building opened[36]:

> The ornamental iron in the Rookery was furnished by the Hecla Ironworks of New York, Poulsen and Eger, proprietors, who were the pioneers in this branch of industrial art work in this country, being the first to practically electro-plate and Bower-Barff ornamental iron work for buildings. The work in the building was largely the result of their intelligent cooperation with the architects. With the exception of two other buildings in Chicago which had shortly before this been furnished with a high grade of ironwork by the same firm, ironwork in this city had been of the commonest description—so much so that it would have been a misnomer to call it "ornamental." Owing to the immense quality of ornamental iron in the Rookery it was let by the architects as a separate contract and this was really the beginning of this practice, which had since become general. It was the largest contract for ornamental iron ever let up to that time, and one of the largest on record up to the present time. The oriel stairs extending to the top of the building were the first ever built in this form of iron. Oriels of a similar nature have since been built by the Winslow Brothers Company in the Mills building, San Francisco; Hotel Midland, Kansas City; and Northern Hotel, Chicago [all by Burnham and Root].

> The immense skylight in the court over the first floor was also a novel feature being of ornamental cast-iron construction. The lights are in small sizes and set in elaborate cast iron frames. A very striking feature is the grand staircase in the court to second floor balcony, with imposing cast iron newels and electroliers at the start. If these electroliers were to be made at the present time the treatment of the iron would be different in some respects but at that time they represented a great stride in the art of ironwork in this country. Hammered leafwork, which is so beautiful and appropriate for purposes of this kind had its inception in this country at a later date. The round pipe uprights in the elevator enclosures were a new feature, and proved to be of such excellence both as to appearance and practicality that their use has continued to this day. More expensive constructions for elevator enclosures are coming into use now for the lower stories especially broad lightly ornamental cast pilasters, but the round upright has had a long and honored service. Although it was not customary to give ironwork more than a six hour electroplating, instead of twenty-two hours as at present, the electroplating in the Rookery has successfully withstood eight years handling by the public, and the rubbing and scouring of the janitor. What impresses one very forcibly in looking at the Rookery work, is that although other buildings erected at the same time or shortly before are now old fashioned, the Rookery is still one of the most modern of modern office buildings.

According to *Ornamental Iron*, the Rookery's ironwork represented the largest contract for ornamental architectural ironwork for its time. In addition, the application of the Bower-Barff finish would have been among the earliest (only five years after its patent) and certainly the largest application of the technique for architectural work. Used in conjunction with patinated and polished copperplated cast-iron elements, the effect was both durable and elegant. This complex treatment of the Rookery's ornamental ironwork was described almost exactly by Poulsen and Eger in their *Illustrated Catalogue [of the] Hecla Iron Works* (1890):

> We electro-plate in bronze, brass and antique brass the railings, newels, and ornamental parts, and as much as possible design our work so that all high relief lines and mouldings are polished to afford artistic contrast. The electro-plate is thoroughly substantial and durable. The Bower-Barff Rustless Process imparts to the surface of the iron a blue-black color on rougher castings, while on polished work the iron retains its lustrous appearance with a beautiful steel-blue shade that harmonizes with electro-plated work, where used in combination.

Although concealed by layers of subsequent overpaint (Figs. 4, 6), nearly all of the Rookery's interior ornamental ironwork, with the exception of the painted and gilded skylight, was originally finished by patinated electroplating and in the Bower-Barff manner with surfaces ranging in color and luster from matte black to dark reddish-brown bronze to light gold-colored highlights (Figs. 6, 7). This was confirmed through examination and analysis of representative samples of metalwork, as detailed in the Appendix (Lins and Meighan 1991), as well as the discovery of Root's

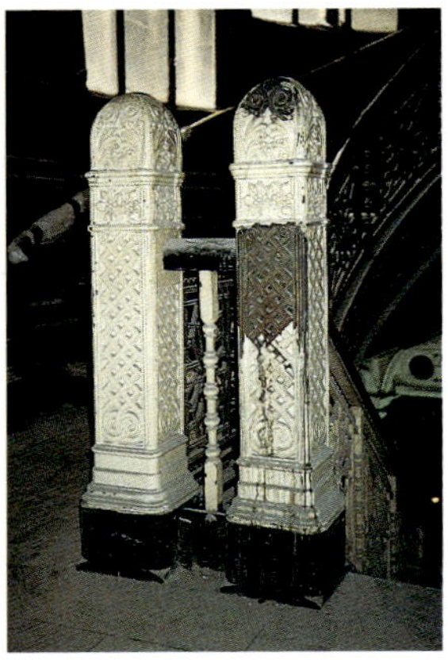

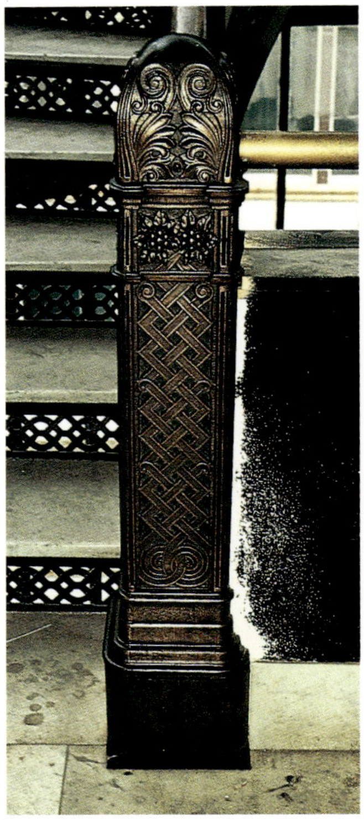

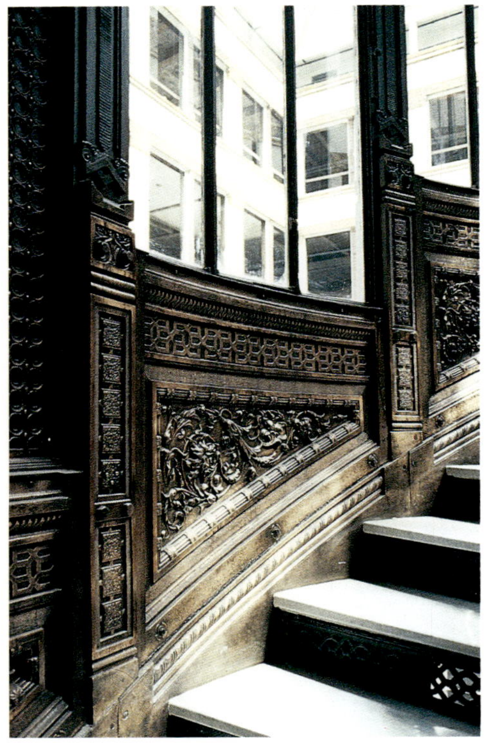

FIGURE 6. *The Rookery, oriel stair newel posts, above, overpainted condition before restoration.*

FIGURE 7. *The Rookery, oriel stair, near right, eleventh-floor newel post and stair risers cleaned and lacquered to reveal original patinated copperplate on the newel post black and Bower-Barff finish on the stair risers.*

FIGURE 8. *The Rookery, oriel stair wall ironwork, above far right. Cleaning tests to reveal original surface finishes.*

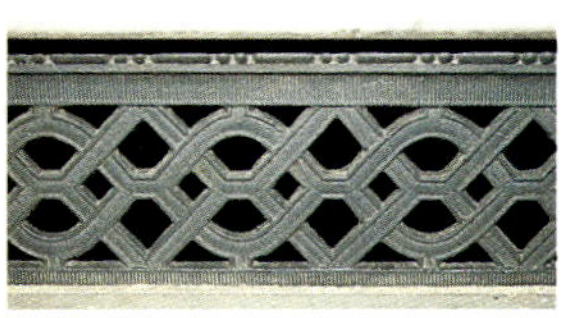

FIGURE 9. *The Rookery, oriel stair riser. Cleaned original Bower-Barff finish.*

notations on his drawings of the stair metalwork. Root describes "the finish of the stairs to be of dark and gold bronze" and "the leaves to be touched with gold bronze not too freely used and rest of dark bronze."[39]

In general, textured flatwork with little or no relief such as the pierced stair risers and window muntins of the oriel stairs were Bower-Barff finished to a matte black or a lustrous blue-black as seen on the smooth surfaces of the window and door hardware (Fig. 9). Metallographic examination and analysis (Lins and Meighan 1991)[40] of an original window handle (Appendix: Sample 1) identified the hardware as grey cast iron with rosettes of flake graphite and a surface patina approximately 25–30 μm thick, consisting of two clear zones of equal thickness of dark grey to black material (Fig. 2). The inner zone clearly displays metallurgical features, such as the remnants of the cast-iron microstructure and voids, which are continuous with the bulk metal. This zone appears to be the cast-iron substrate, which was oxidized by the superheated steam of the Bower-Barff process. This zone may also be described as a depletion zone from which iron ions moved to the outer zone, an oxidation layer grown on the original finished surface of the cast iron during the process. On the uppermost surface of the outer zone there is generally a very thin (1–2 μm) grey-black layer, often separated from the underlying zone by voids. This may represent the alteration of the finish over time or from an initial quenching. No intentional organic coating or residue from an oil quenching was detected using high-power UV microscopy (×1000). Identification of the patina as magnetite using X-ray diffraction was conclusive, although both the sample and the standard appear to be either cryptocrystalline or poorly crystallized (Appendix: Table 1).

High-relief work such as the oriel stair wall panels and stair railing appears to have been copperplated and chemically patinated to a rich dark-brown finish. This may have been somewhat highlighted by the selective removal of the patina; however, the evidence of the degree of removal is unclear. Areas that clearly show intentional mechanical abrading and polishing of the plated surface for highlighting can still be observed on the flat fillet borders and foliate borders of the oriel stairs, wall panels, and stair newel posts as well as on the curvilinear details of the stair-railing balusters and stringers (Fig. 8).

A cross section of a typical oriel stair baluster (Appendix: Sample 3) shows the microstructure to be grey cast iron, with coarse-flake graphite in rosette patterns (Fig. 10). A distinct layer of copper is evident on the surface, varying in thickness from 10 to 20 µm. Analysis of the copper by emission spectroscopy revealed copper with no tin and only a trace of zinc (less than 10 ppm). The plating is therefore copper and not brass or bronze. No evidence of original organic coatings was found using ultraviolet microscopy, suggesting that chemical patination, not resolvable by the analytical means available, probably provided the original final finish.

Root's drawings of the ironwork, as well as period photographs of the original interior, indicate that a combination of finishes was designed and executed as planned. Examination and analysis of a detail of the foliate design of the pierced grille of the light court's main stairs (Appendix: Sample 2), annotated as "gold bronze" on Root's original drawings, suggest a surface treatment different from the "dark bronze" and polished highlighting observed on the elements described above. Here, the surface treatment of the grey cast iron is relatively complex in cross section (Fig. 11). A thin, highly discontinuous layer containing copper was observed on the surface of the iron, which is relatively rough for plating. This copper is most probably a very thin electroplate layer or "strike" coat. Directly above this is what appears to be the original surface-finish coating of a colorless layer approximately 50–75 µm thick, with golden reddish-brown particles in its uppermost levels, suggesting that the method of application may have been by dusting dry colorants onto a warmed or fresh varnish. Fourier transform infrared (FTIR) spectroscopy analysis of a microtomed cross section of the layers above the metal indicates that both the clear and golden reddish-brown layers are natural (tree) resins. Areas of green staining observed within this layer are probably attributable to corrosion of the copper.

FIGURE 10. *Photomicrograph of cross section, below left, of an oriel stairs baluster (Appendix: Sample 3) taken under visible light at ×400. Note the distinct layer of copperplating (1).*

FIGURE 11. *Photomicrograph of cross section, below right, of the ornamental grillework of the main stairs carriage (Appendix: Sample 2) taken under mixed ultraviolet and visible light at ×200 (Leitz Laborlux S ultraviolet microscope with filter block D, BP 355–425). Note the thick resinous layer (3) above the copper film (2) and cast-iron substrate (1), believed to be an original varnish highlighting.*

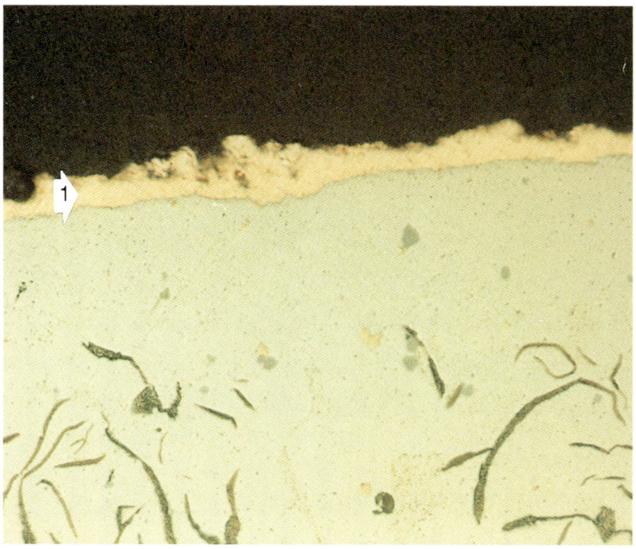

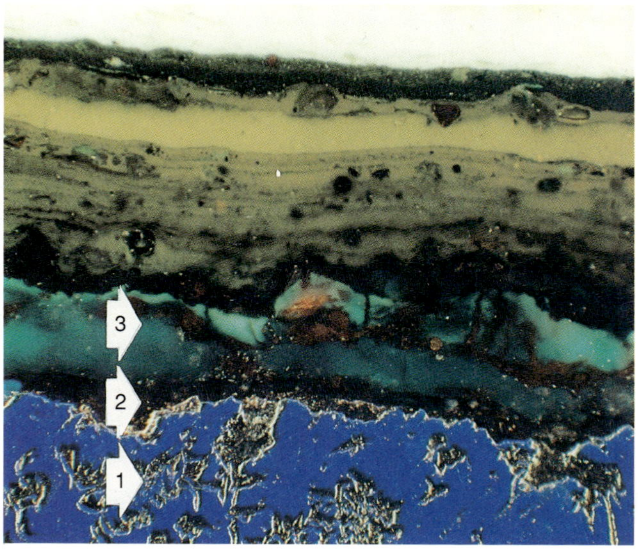

Although it is possible that the thin cuprous layer found above the iron and below the resinous layer is fallout from cuprous "bronzing" powders used in the overlying varnish, it is more likely that the highlighting designated as "gold bronze not too freely used" on the drawings was additional enhancement with tinted varnish of the overall patinated copper electroplate. Such practices employing varnishes, cuprous bronzing powders, or both on both plated and painted iron surfaces were frequently practiced as described in the trade manuals.[41]

This range of integral black and metallic bronze finishes with polished and applied highlights provided a subtle yet dramatic contrast against the white marble walls and white painted and gilded skylight of the Rookery's principal public spaces. Although it appears that this combination of finishes was employed by Root in earlier commissions, the Rookery was by far the most successful and dramatic of his interior designs and his most ambitious use of ornamental iron at that time. This exuberant use of darkly finished Bower-Barffed and richly patinated copperplated ironwork became a requisite feature for succeeding generations of high-rise commercial office buildings throughout the country. It was documented in countless examples in the pages of *Ornamental Iron* and architectural periodicals, such as the following from the January 1885 issue of *Carpentry and Building*:

> It is mainly in city buildings of a business character . . . that the resources for architectural purposes are best illustrated. While iron enters largely into the purely constructive features of these loftier and more substantial erections, it is also made contributory to decorative treatment. . . . Iron balustrades for stairways date far back, but . . . now made in arabesque and other open patterns, they add greatly to interior structural beauty. Architects have turned handsome balustrade designs to good account in bordering successive landings in the spacing and deep wells that furnish interior light to some of our city buildings. Coated with color and picked out with gilt, the designs are sometimes rich and intricate, like interlaced ornamentation; in others, they may be likened to a succession of central bosses, from which radiate curved and spiral forms; occasionally they are a certain arabesque style, with graceful curvilinear lines set between slender shafts.

In addition to their great beauty and richness, these finishes provided extremely durable and maintenance-free surfaces predicted by Barff ten years earlier. As noted in *Ornamental Iron* (1893:[1]97–98), "the electro-plating in the Rookery has successfully withstood eight years handling by the public, and the rubbing and scouring of the janitor." Certainly the predictions of durability have proven true for the surface finishes as well as for the electroplated finishes revealed after the recent paint removal and cleaning.

The application of new technology, fine craftsmanship, and innovative design all contributed to the realization and appreciation of Root's new architectural sensibility. The completed building was widely published in its day and continues to be revered as a work of great creativity, "an improvisation in tones and harmonies both delicate and strong, spontaneity of expression with profundity of thought, all fashioned with the inspiration of genius" (Tallmadge 1941:151). The Rookery's structural and ornamental ironwork played a critical role in the building's acclaim and contributed toward the establishment of architectural ironwork as principal structural and decorative components of the modern commercial high rise.

Appendix

EXAMINATION REPORT: THREE EXAMPLES OF METALWORK FROM THE ROOKERY

209 South LaSalle Street, Chicago, Illinois
Original architects Burnham and Root (1885–88)
Alterations by Wright (1905–07) and Drummond (1931)
P. Andrew Lins and Melissa S. Meighan

Three samples from different areas of original metalwork in the Rookery were submitted for metallographic evaluation, with particular emphasis on surface finishes. The three samples are the following:

1. Window hardware, consisting of the handle proper, a long, biconical "bead," the rod which passes through the bead and which has threaded, decorative spiral knobs at either end, and the two mounting brackets which screw to the window frame and which have projecting arms which are pierced to accommodate the rod. Overall dimensions assembled are 6⅝ inches wide, 1⅞ inches deep, and 2 inches high. The finish, a dense, smooth, lustrous black, is assumed to be the original surface and to have been produced by the Bower-Barff process.
2. A section of the ornamental screen or grate, which partially encloses the space under the main staircase in the light court, basically a one-sided, stylized stem with curling leaves and some cast in decorative patterns. Dimensions are 19½ inches long, 4¾ inches wide, and 11/16 inches thick at the maximum. The sample has several layers of overpaint, the last being a cream-white color.
3. A relatively simple baluster from the stair railing, consisting of square upper and lower portions, 5½ and 7¼ inches long, with two of four surfaces having vermiculated decoration, and a plain, central shaft of alternating balls and lengths of rod with flaring ends. Overall dimensions are 26¼ inches long, 1⅜ inches square at top and bottom. The baluster has been overpainted black.

Each sample was cross-sectioned using a Leco diamond saw (Vari/cut VC-50) and mounted in Leco epoxy. Each sample was polished to 600 grit with SiC using water or Shelsol, followed by 6 μm and 1 μm diamond using oil as lubricant. The sections were examined at magnifications up to ×500 on a Zeiss metallographic microscope and to ×1000 on a Leitz Laborlux S microscope with a LEP 50W Hg vapor as well. Due to the range of surface treatments employed on the samples, a variety of other analytical techniques were utilized in our analyses, as described below.

SAMPLE 1. WINDOW HANDLE WITH BOWER-BARFF SURFACE TREATMENT

The cross section of this sample (Fig. 2), taken from the bracket, showed typical grey cast-iron structure, with rosettes of flake graphite (maximum flake length approximately 65 µm, with many flakes 10 µm or less). The castings, probably sand casts, are of good quality, with relatively few flaws, and well finished. The surface patina, approximately 25–30 µm thick, consists of two clear zones of approximately equal thickness, of dark-grey or black material. The inner zone has metallurgical features, such as remnants of the cast-iron microstructure and voids, which are continuous with the bulk metal. This zone appears to be the cast-iron substrate that reacted with (was oxidized by) the superheated steam of the Bower-Barff process. This zone also may be described as a depletion zone, from which iron ions moved to the outer zone, an oxidation layer grown on the original finished surface of the cast iron during the Bower-Barff process. On the uppermost surface of the outer zone there is commonly a very thin, approximately 1–2 µm thick maximum, grey-black layer, often separated from the underlying zone by voids, which may represent the alteration of the finish over time or from an initial quenching. Examination under the UV microscope up to ×1000 indicated that there was no intentional organic coating, or residue from an oil quenching, on the finished Bower-Barff surface.

The patina is relatively hard (Mohs 6–6½). It was sampled by scraping with either a glass rod in areas where there was cleavage between the two main layers noted above or with a quartz crystal. Samples were analyzed by X-ray diffraction (Phillips PW 1729 and 1840) in a Gondolfi camera, by emission spectroscopy (Baird Assoc. emission spectrograph, and Fourier transform infrared spectroscopy (FTIR—Nicolet 510P FT-IR Spectrometer, with a Nic-Plan IR microscope). High-purity iron (Puratronic, iron powder, grade 1), Fe_3O_4 (Johnson Matthey, Puratronic, iron [II,III] oxide, 99.999% metals basis), Fe_2O_3 (Johnson Matthey, iron [III] oxide, grade 2), and FeO (Johnson Matthey, iron [II] oxide, 99.9% metals basis) were purchased and used as standards for comparison.

The patina sample and the standard were not readily analyzed and they appear to be either cryptocrystalline or poorly crystallized. The Fe_3O_4 standard produced eight relatively weak lines, seven of which matched d-spacings for JCPDS pattern 19–629, or magnetite. The sample produced nine weak lines, seven of which matched the most significant lines for magnetite, JCPDS 19-629. The patina is definitely magnetite, with minor impurities (Table 1).

The sample and several standards were tested with a magnet. The Fe_2O_3 (red) was only very slightly magnetic. The FeO (wustite, black) was magnetic. The Fe_3O_4 (black) standard and the sample, both identified by X-ray diffraction as magnetite, were—not surprisingly—highly magnetic.

Emission spectroscopy was run on approximately 2 or 2.5 mg of sample; each of the standards noted above mixed with equal amounts of graphite in order to burn more thoroughly. This analysis indicated the presence of small amounts of silicon in the patina (analysis for sulfur or phosphorus was not undertaken). A cross section was analyzed by Nick Spencer of the Research Laboratory of W. R. Grace, Baltimore,

TABLE 1. *Data from X-ray diffraction analysis of Bower-Barff patina on a window handle from the Rookery.*

d-spacings (I/II$_1$) sample from handle	*d-spacings standard Fe (II,III) oxide Fe_3O_4 99.999% metal basis*	*d-spacings JCPDS 19-629 (Fe_3O_4)56F Magnetite*
	4.90 (VW)	4.85 (8)
3.50 (VW)		
2.98 (W)	2.96 (M)	2.97 (30)
2.55 (S)	2.52 (S)	2.53 (100)
	2.43 (VW)	2.42 (8)
2.10 (VW)	2.10 (M-W)	2.10 (20)
1.71 (VW)	1.71 (VW)	1.71 (10)
1.62 (M)	1.62 (M)	1.62 (30)
1.53 (VW)		
1.49 (M)		1.48 (40)
	1.38 (M)	(6 lines I=10 or below)
1.09 (VW)		1.09 (12) (11 lines I=8 or below)

by means of a scanning electron microscope and energy dispersive analysis. Sulfur and silicon were identified both in the bulk of the sample and in the Bower-Barff layers. The data on his analyses will be forthcoming. These elements may be responsible for the unidentified lines in the X-ray diffraction pattern.

The IR analysis in the 200–600 cm^{-1} region, coordinated by Beth Price and carried out on a Nicolet 800 series FTIR instrument at the Chemistry Department of Princeton University by John Eng, gave information in the negative sense, since the sample showed an absence of strong absorption bands in the far IR, comparable to the spectra exhibited by the Fe_2O_3 and Fe_3O_4 standards.

SAMPLE 2. UNDERSTAIR ORNAMENT

The metallographic section of this piece of ornamental metalwork revealed a typical grey cast-iron structure, with the graphite flakes (maximum length approximately 150 μm, most 50 μm) in random and rosette configurations (Fig. 10). The surface treatment of the metal was relatively complex. A thin, highly discontinuous layer containing copper (average thickness 2–5 μm, with maximum thickness 10–15 μm) can be seen on the surface of the iron, which is relatively rough for a plated surface. The copper (not analyzed due to difficulty in sampling) is probably a very thin electroplate layer or "strike" coat, but could be copper or brass particles that have settled out of the overlying varnish layer. Above the copper is what appears to be the original surface-finish coating, a colorless layer approximately 50–75 μm thick, with golden red-brown particles in its uppermost and lowermost levels, suggesting that the method of application may have been dry pigments brushed onto a warmed (stoved) lacquer. Areas of green staining can be seen within this layer and are probably attributable to the presence of copper. It is impossible to state if this green component is the result of corrosion of the copper discussed above or from degraded copper particles or copper-based pigments intentionally added to the colored coating. Finally, it must be pointed out that if a chemical treatment or patina had been applied to the copper to alter its appearance before the application of the colored coating, such a layer would only be nanometers thick and, even if unaltered, would not be resolvable with the instrumentation employed.

FTIR analysis on a microtomed cross section of the layers above the metal indicates that both the clear and golden red-brown layers are natural (tree) resins, not cellulose nitrate, shellac, or wax. Examination of the clear layer under the UV microscope indicates by the strong autofluorescence that the material is a resin.

Above this original coating are numerous other layers. Most interesting is the zone directly above, approximately 60–150 μm thick, presently a very dark brown in visible light and fluorescing ochre in ultraviolet, which is built up of 8 to 15 applications of what appears to be a pigmented, resinous material. This layer was not analyzed, but it can be suggested that this labor-intensive coating might be associated with the Wright renovation.

SAMPLE 3. BALUSTER

The cross section of the baluster shows the microstructure to be grey cast iron, with coarse-flake graphite (100 μm long, on average, with the longest flakes approximately 250 μm) in rosette patterns. In this example there is a very distinct layer of copper on the surface, varying in thickness from 10 to 20 μm (Fig. 11). The copper layer was sampled and analyzed by emission spectroscopy to evaluate its purity; that is, to determine whether it might be brass or bronze plating. Approximately 2.5 mg of the sample and of several standards were mixed with equal amounts of graphite to promote more thorough burning. The comparison standards were: Aesar zinc powder, −100 mesh, 99.9999%; Morton Thiokol copper (I) oxide, 99.95% metal basis; Morton Thiokol tin powder, −20 mesh, 99.9%. No tin was detected, and only a trace of zinc (less than 10 ppm) was found. The plating was found to be copper, not brass or bronze. Such low amounts of zinc could be a contaminant from a variety of sources, such as overpaint. The match with copper was excellent, with some traces of silver.

The examination of the cross section on both the metallographic and ultraviolet microscopes showed no organic coatings that could be associated with the original surface. This suggests that chemical patination, not resolvable by the analytical means available as mentioned above, originally provided the final finish.

Acknowledgments

The author wishes to thank Andrew Lins, Melissa Meighan, and Beth Price of the Conservation Department of the Philadelphia Museum of Art; and Roger Milton, Michael Peters, and

Nick Spencer of W. R. Grace for their generous assistance in the metallographic examination and analysis of the Rookery ironwork. Their work was crucial to the understanding and interpretation of the materials and techniques under consideration.

Notes

1. For example, see Condit 1964.
2. Structural economy lent impetus to studies of the strength of materials—the most notable being that of the English engineer T. Tredgold and his publication of 1822, *Practical Essay on the Strength of Cast Iron*—and the mathematical theory of design. This, together with early scientific speculations on crystallization and solidification and an understanding of the relationship between microstructure and performance, resulted in improvements in production (see Smith 1981:153–61).
3. "After 1840, blast-furnace practice gradually improved, [but] it was not until about 1850 that 150 tons of pig iron could be produced in a week. In 1875 several furnaces existed which could produce 700 tons of pig iron in a week, in 1880 several producing 1,000 tons per week, in 1890 one furnace could make 502 gross tons/day. By 1886, the United States was the largest steel-making nation in the world" (Swank 1892:535).
4. Among the earliest printed trade catalogues were those for metalwork and hardware published in England during the late eighteenth century. By the mid-nineteenth century, all types of metalwares and building components, especially in cast iron and zinc, were available.
5. Many of the ironwork catalogues of the period contain a great many elaborate designs for elevator grilles and enclosures in a wide variety of styles and finishes. For example, see *Illustrated Catalogue [of the] Hecla Iron Works* (Poulsen and Eger n.d.[ca. 1890]).
6. These coatings were usually shellac or copal varnish and, if tinted with colorants, were referred to as "changing varnishes" (Gardner 1888:64).
7. For a complete discussion, see Hawkes 1980.
8. Because it forms a lead linoleate soap with linseed oil, red lead was preferred because it provided the more protective, tough, durable, and elastic coating, and its ability to enhance "drying" favored its use as a shop- and site-applied coating. Its major drawbacks were expense, poor hiding power, light instability, and toxicity. In addition, its adulteration with brick dust was cited as one reason for its decline in reputation (Condit 1882:145).
9. See *739 Paint Questions Answered* 1904:227.
10. See Experiments on the preservation of sheet-iron from rust in India. *Journal of the Franklin Institute* 20 (October 1835):280–82.
11. See *739 Paint Questions Answered* 1904:188.
12. "The gilding of bronze and brass castings is performed, in the dry way, by making the surface perfectly smooth, then brushing it over with an amalgam of gold, and dissipating the mercury by heat, which leaves a durable film of gold over the surface. This surface may be burnished or deadened. The amalgam is made by heating one part of gold, in thin laminae, in a crucible, and when it becomes faintly red, pour over it eight parts of mercury, pour the combined gold and mercury into cold water, and squeeze the surplus

mercury out. The amalgam is then enclosed in canvas or chamois leather, and some more mercury pressed out; the remainder will contain one part of gold to four parts of mercury. This amalgam is rubbed over the objects to be gilded: it may be had in its true composition from the gold mines of Virginia, and of the best quality from North Carolina. It is advisable to brush the brass over with a thin solution of nitrate of mercury and some free nitric acid, as this facilitates the adherence of the amalgam. The gilt and burnished articles may be coloured by a simple process to any shade from a bright and crimson red to a violet and deep blue, by being submerged in a bath of caustic potash in which some metallic oxide is dissolved" (Field and Weill 1951:243–44).

13. "Metals are bronzed either by chemical agents or by the application of bronze powders. The latter are brass powders of varying shades applied to coats of drying varnish, pigmented or unpigmented by means of badger brushes and then coated with a transparent varnish" (Brannt 1896:148). "Bronze powders are alloys of varying composition of copper and zinc, with small amounts of iron, silver, [and] tin producing colors yellow, pale green, lemon, copper-red, orange, pale yellow, and crimson" (Brannt 1896:154).
14. "Copper Bronze, Red Bronze, [and] Green Bronze [are] produced by varying the color of the base lacquer followed by dusting of the powder" (Brannt 1896:149).
15. These techniques are described in various sources, as follows:

> "Iron railings should first be painted a deep bronze green. The green is made by mixing chrome yellow with lampblack to the desired color, with carriage-rubbing varnish and turpentine. Then, before the paint is dry, take the velvet dauber, and dipping it into the bronze rub it gently over the most prominent parts, or those which project from the surface of the tips of the rods or spearheads, etc." (Gardner 1888:83).
>
> "In bronzing iron it is very advantageous to first coat it, without regard to the succeeding colors, with red oxide of iron, rubbed up in varnish, and then apply the actual colors. A very neat bronzing is produced with a paint of graphite and varnish, the last coat of which is dusted over before becoming entirely dry with aluminum powder, the whole acquiring thereby the appearance of polished iron" (Brannt 1896:149).
>
> "Cast iron may be bronzed by dipping it in a thin solution of sulphate of copper, or muriate of copper, and when sufficiently covered with copper, it is washed and painted with oil varnish. All objects to be bronzed may, however, be painted of any colour, either a shade of green, from the faintest to an almost black green, or of a blue or bluish green. The paint cover should be coated with pure varnish, and when that is nearly dry, a metallic powder is dusted over it by a dusting-bag, or rubbed on by the fingers, a linen pad, or a paint-brush. The metallic powder is generally mosaic gold, which is made of almost every shade, and is of great beauty; or it may be copper in powder, gold leaf, silver lead, and similar articles; dry paint of a convenient shade may also be used. The highest parts of the articles are generally bronzed so as to appear as if rubbed and worn by use. Over the whole of these, a last coating of spirit varnish is laid on" (Overman 1872:242–43).

16. During this period simultaneous experiments in electrochemistry revealed the potential for commercial application. In 1839 T. Spencer of the Liverpool Polytechnic School

described the electrotype process, and in 1840 de Roulz patented a process for electrogilding and silvering and electrodeposition of copper, lead, tin, cobalt, nickel, and zinc. By 1847 George Elkington established at Birmingham, England, the first plant for practical electroplating. For a chronicle of the specific discoveries and events (see Goodale and Speer 1920).

17. Construction of the first dynamo for electroplating in 1875 by Edward Weston led to the firm establishment of the plating industry in America (Burns and Bradley 1967:56).
18. "We are thoroughly equipped for the special finishes of Electro-plating in Bronze, Brass, Copper, and Silver, The Bower-Barff Process, and Electro-graphic or Galvano-plastic work. Duplex electro-bronze work, for outdoor purposes—Store Fronts, Front railings, Grilles, Etc." (Poulsen and Eger n.d. [ca. 1890]).
19. "Electro-bronzing is a process of preserving and beautifying iron heretofore much neglected. Its possibilities in a decorative way are without limit, since it may be applied wherever desired, and in a number of shades. In combination with other metals, the variety of colors is limitless" (*The Inland Architect and News Record* 32 [September]1889:19).
20. In his initial design for the Schlesinger and Mayer Store (now Carson Pirie Scott) in Chicago (1899–1904), Louis Sullivan originally intended the elaborate ground- and second-story metalwork to be cast "statuary bronzework"; however, this was later changed in the specifications of 1898 to be of electroplated cast iron. Ironically, even this was not done for lack of money, and instead the ironwork was painted with a primer of asphaltum and alternating layers of red and green to simulate the appearance of patinated bronze (see Siry 1988:178).
21. The ornamental ironwork of Sullivan's Schlesinger and Mayor Store (now Carson Pirie Scott) was specified to be cast iron "Duplex electroplated in copper bronze" by immersing the metal in an acid bath for 36 hours longer than the standard 22 hours used for electroplating interior metalwork (*Specifications* Section F [December 10] 1898:4).
22. In recounting the commercial introduction of exterior electroplating in America in 1889–90 for use on the ironwork of the rehabilitated Chamber of Commerce Building in Chicago, *Ornamental Iron* states: "The Duplex-Bronze process, original in its application to artistic metal work, with this company [Winslow Brothers], was here applied practically for the first time after repeated and costly experiments. It has fully met the requirements in affording an imperishable envelope to the metallic base, and a surface finish, equal in appearance to solid bronze, fully as durable, and at a greatly reduced cost . . . as perfect now as when was first put in place" (*Ornamental Iron* 2 [March]1895:79).
23. For patination recipes, see Overman 1872:241–42.
24. Brannt describes these methods in various passages from chapter 6 of his "Bronzing and Coloring" as follows ([1890] 1896):

> From page 154:
> Bronzing of Cast-iron.—To give cast-iron the appearance of bronze coat the polished iron with a thin layer of linseed oil or linseed oil varnish and thoroughly heat it in the air to bring about the oxidation of the metal. The temperature must be higher or lower according to whether a pale yellow or dark brown coloration is to be produced. The so-called Tucker bronze is obtained by greasing the polished iron and exposing it for 2 to 5 minutes to the action of vapors produced by a bath composed of equal parts

concentrated nitric and hydrochloric acids, then coating the iron with vaseline and heating until the latter commences to decompose.

To Bronze Cast-iron.—First clean and smooth the surface and then coat it uniformly with a layer of vegetable oil, for instance a poor quality of olive oil. Then heat, without, however, raising the temperature to the burning point of the oil. In this manner the cast-iron at the moment the decomposition of the oil takes place absorbs oxygen and a brown surface of oxide is formed on the surface which adheres very firmly and acquires a good polish, so that the surface of the cast-iron assumes a bronze-like appearance.

Bronze-like Surface on Iron or Steel.—A very fine bronze-like surface, and one that will prevent rusting, may be produced on iron or steel as follows: The object to be acted upon must be cleansed, so as to take off all oxidation or other impurity. It is then exposed for two or three minutes to the action of the vapors of a heated mixture of hydrochloric acid and nitric acid in equal proportions, at a temperature ranging from 550 to 650 degrees F. After the objects have cooled, they are to [be] rubbed over with vaseline, and again heated until the vaseline begins to decompose. This treatment with the vaseline has to be repeated once. Should a lighter coloring than bronze be desired, it can be produced by mixing acetic acid with the other acids. In using the above-named chemicals great care should be observed in handling and preserving.

From page 166:
A very beautiful and durable black color is produced upon articles by dipping or sponging them with platinum chloride solution . . . [described as] "dead black." Lustrous black color is obtained by polishing with a soft piece of leather moistened with oil, and a lustrous grey-black color by polishing with the burnisher or burnishing stone. The color, especially when polished, is very durable, since platinum is not changed by the action of the air.

Steel and iron articles could be given a fine blue color with a coating of a thin film of a lead sulphide.

To Blacken Small Iron Articles in bulk.—A deep black upon small iron articles is produced by heating them in bulk in connection with oiled saw-dust. . . . The coloring agent consists of an intimate mixture of 10 parts of dry saw-dust and 1 part of linseed oil. The saw-dust thus oiled is shaken in the drum together with the articles to be blackened and exposed to the heat while revolving the drum. The saw-dust undergoing combustion evolves a thick smoke which cannot immediately escape from the drum, it passing out slowly through the aperture in the funnel. This smoke coats the articles with a firmly adhering black color.

From page 18:
Lustrous Black on Iron is obtained by the application of a solution of sulphur in spirits of turpentine prepared by boiling upon the water bath. After the evaporation of the spirits of turpentine a thin layer of sulphur remains upon the iron, which, on heating the article, intimately combines with the metal.

> By another method the cleansed and pickled iron articles are coated when dry with linseed oil and heated to a dark red. If pickling is omitted the coating with linseed oil and heating have to be repeated twice or three times. . . . According to Meriten a lustrous black on iron is obtained by placing the articles as anode in distilled water heated to 158 degrees F and using an iron plate as cathode. A layer of ferroso-ferric oxide is formed which, however, can only be obtained in a firmly adhering state upon wrought-iron. The lustre appears by brushing with a soft waxed brush. The current conducted into the bath must only be strong enough to decompose the water without perceptible development of gas.
>
> Brown-black Coating with Bronze Lustre on Iron.—Heat the bright iron objects and brush them over with a concentrated solution of potassium bichromate. When dry heat them over a charcoal fire and wash until the water running off shows no longer a yellow color. Repeat the operations twice or three times. A similar coating is obtained by heating the iron objects with a solution of 10 parts of sulphate of iron (green vitriol) and 1 part of sal-ammoniac in water.

In his chapter 10, "Decorating, Enamelling, Engraving, Etching," in the same work, Brannt provides the following information:

> From page 196:
> Iron and steel tools protected by an acid-resistant lustrous black coating which resists acids by placing cleaned tools in a box by burning crushed blacksmith's coal and coating tools.
>
> Ward's Inoxidizing Process.—Cast or wrought iron articles are coated by means of a brush or by immersion with a silicate, heating, and after cooling forming a dense uniform coating of a dead-black color, which does not change by the action of the atmosphere nor crack off from the article. Could be colored as well.
>
> From pages 296–97:
> Inoxidizing Process for Cast-iron.—The cast-iron articles, entire gas-chandeliers, water pipes, ornamental pieces, railings, kitchen-pots, etc., are placed upon an iron-carriage and first exposed in a reverberatory furnace of special construction at a temperature of from 112–1292 degrees F for 15 minutes to the oxidizing, and then for 20 minutes to the reducing action of gas generators. After removal from the furnace and cooling, the inoxidized articles acquire a uniform slate color; they may, however, be enamelled and decorated in any manner desired. . . . Various articles exposed in the open air for two years to all kinds of atmospheric influences showed no trace of deterioration. Numerous experiments have demonstrated that the tensile strength of the iron is but little decreased by inoxidation, in fact, no more than by annealing. As regards the durability of the surface layer with a high load, it has been determined that from wrought-iron bars it cracked off in the form of small lamina with an average tension of 4,025 lbs., but from cast-iron it did not crack off even with a breaking load.

25. Professor Barff's process was reported in the *London Times* of March 6, 1877. He delivered a paper describing it in detail before the Society of Arts, London. The paper, titled "Zinc

white as paint, and the treatment of iron for the prevention of corrosion," was delivered on Wednesday, February 14, 1877 and appeared in *The Journal of the Society of Arts* on February 16, 1877, The lecture was reprinted in its entirety for the American audience in *The Iron Age* on April 26, 1877:5–7, and notices of the discovery were published in *The Journal of the Institute of Iron and Steel* in 1877.

26. Current practice indicates that exposure to superheated steam at 500 °C and upward produces an adherent black coating that varies in composition, thickness, and corrosion resistance with the steam temperature. Resultant films can be 2–3 µm at 500 °C to over 10 µm at 600 °C. Cast iron subjected at 700 °C produces a film of 37 µm in 60 minutes. A magnetite film of 25–30 µm was observed on the Bower-Barffed hardware at the Rookery. See Fishlock 1962:229.
27. At least one member of the Institute of Iron and Steel took exception to the unique credit given to Barff for the alleged "discovery" of the magnetite formation on ferrous metals (see Percy 1877).
28. Barff published articles on the process in *Van Nostrand's Eclectic Engineering Magazine* 16(April 1878):300–2; and 18(April 1879):350–54.
29. The paper was subsequently published as "The corrosion of iron." In *Royal Institute of British Architects Sessional Papers* 1878:265–76. London: RIBA.
30. Published as "The Barff process for the protection of iron," *The Iron Age* (January 22) 1880:7.
31. These proceedings were published in *The Journal of the Institute of Iron and Steel* 2, 1881:167–70.
32. "In this process the parts to be treated are heated to nearly 900 degrees C in a closed retort; superheated steam is then led in for twenty minutes and a coating consisting of a mixture of the black and red oxides of iron is formed. Producer gas is now substituted for the steam and passed in for about the same length of time. If the coating thus formed is not sufficiently thick, these operations are repeated as often as necessary. Paraffin or some other oil is afterward applied to the parts, whereby a fine black color is obtained and additional protection afforded. As regards cost, this process is expensive, the price ranging from $5 to $20 per ton, but it can be applied to almost anything which will not be injured by the heat and is small enough to be placed in the furnace, and it affords very efficient protection, especially against sea-water, acid fumes, and like influences. Paints and enamels can be applied and adhere strongly. Parts given this treatment will stand almost any degree of heat without injury, but they may not be bent or machined. Of course, the process is not applicable to tools or tempered pieces. Also, the parts are enlarged slightly so that those which must maintain close limits on dimensions should not be given this treatment" (Wilson 1915:48–49).
33. This was presented in response to an article published in *Iron Age* and reprinted with commentary in *Ornamental Iron* (Art iron at the fair 1893:19–20).
34. "The durability of these finishes has been well established, and they have had such substantial recognition by our leading Architects that we need not enlarge upon them." "Announcement" for the Winslow Brothers Company Artistic Metal Work in *Illustrated Catalogue [of the] Hecla Iron Works*, ca. 1890.
35. Hecla Architectural Bronze and Iron Works, owned by Poulsen and Eger, was located between North 10th and 11th and Berry streets in Brooklyn, New York, and employed the

first furnace for the application of the Bower-Barff process in the United States (*Ornamental Iron* 1[November]1893:94–98).

36. Winslow Brothers Company of Chicago, Illinois, had their city office and exhibit rooms as well as their publication office for *Ornamental Iron* in the Rookery beginning in 1893 (*Ornamental Iron* 1[November]1893:94–98). According to *The Architect and Contractor*, the firm had the largest plant for the manufacture of ornamental ironwork in the country and was the recipient of eight medals awarded at the Columbia Exposition, including "most successful application of the Bower-Barff Rustless Process" and the most "successful electroplating of bronze on iron . . . and on wood" (*Ornamental Iron* 1[April–May]1894:165).
37. Design no. 2044—railing panels of the light court mezzanine and grand stairs; no. 2048 (Design A)—railing panels and turned balusters of the light-court double-cantilevered stairs and oriel staircase, and (Design B)—skylight muntin pattern; no. 2049—stairs and mezzanine newels and stringers; and no. 3563—electrolier and column bases of the light court, in Poulsen and Eger, *Illustrated Catalogue [of the] Hecla Iron Works*, n.d. [ca. 1890].
38. Root's geometric designs for the ornamental incising of the marble, the floor mosaics, exterior terra-cotta and stone, and the metalwork can all be traced to plates of designs of "Arabian," "Turkish," and "Moresque Ornament" in Owen Jones's *The Grammar of Ornament*, published in London in 1856 and later in America in 1880. Of the metalwork, the oriel stair railings are very similar to the square diaper pattern of "Moresque no. 4" (Plate 42, no. 6) and the risers and stringers to the interlaced patterns of "Arabian no. 5" (Plate 35) and "Moresque no. 1."
39. Burnham and Root, Drawing no. 73, "Grand Staircase, Rookery Building," revised October 1887 (Burnham Library, Art Institute of Chicago).
40. All technical descriptions and observations are verbatim from Lins and Meighan 1991 (see Appendix).
41. "Cast iron may be bronzed by dipping it in a thin solution of sulphate of copper, or muriate of copper, and when sufficiently covered with copper, it is washed and painted with oil varnish" (Overman 1872:242).

> Gold Bronze of Great Lustre on Iron.—Dissolve 3 oz. of finely powdered shellac in 1¾ pints of spirit of wine. Filter the varnish through linen and triturate a sufficient quantity of Dutch gold with the filtrate to give to it a lustrous appearance. The iron, previously polished and heated, is brushed over with vinegar and the color applied with a brush. When dry the article may be coated with copal lacquer to which some amber lacquer has been added (Brannt 1896:154).

References

ART IRON AT THE FAIR

1893 *Ornamental Iron* 1(September):65.

BARFF, F.

1877 Zinc white as paint, and the treatment of iron for the prevention of corrosion. *The Journal of the Society of Arts* (February):255–56.

BRANNT, W. T., ED.

[1890] 1896 *The Metal Worker's Handy-Book of Receipts and Processes*. Philadelphia: Henry Carey Baird (1890); London: Sampson Low, Marston, Searle and Rivington (1896).

BURNS, R. M., AND W. W. BRADLEY

1967 *Protective Coatings for Metals*. New York: Reinhold Publishing.

1885 *Carpentry and Building* vol. 7 (January):2.

CONDIT, C. W.

1964 *The Chicago School of Architecture*. Chicago and London: University of Chicago Press.

CONDIT, C.

1882 *Painting and Painter's Materials*. New York: The Railway Gazette.

FIELD, S., AND A. D. WEILL

1951 *Electro-plating*. 6th ed. New York, Toronto, and London: Pitman Publishing.

FISHLOCK, D.

1962 *Metal Colouring*. Teddington, Middlesex: Robert Draper.

A FRONT OF BRONZE

1893 *Ornamental Iron* 1(July):22.

FRYER, W. J.

1876 *Architectural Iron Work: A Practical Work*. New York: John Wiley & Sons.

GARDNER, F. B.

1888 *Everybody's Paint Book*. New York: M. T. Richardson.

GOODALE, S. L., COMP., AND J. R. SPEER, ED.

1920 *Chronology of Iron and Steel*. Pittsburgh: Pittsburgh Iron and Steel Foundries.

HAWKES, P.

1980 Paints for cast iron. Master's thesis, Columbia University.

HOFFMAN, D.

1973 *John Wellborn Root*. Chicago and London: University of Chicago Press.

1889 *The Inland Architect and News Recover* vol. 32 (September):18–19.

JONES, O.

[1856]1880 *The Grammar of Ornament*. London: J. W. Bouton.

LEE, A. J.

1983 Cast iron in American architecture: A synoptic view. In *The Technology of Historic American Buildings*. H. Ward Jandl, ed. 100–1. Washington, D.C.: Foundation for Preservation Technology.

LINS, P. A., AND M. S. MEIGHAN

1991 *Examination Report: Three Examples of Metalwork from the Rookery*. Philadelphia: Philadelphia Museum of Art.

ORNAMENTAL IRON

1893 Vol. 1 (November):94–98.

1895 Vol. 2 (March):79.

OVERMAN, F.

1872 *The Moulder's and Founder's Pocket Guide*. Philadelphia: Henry Carey Baird.

PERCY, J.

1877 On the protection from atmospheric action which is imparted to metals by a coating of certain of their own oxides, respectively. In *The Journal of the Institute of Iron and Steel*, 456–60. London: E. & F. N. Spoon.

PORTER, J. J.

1915 Manufacture of pig iron. In *The ABC of Iron and Steel*. A. O. Backert, ed. 78–79. Cleveland: The Penton Company.

POULSEN AND EGER, COMPS.

n.d. [ca. 1890] *Illustrated Catalogue [of the] Hecla Iron Works*. New York: Poulsen and Eger.

739 PAINT QUESTIONS ANSWERED

1904 New York: The Painter's Magazine.

SIRY, J.

1988 *Carson Pirie Scott, Louis Sullivan and the Chicago Department Store*. Chicago and London: University of Chicago Press.

SMITH, C. S.

1981 The early history of casting, molds, and the science of solidification. In *A Search for Structure*, 127–73. Cambridge and London: MIT Press.

SPRETSON, N. E.

1878 *A Practical Treatise on Casting and Founding*. London and New York: E. & F. N. Spoon.

SWANK, J. A.

1892 *History of the Manufacture of Iron in All Ages*. Philadelphia: American Iron and Steel Association.

TALLMADGE, T. E.

1941 *Architecture in Old Chicago*. Chicago: University of Chicago Press.

TREDGOLD, T.

1822 *Practical Essay on the Strength of Cast Iron*.

WILSON, L. C.

1915 *The Corrosion of Iron*. New York: The Engineering Magazine.

WORRILOW, W. H.

1949 Foundries—past and present: Their transition from manual to mechanical operation. *Journal of the Franklin Institute* (May):478.

Biography

Frank G. Matero is associate professor of architecture in the Graduate Program in Historic Preservation of the University of Pennsylvania and director of The Architectural Conservation Laboratory. He is also a lecturer at the International Center for the Study of Preservation and the Restoration of Cultural Property in Rome. From 1981 to 1990 he was assistant professor of architecture and director of the Center for Preservation Research in the Graduate School of Architecture, Planning and Preservation of Columbia University. His teaching and research are focused on the history of building technology and the conservation of historic building materials, particularly stone and architectural surface finishes. He is a professional associate of the American Institute for Conservation of Historic and Artistic Works and a member of the Advisory Committee of the International Masonry Institute.

Techniques of Mercury Gilding in the Eighteenth Century

MARTIN CHAPMAN

The art of gilding flourished to an unparalleled degree during the Age of Enlightenment. As Diderot's *Encyclopédie ou Dictionnaire Raisonné des Sciences, des Arts et des Métiers* claimed in its section on gilding,

> *"Cet art n'etoit point inconnu aux anciens, mais ils ne l'ont jamais poussé à la même perfection que les modernes"* [This art was not unknown to the ancient world, but it never reached the same level of perfection as that of today](Diderot and d'Alembert 1751–57, vol. 5:57).

In other words, though gilding was performed by techniques that had been practiced for centuries, the richness and refinement in the finish of eighteenth-century gilding raised it far above that of its predecessors.

From the seventeenth century, gilding had been in great demand for the decoration of interiors. In their treatise on japanning, Stalker and Parker declared in 1688, "We are all great admirers of Gold and by consequence must be enamoured with Guilding, which is so nearly related to it." Gilding was highly desirable in adding that necessary touch of richness to the Parisian luxury interior. The trades of Paris practiced the many traditional techniques of gilding on wood, plaster, and metal to respond to this demand.

Of all methods of gilding practiced in the eighteenth century, that done by the mercury process was the most remarkable. This technique, known as early as the ancient world, was developed to its highest degree during the eighteenth century in the hands of skilled Parisian craftsmen. Mercurial gilt bronze became an essential and conspicuous part of Parisian interior decoration from the early eighteenth century, partly because of its rich, metallic glow, and also because of its durability, especially when compared to gilding on wood or plaster. Gilt bronze was applied in the form of mounts to furniture and to any furnishings that formed the focal point in a room. Elaborate mounts were applied to the most expensive pieces of furniture and decorative objects. They virtually overwhelmed the primary material—whether it was wood, stone, or porcelain—with their brilliance and often assumed the leading theme of ornament. Gilt bronze was favored for any purpose associated with artificial light—such as the chandelier, candlestick, or fire dog—where the flames would be attractively reflected in the rich, gilded surfaces. Gilding was also used for clock cases hung on the wall, on a piece of furniture, or on a chimneypiece.

This predilection for gilt bronze led to manufacturing methods that became more sophisticated as the century progressed and richer, more refined finishes were sought. At the beginning of the century, the gilding trade was geared to a perfunctory execution of the process more or less as it had been practiced for centuries. By the 1780s, however, the final stages of mercury gilding were refined to a degree that ensured the surface color and richness of fine gold and the most subtle contrasts of light and texture. These later developments were closely guarded trade secrets, but some of them were derived from a variety of contemporary accounts, both French and English.

The Gilding Process

The process of mercury gilding in the eighteenth century was, in essence, the same as that described by Theophilus in the twelfth century in his *Treatise on Divers Arts*. Briefly, the method consisted of forming an amalgam of gold and mercury and depositing it in prescribed amounts on the metal to be gilded, then volatilizing the mercury by heat application to leave the gold alloyed to the surface of the metal. This could be done with silver as well as bronze.

The best-known contemporaneous source for discussing the mercury gilding process is Diderot's *Encyclopédie* (1751–57). This is an important compendium for details of techniques used in the eighteenth century, but it is by no means the first or indeed the fullest account of mercury gilding. The version quoted by Diderot under "Dorure," the section of the *Encyclopédie* devoted to gilding, is lifted straight out of Chamber's *Cyclopaedia*, published in London in 1728, and to a lesser extent Stalker and Parker's *A Treatise of Japanning and Varnishing* of 1688. Diderot's description of the process outlines the general principles rather than providing an eyewitness account of the contemporary practice used in Parisian workshops of the mid-century.

If Diderot's text is somewhat perfunctory, his illustrations are not. They are, in fact, the first visual accounts of Parisian trades in operation. In the section "Doreur, sur Metaux," he shows the interior of a gilding workshop where workers are performing different techniques. His illustration (Fig. 1) shows the two processes of gilding on metal—mercury gilding and leaf gilding—which were used extensively in the eighteenth century, sometimes in combination (Verlet 1987:174).

Leaf gilding was simpler (and probably cheaper) to perform but was less durable than mercurial gilding. Also, the leaf process was inappropriate for chased work, so it tended to be used only when brilliantly polished effects were required. The workman in the back center of the scene is gilding a crucifix using the leaf process. The gilder is shown lifting the gold leaf with a gilding knife and applying it to the cross.

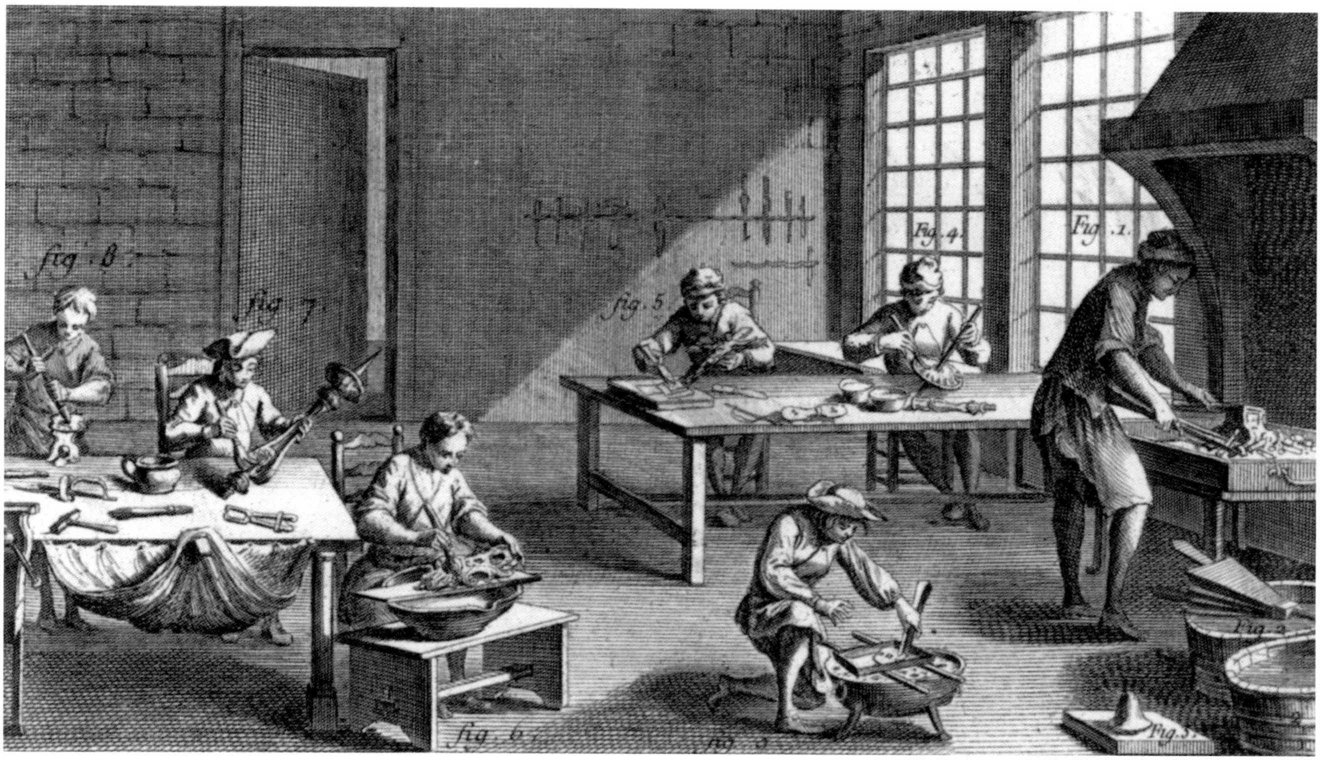

FIGURE 1. *The metal gilder's workshop, detail of plate I from Diderot's* Encyclopédie *(Diderot and d'Alembert 1751–57).*

Kneeling in the foreground, a worker in a tricorn hat is blueing a piece of bronze by heating it over a brazier, which was the necessary preliminary step in leaf gilding. The person at right, near the window, is burnishing part of a candlestick, which was also gilded by the leaf process.

The remaining workmen in the scene are performing various steps of the mercury-gilding process. (The earlier stages, which involved the preparation of the gold, are not illustrated.) First the gold would have been obtained from the gold beater. According to Diderot, ready-ground *or moulu* (powdered gold) cost 104 livres an ounce in the 1750s compared with leaf gold, which cost 90 livres (Diderot 1751–57:737). The term *or moulu* meant more than just ground gold; by association it came to describe the finest type of gilding. It was assimilated into the English language as the term *ormolu* about 1760.

Next, the amalgam was formed by putting the gold into a crucible onto a charcoal fire. Then mercury was added in proportions of about one part gold to eight parts mercury.[1] When heated together the two metals readily combined to form a silvery amalgam the consistency of putty. The amalgam was poured into a vessel of water and then squeezed through chamois leather or muslin to remove the excess mercury. This excess was often kept for further use as it inevitably contained some residual gold.

Preparation of the Casting

The casting arrived in the gilding workshop already chased in the finest details. Though the term *gilt bronze* is generally used, the casting was actually made of brass; that is, a copper-zinc alloy rather than the copper-tin alloy usually associated with bronze. The terms *bronze* and *brass* were somewhat vague in both England and France at this time, as was the composition of the metals; the proportions of copper and zinc used for casting varied from foundry to foundry. Matthew Boulton, the Birmingham ormolu manufacturer who turned his hand to producing gilt bronzes in the 1760s and 1770s, recognized that a high-copper content was necessary to provide the best foundation for gilding (Goodison 1974:70, 73). Indeed, what is described as gilding metal today is what Boulton advocated: about 95% copper, 5% zinc. However, as late as 1818 the chemist Jean Pierre Joseph d'Arcet, after much research into the question, recommended a rather different alloy comprising 82 parts copper, 18 zinc, 3 tin, and 1½ lead (1818:vii). Investigations of the collection of gilt bronzes in the Victoria and Albert Museum show a wide fluctuation in the composition of the alloy.[2] This is undoubtedly due to the haphazard foundry practice of throwing old bronzes into the melting pot. By tradition and superstition it was thought that old bronzes improved the consistency of the alloy, with the result that the material contained a curious mixture of different metals that were recycled from one generation to the next (Hachet n.d.:36).

In preparation for receiving the gilding, the casting was cleaned with nitric acid. This process was known as *dérochage*. The workman at the low bench in the foreground of Diderot's workshop (Fig. 1) is working the surface of what appears to be a large furniture mount with a scratch brush made of metal wire dipped in nitric acid. If the piece were especially dirty, it would be left to soak for half a day in *eau seconde*, a weaker solution of old nitric acid that was kept in the tubs shown at the extreme right in the illustration. Apparently this solution had the added advantage of lending the brass a reddish color, which was an appropriate foundation for gilding. The purpose of this treatment was not only to clean the surface of the brass, but also to open up the metal's pores to receive the gilding. As the *Dictionary of Arts and Sciences* suggests:

> The use of aquafortis or nitric acid . . . is not, as is generally supposed, confined to merely cleansing the surface of the metal to be gilt . . . ; but it also greatly facilitates the application of the amalgam to the surface of the metal, probably in the following manner: it first dissolves part of the mercury of the amalgam; and when the solution is applied to the copper, this latter metal, having a stronger disposition to unite with the nitrous acid than the mercury has, precipitates the mercury upon the surface . . . When the metal to be gilt is thus covered with a thin coat of the precipitated mercury, it readily receives the amalgam (Gregory 1806–7:843–49).

Whether this interpretation is correct I leave to the chemists to determine.

The amalgam, kept in a pot like the one shown in Diderot's second plate (Fig. 2), was then applied to the casting with a stiff brush made of brass wire, such as the one shown in Figure 3 (fig. 10 in illustration). Sometimes a layer of pure mercury was applied first to facilitate adherence of the amalgam. The gilder first dipped the wire brush into water, then into the amalgam, which he deposited across the surface of the casting, working in small areas to ensure that the entire exposed surface was covered. Invisible areas on the back of a furniture mount, for example, were left

FIGURE 2. *Tools and instruments, such as those used by workers in Figure 1. Detail of plate I from Diderot's* Encyclopédie *(Diderot and d'Alembert 1751–57).*

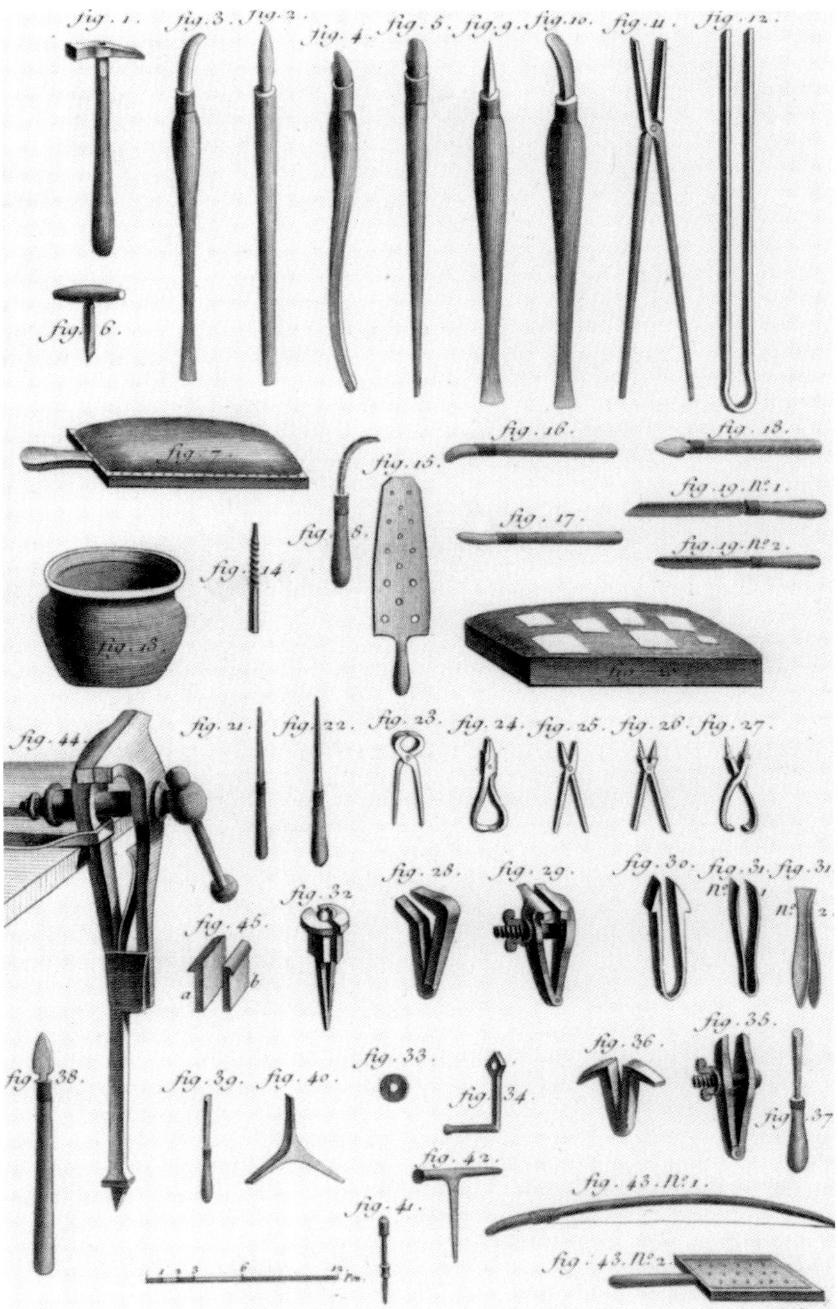

untouched by gilding. It is frequently a sign of later electrogilding if the back or underside of an object, which was not intended to be seen, is covered with gilding.

The amalgam-coated casting was then put onto a hot charcoal fire, the heat of which, Diderot says, can only be judged by experience (1751–57, vol. 5:59). This temperature was crucial, as the amalgam would run if it were too hot (Goodison 1974:69). When the casting reached a certain temperature, which was usually judged by the workman spitting on the piece, the mercury was volatilized in clouds of vapor, leaving the gold alloyed with the metal. The workman standing at the far right in Diderot's illustration (Fig. 1) is performing this part of the process. This was the most dangerous part of the operation, as the fumes were excessively toxic. Gilders had a short life expectancy, suffering most commonly from nervous system disorders such as the shakes, paranoia, and eventually madness. The expression "mad as a hatter" derives from the unfortunate effects of mercury on workers who came into

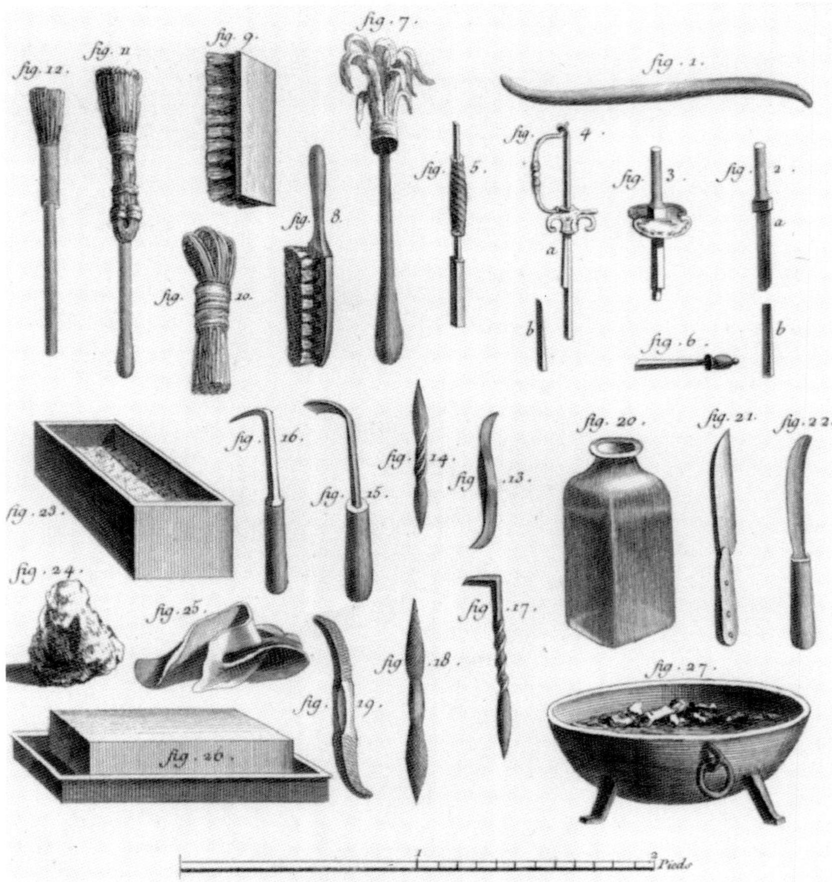

FIGURE 3. *Tools and instruments, detail of plate II from Diderot's* Encyclopédie *(Diderot and d'Alembert 1751–57).*

contact with it in England's hat trade during the nineteeth century. Ravrio, a successful manufacturer of gilt-bronze, left a legacy of 3,000 francs in his will to be awarded by the Academie des Sciences to anyone who found a method of gilding that avoided the hazards of mercury. It was awarded to the chemist d'Arcet in 1818 for his adaptation of the muffle kiln to improve chimney drafts and conduct fumes away from the workers. However, the only precautions visible in Diderot's workshop are the hood above the fireplace to collect rising fumes and the door at the back which is left ajar for ventilation (Fig. 1). D'Arcet remarks that more workers were poisoned in winter than in summer, as the doors and windows which created drafts and allowed fumes to escape were often closed in cold weather (d'Arcet 1818:ch. 13).

Finishing the Object

After the mercury had evaporated, the casting acquired a yellow (straw) color with a dull finish. In order to achieve the appearance of fine gold, further processes were used to modify and improve the appearance of the surface. The Parisian gilders refined these methods to an art, drawing on the experience of the past and employing the traditional methods used by goldsmiths in the gilding of silver.

With a huge tool, the worker at the far left in Diderot's illustration (Fig. 1) is burnishing a mount held in a vise. Burnishing was the process of polishing a surface by rubbing it with a hard material until it became mirrorlike. In some cases the burnishing would be enough to finish the piece, especially if there were no modeled or chased areas. The curved tools in the center row of Diderot's plate II (Fig. 3) are burnishers. Several are tipped with *pierre sanguine*, a red hematite stone imported from Germany. Diderot mentions the *ferrette d'Espagne*, a similar stone brought back from Spain by pilgrims, presumably from their journey to Santiago de Compostella. Agates and hounds' teeth are often erroneously mentioned as burnishers in accounts of mercury gilding, but these materials were better suited for softer leaf gilding and gilding on wood.

The process that gave gilding its richness of color was known as *mise en couleur de l'or*. According to Diderot, this was the gilders' secret. Diderot provides no detailed account of its composition under his section on gilding. He only mentions the process used in minting coins, which involved common salt and *tartre de Montpellier*, the latter derived from the crust formed inside wine casks during fermentation. In other sections of the *Encyclopédie* Diderot describes a concoction of tartar, sulfur, and salt mixed with water, in which the object is left to soak until the desired color is achieved (1751–57, vol. 5:59). Diderot lifted this recipe straight from the pages of Stalker and Parker's account of 1688, and therefore it was hardly a secret in the 1750s. Some recipes for coloring the gold simply called for the addition of a yellow coloring agent, such as crushed licorice, turmeric, or saffron. The workman second from left in Diderot's workshop (Fig. 1) is painting the stem of an altar candlestick, probably with this type of varnish in the small pot beside him.

Stalker and Parker discuss another way to heighten the color of the gilding. The recipe involves sal ammoniack (ammonium chloride), saltpeter (potassium nitrate), sandiver (glass gall, "a liquid saline matter found floating over glass after vitrification," as defined by the *Oxford English Dictionary*), verdigris (cupric acetate), white vitriol (zinc sulphate), and green vitriol (either cupric or ferrous sulphate). The process is described as follows: "Grind them together with white wine vinegar, which lay all over your work, then lay it on a fire, and give it a small heat that it may make it smoak, then take it off and quench it in urine" (Stalker and Parker [1688] 1971:66).

A similar recipe is quoted by Goodison in his book on Boulton, the English ormolu manufacturer. Boulton was particularly anxious to learn "the secret whereby the French give that fine colour to their sconces," and he asked his agent, Solomon Hyman in Paris, to investigate it for him there (Goodison 1974:73). Some of the recipes that have come down to us in Boulton's notebooks may therefore have been obtained from Parisian workshops. For coloring the gilding a "pale, middle or clearing colour," a recipe such as this would be tried, as quoted by Goodison (1974):

	oz	*dwt (pennyweights)*
Sal ammoniack	1	10
Green copperas	1	10
Salt	1	—
Verdegrees	0	15
Salt petre	0	15

Goodison analyzes the process as follows: "Copper was introduced into the solution in the form of cupric acetate ('verdigris') . . . and was induced to precipitate with the gold by the use of iron or zinc in the form of ferrous sulphate or zinc sulphate ('copperas')" (1974:71–72).

More than twenty-five recipes for coloring the gilding are quoted by Godfrey Smith in his book, *The Laboratory*, first published in 1755. Many of these formulas are variations on the concoctions already mentioned. One, called "Nuremberg gilding wax" (Smith [1755] 1770:21), seems to follow the same principles but uses beeswax as the medium. This method was undoubtedly derived from the German silversmithing trade, which had a high reputation for gilding, even in France.[3] Diderot gives recipes for two of these Nuremberg waxes under his section on *or moulu*. Both recipes seem to employ the formula of verdigris, white vitriol, borax, and *sanguine* (red chalk) mixed with yellow wax (Diderot 1751–57, vol. 11:529).

Stalker and Parker ([1688] 1971) provide the ultimate solution to coloring the gilding by resorting to the hopelessly expensive expedient of layering it over and over with gold amalgam. They suggest, "You may again, quicken it with quicksilver [mercury] and aqua fortis [nitric acid] and guild over and over again after the former method [i.e., the application of the amalgam already described] and repeat it so often as you please, till your gold lies thick as your nail upon the metal." This of course is cheating. By gilding the bronze over and over again the right effect of fine gold would be achieved, but at what cost? This last process was possibly used in the eighteenth century, but probably not by economically minded gilders who wanted to survive in a fiercely competitive business.

Matt Gilding

The last refinement in the gilding process is that of matt gilding.[4] Diderot does not discuss this process, probably because the technique was not introduced to the Parisian gilt-bronze trade until the last quarter of the eighteenth century. Matt gilding is actually another form of coloring; it imparts a dull sheen by breaking up the surface texture, providing a pleasant contrast with the high polish of the burnishing. Matt gilding produces a more delicate surface than can be achieved mechanically by chasing, which breaks up the surface by the application of patterned punches.

The detail of the foot and handle of a vase in the Victoria and Albert Museum, dating from the 1780s, demonstrates the subtlety of this finish (Figs. 4, 5). The modeled areas of the handle are matt gilded, enhancing the naturalistic qualities of the sinewy ivy branch and the arrangement of the leaves (Fig. 4). The lions' heads on the foot of the vase (Fig. 5) are matt gilded to enhance the sculptural quality of the lions' angry, growling expressions. Contrast is achieved through mirrorlike burnishing of the plinth. On a tiny scale, these mounts show the use of the most delicate, satiny matt gilding as well as the most brilliant burnishing, which together enliven the vase's sculptural detail.

According to Goodison, matt gilding is achieved by the action of chlorine etching the surface of the gold. One of the recipes in Boulton's notes for achieving what he called the "dead colour" contains "2 oz sal petre, 1 oz allum, and 5 pennyweights salt." Goodison interprets this as the salt (sodium chloride) and the "sal petre" (potassium nitrate) dissolving the gold and then redepositing it by "allum" (potassium aluminum sulfate; 1974:71), the essential role being played by the chloride.

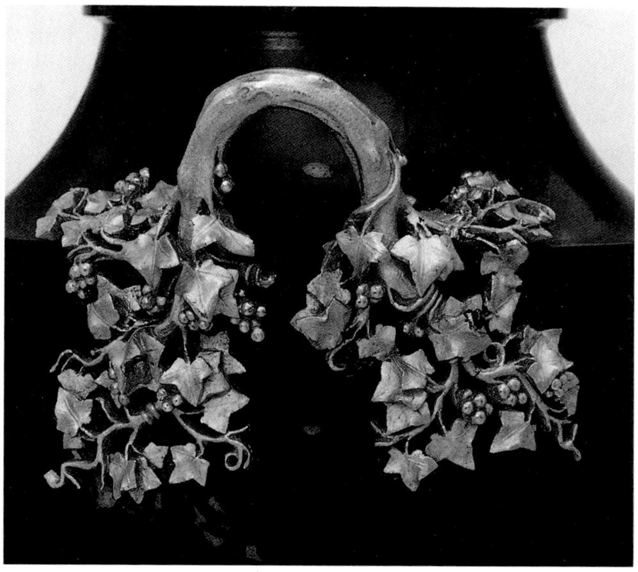

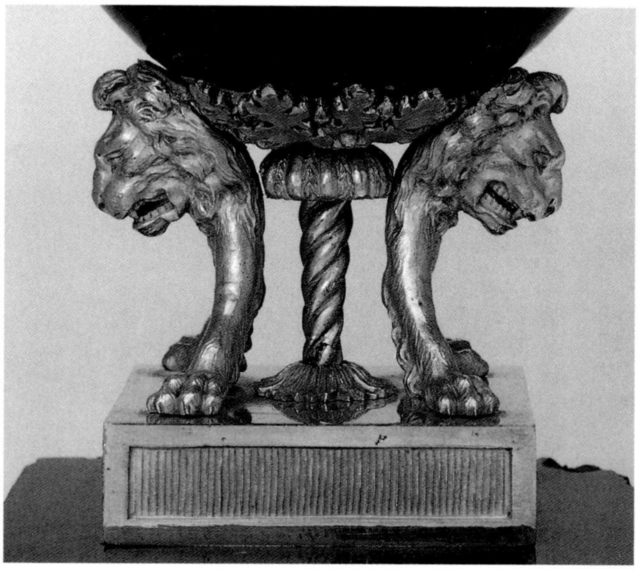

FIGURE 4. *Handle of a vase, detail, above left. Black marble and gilt bronze, France, ca. 1780. This magnified detail shows the use of matt gilding to enhance the naturalistic qualities in the modeling of the ivy branch and leaves. Jones Collection, Victoria and Albert Museum, London, inv. no. 1144-1882.*

FIGURE 5. *Foot of the vase shown in Figure 4, detail, above right. Note both matt and burnished gilding; matt gilding is applied to the highly modeled faces of the lions, while mirrorlike burnished gilding is employed on the plinth.*

D'Arcet learned in 1818 that a similar substance known as *mat* was sold ready-mixed to Parisian gilders. He found it to contain 40% saltpeter (potassium nitrate), 25% alum (potassium aluminum sulfate), and 35% sea salt (sodium chloride; d'Arcet 1818:65[i]). In the early nineteenth century matt gilding was widely used by the large bronze-making firms, and it is present on many so-called Empire bronzes. However, the finish on these objects lacks the vitality observed in the finest pieces dating from the late eighteenth century. The tendency toward a more mechanical finish was augmented by the invention of electrogilding in the 1840s, which produced a perfectly even and therefore rather bland surface.

Conclusion

We look back to the work of the Parisian gilders of the late eighteenth century for the most beautiful gilding effects. As Diderot said, most of these were achieved by drawing on traditional techniques. It was perhaps less the invention of these diverse methods and more the strenuous application of them that produced the most beautiful objects of the period. Most of the chemistry, whether it was understood or not, had been already devised by previous generations dating as far back as the ancient world. But the Parisian gilders were inquisitive and resourceful enough to gather up all these processes and use them to their greatest advantage in producing the most delicate and sophisticated gilt bronzes.

Notes

1. Proportions vary from account to account, but this seems to be the most frequently cited figure. See d'Arcet 1818:viii.
2. By X ray performed by Graham Martin, all pieces tested were copper-zinc alloys.
3. While German gilding was acknowledged as being more brilliant, the silver was considered inferior, as it was alloyed with more copper than French silver. See Diderot 1751–57:vol. 9:529.

4. Until quite recently it was often claimed that matt gilding was the invention of the great *bronzier* Pierre Gouthiere (1743–1813), who was at the zenith of his career during the 1770s and early 1780s (Verlet 1987:174). Though he frequently exploited the beautiful effect of matt gilding, the technique seems to have existed earlier in the repertoire of the medalist and goldsmith.

References

D'ARCET, J. P. J.

1818 *Memoire sur L'Art de Dorer le Bronze.* Paris: Veuve Agasse.

CHAMBERS, E.

1728 *Cyclopaedia: or, an Universal Dictionary of Arts and Sciences.* 2 vols. London: Printed for W. Innys et al.

DIDEROT, D., AND J. L. D'ALEMBERT

1751–57 *Encyclopédie ou Dictionnaire Raisonné des Sciences, des Arts et des Métiers,* vols. 5, 11. Paris: Chez Briasson.

GOODISON, N.

1974 *Ormolu, the Work of Matthew Boulton.* London: Phaidon.

GREGORY, G.

1806–7 *Dictionary of Arts and Sciences.* London: R. Phillips.

HATCHET, S.

n.d. *L'Industrie, Exposition de 1834.* Paris: Henry Dupuy.

SMITH, G.

[1755] 1770 *The Laboratory: or school of arts.* Reprint. London: Printed for James Hodges.

STALKER, J., AND G. PARKER

[1688] 1971 *A Treatise of Japanning and Varnishing.* Reprint. London: Alec Tiranti.

VERLET, P.

1987 *Les Bronzes Dorés Francais du XVIIIe Siecle.* Paris: Picard.

Biography

Martin Chapman is curator of European Decorative Arts and the Gilbert Collection of Gold and Silver at the Los Angeles County Museum of Art. From 1980 to 1990, he was a curator in the Department of Metalwork at the Victoria and Albert Museum, where he specialized in the collection of French gilt bronzes. In 1983–84, he was a fellow in the Department of Decorative Arts at the J. Paul Getty Museum, and from 1974 to 1980 he served as assistant curator in the Department of Prints and Drawings at the Victoria and Albert Musuem. He has lectured and published widely on the subject of French gilt bronzes.

Production and Restoration of Nineteenth-century Zinc Sculpture in Denmark

KNUD HOLM

Danish productivity of cast statuary between approximately 1840 and 1940 is rather remarkable, at least from the standpoint of Denmark's cultural history. During the seventeenth and eighteenth centuries, Denmark, like several other small European countries, had to turn to France whenever a major sculptural project was under consideration and a suitable sculptor and skilled founder needed to be found.

The fact that the work was done by French artists or artisans, however, does not necessarily mean that it was fabricated in France. One of the foremost examples of French sculpture in Denmark is the eighteenth-century equestrian statue of King Frederik V at the Amalienborg Castle by the sculptor Jacques-François-Joseph Saly, cast by Pierre Gor in Copenhagen (Holm 1981). This statue took seventeen years to complete. During that time, Saly became director of the Royal Danish Academy of Fine Arts. Both Saly and Gor subsequently left Denmark. Another well-known French statue is that of King Christian V on Kongens Nytorv in Copenhagen, sculpted by Abraham César L'amoureux (d. 1692), assisted by his brother, Claude. This statue was cast in lead in Copenhagen and gilded.

Early Discoveries in Sculpture Casting

In Denmark today, sculpture casting is performed by only a few small workshops; most sculptures are sent to foundries in Italy for casting. During the period from about 1840 to 1940, however, a number of enterprises accomplished this work in Denmark. Yet only one, the firm owned by Lauritz Rasmussen, extended its activities beyond the death of its founder. This establishment operated for three generations and closed finally in 1967. The first sculpture-casting workshop in Denmark, founded by Jørgen Balthasar Dalhoff (1800–90), is also perhaps the most interesting.

Dalhoff was a goldsmith who took up sculpture casting in bronze, zinc, and terra-cotta. He was very quick to employ new techniques and new materials. During

Dalhoff's time, zinc was considered a rather new material, at least as far as sculpture casting was concerned. Dalhoff's first zinc statues were made in 1841–42.

Zinc played another important role related to the manufacture of statuary: It was an essential element in the discovery of electrical current (c. 1800). The Italian physicist Alessandro Volta demonstrated that piling up multiple, small, galvanic elements based on copper and zinc results in a lasting source of electricity. Since then, zinc has been used as a common component of electrical elements.

One of the earliest technical applications of this discovery led to another discovery: the electrolytic deposition of copper or other metals on a surface through the action of an electric current (1837). The reduction of metals from a solution was known early in the century, but the production of a coherent layer of copper was quite new. This led to the development of electrotyping by M. H. Jacobi, who published his work in St. Petersburg under the German title *Die Galvanoplastik* in 1840. The same year, Dalhoff exhibited some small works made by electrotyping. He did not continue to pursue the process, however. Other workshops took it up, but as far as can be determined, the production of large-scale objects never became their principal business.

A number of statues made in copper by electrotyping can be found in Copenhagen. Many are copies of figures from classical antiquity, such as those situated in Copenhagen's Botanical Garden. These statues were produced in Germany and presented to the city of Copenhagen by the brewer Carl Jacobsen, either as a private donator or through his foundations—the Albertina Foundation (named after Bertel "Alberto" Thorvaldsen) and the New Carlsberg Foundation.

Another new material was found at the beginning of the century: aluminum. In 1825 the Danish chemist and physicist Hans Christian Ørsted (Oersted) was the first to produce a sample of the metal in the laboratory. He was followed by Friedrich Wöhler in 1827. A purer aluminum was obtained by another reducing method devised by Henri Sainte-Claire Deville in 1854 (Rancke-Madsen 1984).[1] This innovation made it possible to produce aluminum utensils and other objects. Aluminum was extremely expensive at first, even more costly than platinum. Nonetheless, Dalhoff made an aluminum helmet for King Frederik VII, which is now in the Chronological Collection of the Danish Kings at Rosenborg Castle in Copenhagen.

The St. Ansgar Statues

The zinc statues cast by Dalhoff in 1841–42 were made by sculptor Gotthilf Borup (ca. 1800–79) to decorate the facade of the new Catholic Church of St. Ansgar in Copenhagen. Five biblical figures—Isaiah, Moses, David, Aaron, and Elijah—stand in individual niches on the upper front of the church, and two kneeling angels are positioned at the base of the roof. These pieces are among the few of Borup's works that still exist. Borup's most important work is said to have been a large pedimental group on the front of the second Christiansborg Castle in Copenhagen, cast in terracotta by Dalhoff. The original sketch for the piece was made by Thorvaldsen. Borup received the commission after the death of another sculptor, Hermann E. Freund,

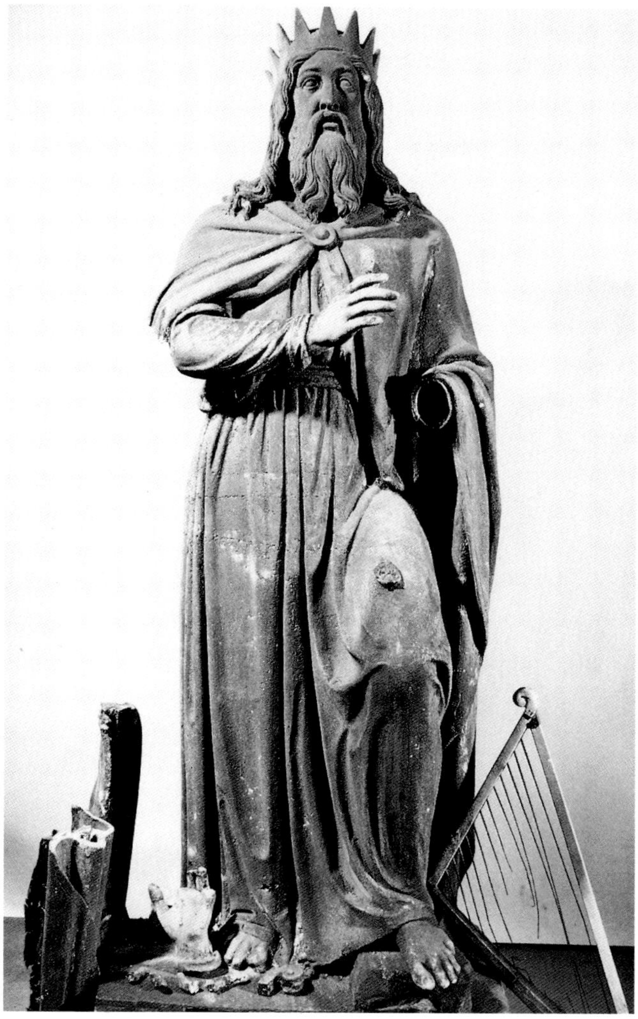

FIGURE 1. *David by G. Borup, cast in zinc by the lost-wax method by J. B. Dalhoff. St. Ansgar Kirke, Bredgade, Copenhagen. Before restoration.*

who was also Thorvaldsen's pupil. Christiansborg Castle burned down in 1884, and only fragmentary remnants of this sculpture remain today. In 1852 Borup left Denmark for the United States and settled in New Haven, Connecticut, where he remained until his death. His work in America, if any, is unknown.

The St. Ansgar figures were cast in zinc by the lost-wax method (Fig. 1), which had been known since prehistoric times. It was developed and perfected during the Bronze Age, as exemplified by Greek sculpture, Nordic lures, and other artifacts. In later years the lost-wax method became the principal technique for casting large sculptures in one piece, such as the many equestrian statues erected in Italy and France during the seventeenth and eighteenth centuries.

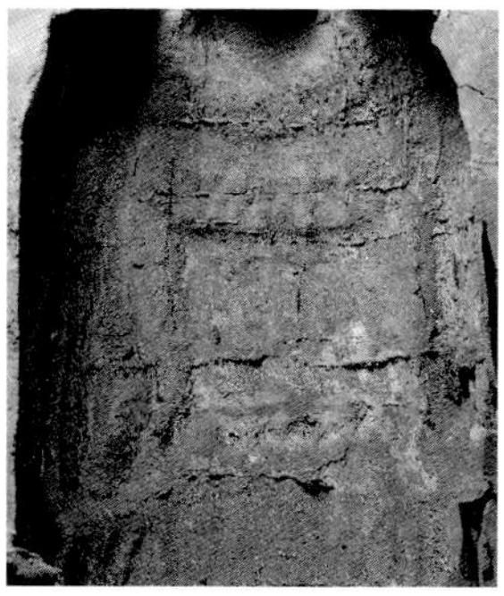

FIGURE 2. *Detail of the inside surface of one of the statues from the church of St. Ansgar.*

Before producing the St. Ansgar statues, Dalhoff traveled to Berlin as well as to Nuremberg, Munich, and Vienna to study casting techniques. He also scheduled a visit to a large foundry in Lauchhammer, Silesia, in east-central Europe. Dalhoff must have known, however, that it would be very difficult to obtain all the technical information he needed, since most foundries jealously guarded their secrets. He evidently did not find out until later that the lost-wax method is not very suitable for zinc casting. Zinc easily develops a number of cracks when it cools and solidifies, since the inner core of the hollow cast prevents the metal from shrinking (Fig. 2). In addition, Borup's models were delivered to Dalhoff in clay, not plaster, so they were only meant to be used once (Dalhoff 1915–16:258). Models are not necessarily

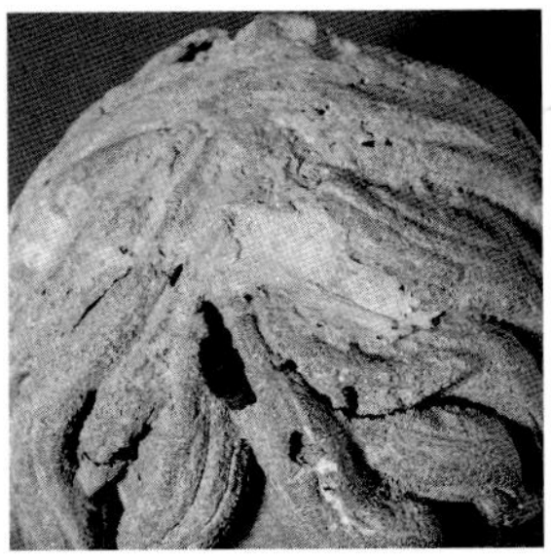

FIGURE 3. *Detail of the head of Elijah showing cracks and holes, church of St. Ansgar.*

destroyed immediately during the process of making molds, but when the lost-wax method is used for large castings, the foundry worker almost always tries to salvage the results of the first casting. Even an extensive repair effort is often easier than starting all over again.

Each figure was cast in two major pieces, an upper half and a lower half. The heads were cast separately, as were the hands and the figures' various symbolic attributes. For these reasons, the casting of the five St. Ansgar figures was far from perfect (Fig. 3). However, because the statues were designed to be painted, nothing prevented Dalhoff from using lead for the bigger repairs and linseed-oil putty for the smaller. Even concrete was used to keep the hand of one figure in place.

Installation of the statues was similarly problematic, since the niches in the church facade were not deep enough to accommodate the cast figures. It was not feasible to make the niches deeper, so the backs of all the figures were cut away along with a portion of their square bases. This operation seriously impaired the stability of the work.

After the St. Ansgar figures, Dalhoff went on to cast a number of bronze statues in Copenhagen, including four colossal statues at Christiansborg Castle (originally placed in niches on the facade of the former castle) and the quadriga on top of the Thorvaldsen Museum.

The Restoration

When restoration was initiated, the St. Ansgar statues were found to be very dirty. Their interiors were partly filled with pigeon dung and nest material. Some parts of the figures were missing. When the statues were removed from the building, special precautions had to be taken to prevent them from falling apart. The statues did not suffer extensive surface corrosion, although they were exposed to exhaust fumes from the city's rather heavy traffic. Their recessed niches sheltered them from wind and heavy (acid) rain.

The statues were cleaned with water using a high-pressure water jet and careful sandblasting. Loose parts were soldered on with lead-tin solder.[2] An inner armature of stainless steel was also tailored to fit each figure. For reinforcement, ribs of zinc were cut to shape from a 7-mm-thick plate and soldered to the inner surface of the zinc at appropriate points. One of the heads was cast so thinly and had so many cracks and holes, that it was necessary to reinforce it by adding a layer of zinc to the inside. This was done using the slush-casting method.

In slush casting, molten metal is poured into a hollow form and then poured out again afterwards, leaving only a thin layer of solidified metal along the walls of the form. To restore the shape of a head that has a number of gaps, it is necessary to cover the head with a shell, which could consist of aluminum foil, clay, and gypsum. The head must be cleaned inside by sandblasting. Before receiving the molten metal, the shell must be dry and the whole head heated to some degree. The old and new metals will not necessarily combine very well, and it is often wise, therefore, to solder the joints from the outside. In cases such as this, a tin solder with a resinous core is advisable. Larger gaps should be modeled before making the shell.

The greatest challenge to the restoration of the St. Ansgar figures lay in reconstructing the proper shapes of two of the figures' attributes: David's harp, which was incomplete, and Aaron's censer, which was missing. The missing parts of the harp had been replaced earlier with a very poor imitation of the instrument; and the censer, which had originally been in Aaron's hands, required an appropriate prototype from which to create a new model. Old photographs did not provide much detail for these objects, so new ones were created and cast in zinc. Since the original statue was made of zinc, there was no reason to introduce other materials for the missing parts. Some existing parts, however, were originally made of other materials. For example, the harp strings were presumably made of steel wire that rusted away long ago; these were replaced with stainless-steel wire. The molds were made with gypsum (Fig. 4).

A few small remnants of paint and a couple of old pictures of the statues indicated that the figures were originally painted green to resemble the natural patina of bronze. To restore their original green color, the statues were painted using a modern acrylic PVC paint called ICOSIT-Dickschicht 5530 (Lechler Chemie GmbH, Stuttgart, Germany). The paint was sprayed on with airless spraying equipment. No primer was used.

The two angels by Borup have not yet been restored. Dalhoff used a different casting technique for these figures, which are made from a number of pieces that were soldered together. Therefore, it appears that Dalhoff adopted the sand-casting method, which became the standard and most readily available technique for producing zinc sculpture, at least when only a few items were needed. Sand casting originated and became widespread primarily for the casting of iron. Eventually sand casting supplanted the need for lost-wax casting of bronze as well, especially in Germany and Denmark. However, lost-wax casting never died out completely, and currently seems to be regaining some degree of popularity. Dalhoff's first zinc castings remain an exception for being cast using the lost-wax method during the 1840s in Denmark.

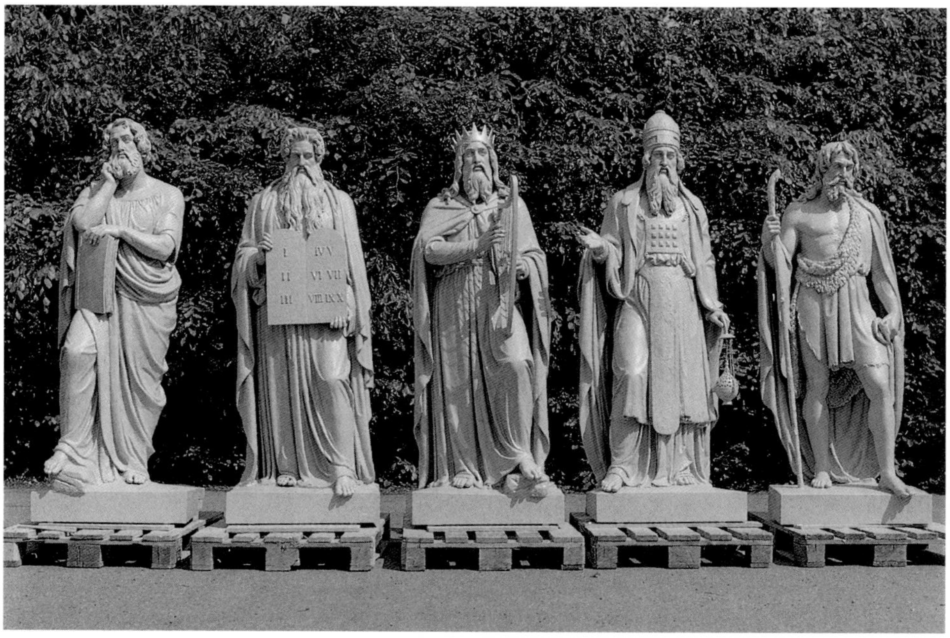

FIGURE 4. *The five statues by G. Borup. From the left: Isaiah, Moses, David, Aaron, Elijah. After restoration.*

Once Dalhoff had introduced the casting trade to Denmark, other foundries began to appear, with sand casting as the principal method. They included those of Thomas Christian Thomsen, Carl Frederik Holm, Lauritz Rasmussen, Vilhelm Bissen, and Oluf Schieltved. Christian Lundqvist is known to have cast at least one statue.[3] All of these foundries cast other objects as well.

The largest example of zinc casting in Denmark was a group of figures created for the new Royal Theater (Fig. 5) by the sculptor Ferdinand Edvard Ring (1829–86). The theater opened in 1874, but the building itself was not completed until 1883. In 1878 Ring's sculpture was placed on top of the facade. The original group of figures was 6.5 m high and about the same width. It depicts Apollo and Pegasus at the Hippocrene spring along with Thalia and Melpomene, the muses of comedy and tragedy. Holm was at first commissioned to have cast this group, but it seems that Rasmussen may have cast it (Clemmensen 1929).[4]

In 1959 the Ring sculpture was dismantled and taken down, since the iron armature inside the casting had become severely rusted. The work was stored in various places and was eventually considered beyond repair. A replica was made, this time in bronze, by the Italian founder Fonderia Artistica Battaglia in Milan. In 1987 the replicated group was returned to its location at the Royal Theater. Fragments of the original statue were sold to finance the replication. Unfortunately, possibilities for restoration of the original were never explored and cannot now be determined.

Another building of interest in regard to nineteenth-century zinc casting is the so-called Marble church, or Frederiks Church, in Copenhagen (Nationalmuseet 1933). This church was planned as a landmark in Frederiksstaden, the new part of Copenhagen named after King Frederik V. The king laid down the foundation stone in 1749, and building was started later the same year. In the following years, plans often changed, including the choice of materials and the architect. In 1770 funds ran

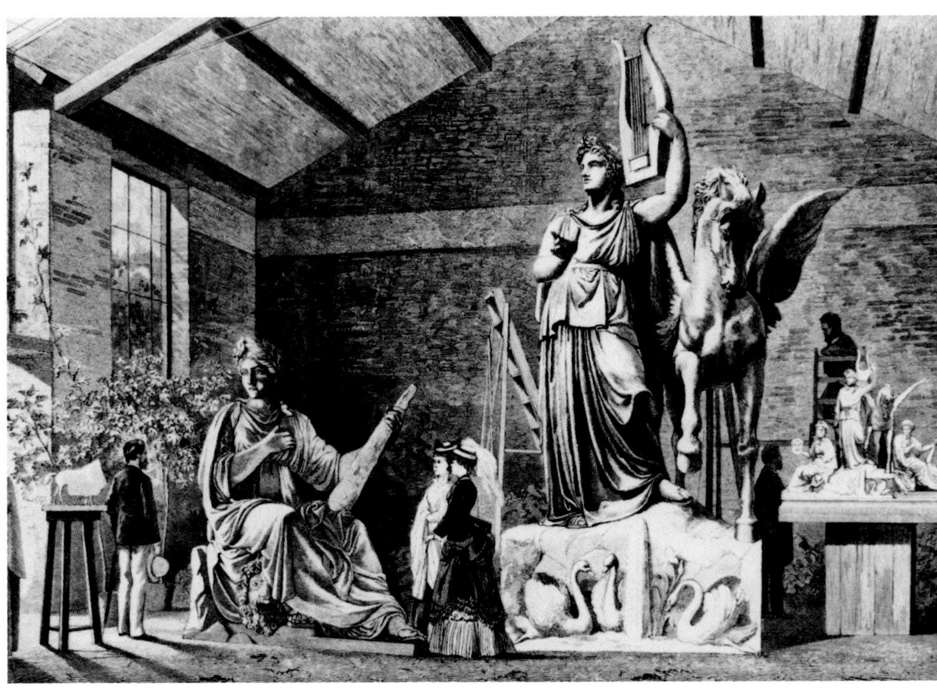

FIGURE 5. *Illustration from the periodical* Illustreret Tidende *(1874) showing the group of figures for the Royal Theater in Copenhagen in the studio of Ferdinand Edvard Ring.*

out and the project was canceled. For more than a century the unfinished building stood as a picturesque ruin in the middle of the otherwise fashionable town quarter. Many proposals for renovation were introduced. Finally in 1874 the ruin was bought by Carl Frederik Tietgen, a finance magnate and bank director who hired the architecture professor Ferdinand Meldahl to build a church on the existing foundations.

A number of statues stand at ground level around the church. However, eighteen large statues made of zinc, which are placed halfway up the building are of prime importance here. Tietgen was very interested in the design and in the choice of figures for the church. He commissioned all the figures and had them produced between 1883 and 1885 by ten different sculptors (Nationalmuseet 1933) and four different founders. The sculptors were Otto Evens (1826–1895), Georg Christian Freund (1821–1904), Johannes Gelert (1852–1914), Aksel Hansen (1853–1933), Johannes Hoffmann (1844–1920), Carl Christian Peters (1822–1899), Ferdinand Edvard Ring (1829–1886), Carl Rohl Smith (1848–1900), August Saabye (1823–1916), Theobald Stein (1829–1901), and Bodelsen and Engelstoft (1947–52).

Gelert went to Chicago in 1887 and to New York in 1898. Rohl Smith emigrated to the United States in 1886 and soon established himself as a sculptor of public monuments. He created a bronze statue of Benjamin Franklin in Philadelphia and another of General William T. Sherman on horseback, which stands in Washington D.C. The Iowa State Soldiers and Sailors Monument in Des Moines and a statue on a street corner in Chicago that commemorates the massacre of American Indians in 1872 are also by Rohl Smith.

The four founders who cast the figures for the Marble church were Holm, Lundqvist, Rasmussen, and Schieltved. The eighteen statues they produced are currently in poor condition and require restoration. Their marked surface corrosion is clear in comparison to the almost uncorroded surfaces of all the solder seams (Figs. 6–9). It is interesting to note that in a written report in 1968 an engineering firm that inspected the statues stated that the figures showed no sign of crumbling, loosening,

FIGURE 6. *Figure of Luther by Theobald Stein, below left, cast in zinc by C. F. Holm. Frederiks Kirke, Frederiksstaden, Copenhagen.*

FIGURE 7. *Back of figure of Athanasius by Carl Rohl Smith, below center, cast in zinc by O. Schieltved. Frederiks Kirke.*

FIGURE 8. *Detail, far right, of one of the figures from Frederiks Church showing tin-solder seams standing out against the corroded zinc surface.*

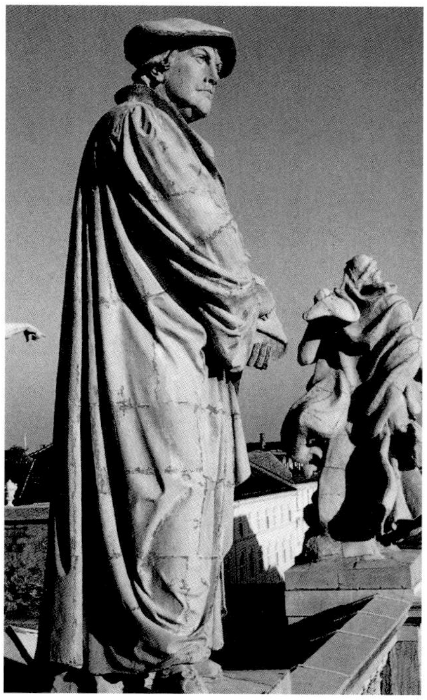

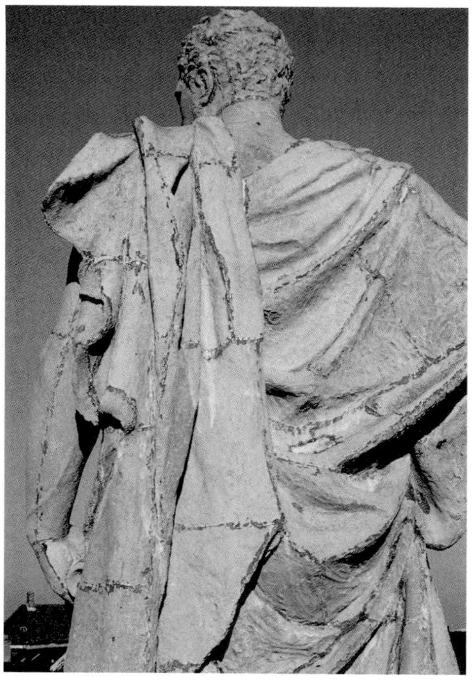

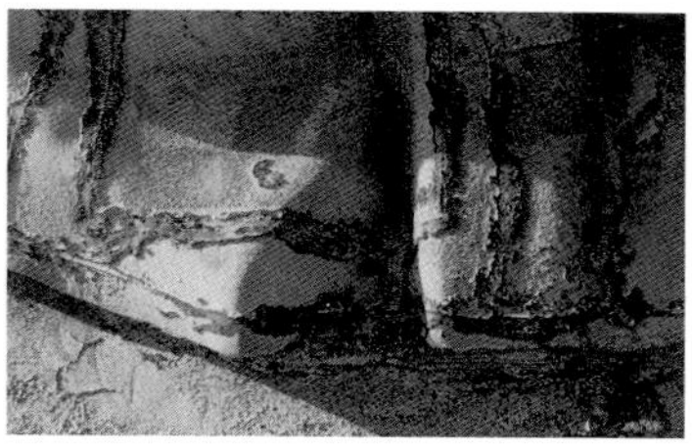

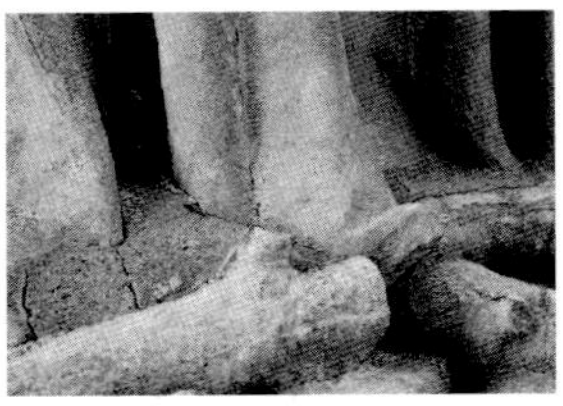

FIGURE 9. *Detail of one of the figures from Frederiks Church showing places where water could be trapped by the shape of the figure. These areas have been filled with mortar to prevent damage from freezing water or to prevent water from seeping in through cracks and holes.*

or development of cracks. The statues still carry traces of paint, indicating that they were finished to resemble sandstone. The most common colors used for this purpose were brown-bronze or yellow or whitish sandstone imitation. Analysis of the patinas has not yet begun, and research is still underway concerning their treatment histories. It is uncertain, therefore, whether these statues have been painted since 1884, but they may have been refinished during an earlier restoration in 1938–39.

The eighteen statues commissioned by Tietgen for the Marble church are unique. In addition to their inherent aesthetic qualities, they stand out in respect to the vast majority of zinc castings that were made in quantity during the same period and could be selected from catalogues.

Conclusion

The many zinc statues cast in Denmark between about 1840 and 1940 now require restoration and repair, work at which the National Museum of Denmark has acquired some experience. The restoration and conservation of outdoor zinc sculpture is in some ways not so problematic as that of bronzes, simply because most of the zinc statues were meant to be painted, and proper maintenance of them has taken that fact into account. However, some zinc statues have been left to form a natural mineral crust, or patina. Generally speaking, zinc does not corrode as much as might be expected from its electrode potential. This is due to the formation of a thin layer of hydroxides and carbonates or, when heated, oxides. Today the increased acidity of the precipitation is an additional threat to those zinc statues that are not protected by a surface coat.

Most of the restoration work that has been carried out has had to deal with structural problems, such as disintegrating inner armatures and the renewal of long-absent surface coatings. Traditional oil paints do not last on zinc. Modern surface-coating materials with better bonding properties on zinc are commercially available, however. The technical and historical study of the Danish sculptures of different periods will continue to contribute to the conservation of each statue.

Notes

1. The English translation of Rancke-Madsen's book (1984) is "The History of the Discovery of the Elements." References to Sainte-Claire Deville and his work can also be found elsewhere in the chemical literature. J. R. Partington mentions it briefly in his *A Short History of Chemistry*, although he fails to mention Oersted in connection with aluminum. Rancke-Madsen describes the method used by Oersted to produce aluminum. This paper was originally published in "Oversigt over det kongelige Danske Videnskabernes Selskabs Forhandlinger, 1824–25," in the proceedings of the Royal Danish Academy of Science and Letters 1824–25. Oersted was the secretary of the academy at this time. A sample of the metal was presented before the academy on April 8, 1825, and is also described in "H. C. Oersted's Scientific Papers I–III" (1920) collected edition, including two essays on his work by Kirstine Meyer, vol. 2:463–66, Copenhagen.

2. One important technique in soldering zinc is the use of a hydrochloric-acid solution (15% in water) instead of an ordinary commercial flux, whether water, resinous, or other type. The solder was lead-tin solder of different composition. The sand used in sandblasting was an ordinary quartz sand. In some other cases, glass beads were used.
3. Thomsen was an apprentice of Dalhoff, Rasmussen of Holm. It is not known where Holm learned to cast statues. He was an apprentice of a brazier by the name of Løchte, but as far as we know, Løchte never made statuary. Bissen was primarily a sculptor. He established a foundry in order to cast his own statues but employed French foundry workers.
4. The group was originally planned to be cast by Holm, according to a contemporaneous press source in Copenhagen (*Illustreret Tidende* 1874: vol. 15, no. 783), but archive documentation fails to confirm this or to provide the name of anyone else who cast this group. A 1929 book (Clemmensen) celebrating the seventy-fifth anniversary of the Lauritz Rasmussen firm, however, mentions that the group was cast by Rasmussen. Although this source is not infallible, it is probably safe to surmise that it is correct in this case.

References

BODELSEN, M., AND P. ENGELSTOFT, EDS.

1947–52 *Weilbachs Kunstnerleksikon*. Copenhagen: Aschehoug Dansk Forlag.

CLEMMENSEN, C. A.

1929 *Lauritz Rasmussen. Kgl. Hof-bronzestøber 1854, 2 November 1929*. Copenhagen: Langkjærs Bogtrykkeri.

DALHOFF, N.

1915–16 *Jørgen Balthasar Dalhoff. Et Liv i Arbejde. Efter hans egne Breve og Optegnelser skildret af hans Søn*. Copenhagen: G.E.C. Gad.

HOLM, K.

1981 *Saly's Rytterstatue of Frederik V. Statuen og lidt om dens bevaringstilstand*. Nationalmuseets Arbejdsmark. Copenhagen: Nationalmuseet.

HOLM, K., AND O. B. JENSEN

1988 *Metalskulpturen i forrige århundrede. Lidt om mænd og materialer*. Nationalmuseets Arbejdsmark. Copenhagen: Nationalmuseet.

JACOBI, M. H.

1840 *Die Galvanoplastik*. St. Petersburg.

NATIONALMUSEET

1933 *Danmarks Kirker*. Copenhagen: Nationalmuseet.

NYROP, C.

1888 *Jørgen Balthasar Dalhoff. Et Afsnit af den danske Kunstindustris Historie*. Tidsskrift for Kunstindustri. Copenhagen: Industriforeningen i København.

RANCKE-MADSEN, E.

1984 *Grundstoffernes Opdagelseshistorie*. Copenhagen: G. E. C. Gad.

ROSTRUP, H.

1945 *H. W. Bissen*. Copenhagen: Kunstforeningen i København.

Further information derived from private and public archives in Copenhagen.

Biography

Knud Holm is chief conservator and former keeper of the Department of Conservation 2 at the National Museum of Denmark, where he has worked since 1951. Holm studied chemistry, physics, and radiography at the University of Copenhagen; metallography at the Technical University of Denmark; and conservation of antiquities at the National Museum. His specialty is metals conservation. He has served as a member of International Institute for Conservation, the ICOM Committee for Conservation, and the Ministry of Cultural Affairs committee reporting on the organization of conservation in Denmark and the establishment of a school for conservation. He has written numerous articles on cultural history, conservation, and the history of technology.

Real-time Survival Rates for Treatments of Archaeological Iron

SUZANNE KEENE

> I there behelde the bones of a man . . . and round about him . . . such nailes were found, wherefore I coniectured them to be the nailes of his coffin. . . . I caused some of the nayles to bee reached vp to mee, and found under the broad heades of them, the olde wood, skant turned into earth. . . . I reserued one . . . but the nayle lying drie, is by scaling greatly wasted.
>
> —John Stowe, *Survey of London,* 1603

Since the 1970s, a series of excavations in London has uncovered a mass of archaeological evidence for life in the capital from pre-Roman times onward. During the short-lived property boom of the mid-1980s, there was an exponential increase in new building and redevelopment in the city of London. The museum was fortunate in persuading many developers to fund excavation on these developments. It was found that progressive land reclamation and revetment along the north bank of the Thames, the main trade route, had resulted in large, stratified dumps of rubbish from the Roman and medieval city. The waterlogged and anaerobic conditions in these sites ensured the survival of the objects contained there, including an unparalleled number and range of objects made of iron.

The Nature of the Problem

The stabilization of archaeological iron presents difficult problems. Iron is a relatively reactive metal, and in the presence of an electrolyte such as ground water, it mineralizes to a greater or lesser extent. The minerals formed will vary depending on burial conditions (Turgoose 1982b). In aerobic conditions, objects often corrode out entirely, although their form can remain embodied within the corrosion products (and can sometimes be recovered by skilled conservators). In anaerobic conditions, corrosion may proceed to a much lesser extent. Such objects often appear extremely well preserved and are sometimes assumed to be stable. Experience has shown, however, that such is not the case.

Whatever the apparent condition of the objects when excavated, they actually consist of a highly complex metal-and-mineral system. Many of the minerals are only stable within the particular burial environment—with its specific redox potential, pH, and so on. When brought above ground, these conditions no longer exist, and different minerals begin to form, due either to changes in the original minerals, or to fresh corrosion at the metal-corrosion interface (Turgoose 1982b).

Because they are so well preserved, these objects are an excellent source of evidence for their manufacture and use. London iron is often decorated by tooling or shaping; it can be tinned or inlaid; tool marks from manufacture or sharpening are preserved in surfaces. But unfortunately for the conservator, most of this detail is contained in the mineral layers, as John Stowe observed nearly 400 years ago. If objects recorrode, fresh mineralization at the metal surface pushes off these delicate surfaces, leaving only the familiar powdery red rust.

Existing Knowledge

The work of Turgoose (1982a, 1982b, 1985) and others (Gilberg and Seeley 1982) has done much to explain the corrosion mechanisms sustained by archaeological iron after excavation. However, developments in treatments to stabilize this material have lagged behind theoretical understanding to some extent.

The "Holy Grail" of iron conservators has been removal of every trace of chloride from the corrosion and metal. Until the 1970s, there was a meager choice of treatments for iron: boiling in purified water, reduction using electrolysis, or soaking in sodium carbonate. These methods were widely found to be unsuccessful; objects treated in this way frequently recorroded within a few years. Other methods have been more recently introduced: hydrogen reduction, chemical reduction, and gas plasma reduction. These are very aggressive treatments, however, and all of them place at risk the delicate corrosion surface of the objects or the metallurgical evidence that they embody, both of which are prime reasons for preserving the objects. Some of them rely on heat, which compromises the metallurgical evidence (Tylecote and Black 1980).

Other measures that do not depend on chloride removal are desiccated storage, storage in an atmosphere of vapor-phase inhibitor (VPI), and storage in atmospheres of nitrogen or with an oxygen scavenger. However, treatments not involving chloride removal, such as corrosion inhibitors, have scarcely been seriously explored.

In the United Kingdom at least, there is a strong body of opinion that treatments for iron are no more effective than simply storing objects at a low relative humidity. Turgoose (1982b) identified 15% RH as the level at which all water will be removed from iron minerals, thus preventing alteration. Therefore, many conservators in the United Kingdom have adopted storage at, or as close as possible to, 15% RH as their preferred procedure for stabilizing this material.

The Reason for This Study

Conservators at the Museum of London were faced with an influx of beautifully preserved but unstable objects, as shown in Figure 1, but no clear professional con-

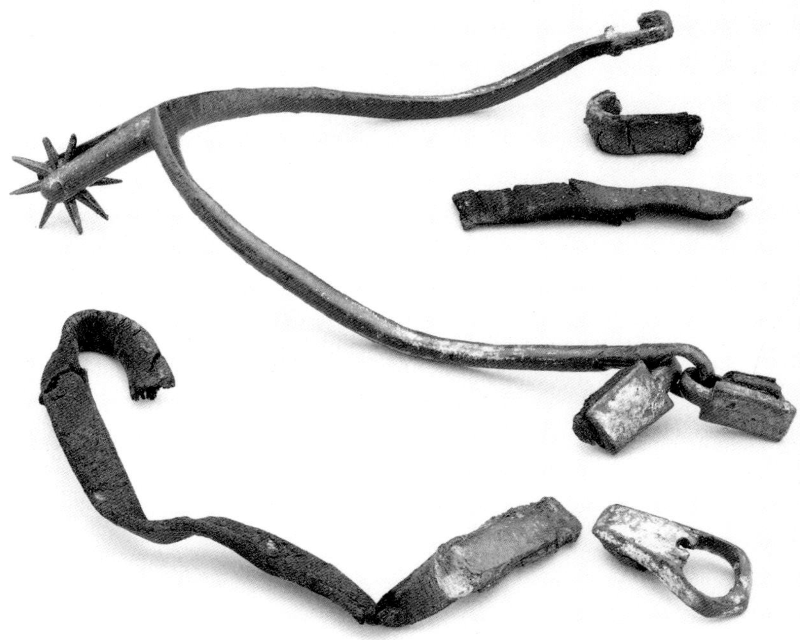

FIGURE 1. *Well-preserved iron objects from Swan Lane, treated using various methods.*

sensus on the most effective treatment for them. Excavation funding included an element for conservation, but this was project-based and only available for a fixed time; decisions could not be deferred.

In 1983–84, in order to gather some objective evidence, a study of iron previously treated by the author was undertaken and published (Keene and Orton 1985). The results of this study suggested that all the treatments that were assessed improved the stability of the objects. Since then, the policy of the Museum of London has been to actively treat iron destined for its collections. The museum also considered it desirable to keep this policy under review. Therefore, in 1983, anticipating an eventual formal assessment, finds from a particular site (Swan Lane, SWA 81) were divided into four batches, and each batch was treated using a different procedure. The objects have since been stored and handled in exactly the same way as are other objects in the museum's collections.

Assessment Design and Procedures

The present study is based on empirical observations of the stability of excavated iron treated in various ways and untreated. The aim was to establish whether any of the treatments that are used confer greater stability than no treatment or storage at RH lower than 15% (desiccated storage).

The sample in the assessment summarized in Table 1 may be described as follows: (a) four batches of iron from the Swan Lane site, treated at one time; (b) iron from the same excavation, not treated and stored in desiccated and ambient conditions; (c) iron from other Museum of London sites, some batches stored in desiccated conditions, some in ambient environments; and (d) iron from completely different areas and types of sites (excavated by the Passmore Edwards Museum), stored using the best practicable means to maintain a low RH.

TABLE 1. *Summary of the data from the assessments of the batches of objects.*

Excavation code and date	*Treatment or storage type*	*Year treated or stored*	*Number in batch*	*Years since treatment or start of storage* 2			3			5			6			7			8			9			13		
				Mis	*Sta*	*Uns*	*Mis*	*Sta*	*Uns*	*Mis*	*Sta*	*Uns*	*Mis*	*Sta*	*Uns*	*Mis*	*Sta*	*Uns*	*Mis*	*Sta*	*Uns*	*Mis*	*Sta*	*Uns*	*Mis*	*Sta*	*Unc*
a. Museum of London, Swan Lane site: treated batches																											
SWA 81	NaOH	1983	24	0	24	0				0	24	0							1	22	1						
SWA 81	Alk. sulf.	1983	19	0	19	0				0	19	0							5	13	1						
SWA 81	Water/VPI	1983	31	7	27	2				4	18	7							3	16	3						
SWA 81	Electrol.	1983	17	6	10	1				3	6	6							2	6	4						
b. Museum of London, Swan Lane site: batches stored without treatment																											
SWA 81	Ambient	1983	31																			0	10	21			
SWA 81	Dessicated	1983	10																			0	3	7			
c. Museum of London, other sites: batches stored without treatment																											
OPT 81	Dessicated	1982	32																			0	23	9			
OPT 81	Ambient	1982	33																			0	13	20			
BWB 83	Dessicated	1984	17													0	9	8									
MOG 86	Dessicated	1986	15							0	2	13															
MOG 86	Ambient	1986	15							0	2	13															
BOY 86	Dessicated	1986	15							0	6	9															
BOY 86	Ambient	1986	20							0	3	17															
d. Passmore Edwards Museum: batches stored without treatment																											
WA/AM 78	Dessicated	1978	164																						0	8	156
BA 85	Dessicated	1985	47										0	10	37												
TF 88	Dessicated	1988	72					0			22		50														
TF 89	Dessicated	1989	26		0			20			6																

Notes: 1. An object is only counted unstable once, and is thereafter omitted from the table.
2. If an object was missing from an assessment but is found at a later one to be stable, it is counted as stable in the earlier assessments.
3. In this table, an object may be considered missing at one assessment but observed as unstable at a subsequent one.
See Table 2 for further details of these points.

The treatment procedures, detailed in Appendix 1, may be summarized as follows:

Water/VPI: Soaking in deionized water with dissolved vapor phase inhibitor (Dichan, dicyclohexyl ammonium carbate, marketed as Shell VPI 260)
Alkaline sulfite: Soaking in changes of alkaline sulfite, followed by changes of deionized water
Sodium hydroxide: Soaking in a dilute solution of sodium hydroxide
Electrolysis and soaking: A short period of electrolytic reduction followed by soaking in water/VPI as above

Objects were allocated to the four treatments in rotation by their accession number; those that were heavily corroded, very delicate, or had tinned or decorated surfaces were not treated with alkaline sulfite nor with sodium hydroxide. Electrolytic reduction was not used as it usually is—to strip away all corrosion to the metal—but rather to loosen corrosion layers in order to make it easier to expose fine detail. In this, it was very successful. The selection for treatment was similar therefore to any typical selection process and was not completely random. A higher proportion of delicate and heavily corroded objects are typically allocated to treatments either by electrolysis or by using water with VPI, although there is no reason to assume that such objects are inherently more or less prone to corrosion. These treatments are summarized in Table 2.

After treatment, the objects were stored at 15% RH in sealed boxes containing silica gel, according to normal practice. Their condition was reviewed in 1985, 1988, and 1991: two, five, and eight years after treatment. The most recent review included untreated objects from the same site stored in the ambient environment but not treated, and a similar sample that had been stored at 15% RH.

To provide further data, objects both from other Museum of London sites and from a completely different context were also examined. A neighboring museum, the Passmore Edwards, has a climate-controlled chamber operated at 18–20% RH. Hygrothermograph records confirm that these conditions were maintained throughout the period of study. Excavated iron has been stored within this chamber in sealed boxes containing completely desiccated silica gel monitored with indicator strips since 1982. The objects have thus been doubly protected. Prior to that time, silica gel in sealed boxes alone was used to create desiccated conditions. The museum is therefore an excellent exponent of the technique of desiccated storage.

In assessing instability, as in the earlier study, an object was graded "unstable" if there was the slightest sign of fresh corrosion or lifting of mineral layers. In the current review of Museum of London iron, microscopic examination was used to settle doubtful cases, and this resulted in regrading a few objects from "unstable" to "stable."

In most cases of treated iron, instability consists of one or two very small flakes or cracks, as shown in Figure 2. Often, however, untreated iron has almost disintegrated, as shown in Figure 3.

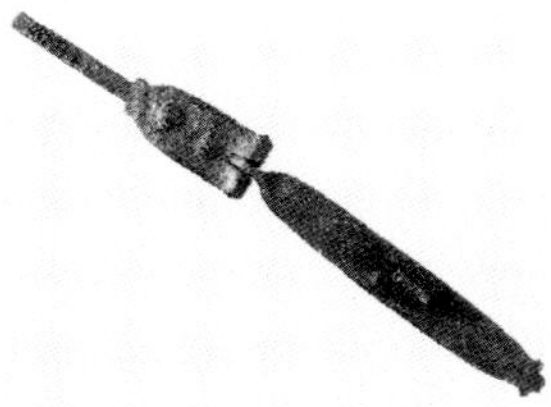

FIGURE 2. *A treated iron object, above, slightly unstable.*

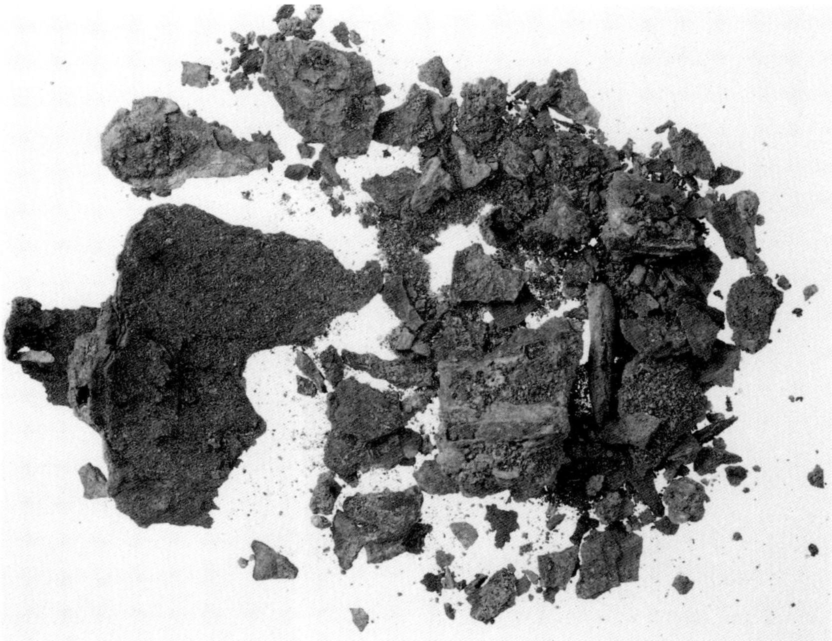

FIGURE 3. *An untreated iron object, right, almost disintegrated.*

Data and Analysis

A total of 588 objects were examined. The data from these assessments are summarized in Table 1. Table 2 shows an example of the data for one batch, which was treated using water and VPI. For each object, the following data were available: year

TABLE 2. *Observations for the batch treated using water/VPI.*

Date of assessment (treated in 1983)	*April 1985*	*October 1988*	*July 1991*
Registration number: SWA 81			
462	m	x	out of trial
463	[o]	[o]	o
468	[o]	[o]	o
470	m	m	x
472	o	[o]	o
474	o	o	o
506	[o]	[o]	o
656	o	m	x
657	o	[o]	o
695	[o]	[o]	o
780	m	x	out of trial
926	o	[o]	o
946	o	m	m
1045	m	x	out of trial
1362	x	out of trial	
1530	x	out of trial	
1591	o	o	o
1739	[o]	[o]	o
1762	[o]	o	m
2230	[o]	o	x
2239	[o]	[o]	o
2240	m	x	out of trial
2313	o	o	o
3068	o	x	out of trial
3079	o	m	x
3091	o	o	o
3273	[o]	o	o
3382	[o]	[o]	o
3388	m	x	out of trial
3863	[o]	[o]	o
3951	m	x	out of trial
Totals:			
Number entering interval (= number from last interval minus number then unstable)	31	29	22
Missing	7	4	3
Newly unstable	2	7	3

Key: m = missing; o = observed stable; [o] = inferred stable from later re-appearance; x = newly unstable of excavation; year treated or storage begun; stable, unstable, or not present on one or more occasions after treatment; years elapsed from start date to examination.

In the precursor to this study (Keene and Orton 1985), the probability of recorrosion and the related parameter of half-life were also calculated. It was not necessary to use this sophisticated technique again, since the new data were considerably simpler. However, complications did arise. On first sight, in fact, it was difficult to determine how to extract all the information these data obviously contained. Batches of objects had been examined at different points in time; some objects were not present for assessment but were found on subsequent occasions when others were missing.

A very similar research situation exists in medicine when clinical trials are undertaken to compare survival rates for different treatments. Patients may survive, or they may die of the disease being treated or from another cause. They may miss checkups, then subsequently return. It was found that the techniques used to assess test results with these kinds of variables could also be applied to the "survival" of archaeological iron. The techniques are described fully in Mould (1981:65–69). Data similar to those available for the iron objects are used to construct "Life Tables." A Life Table for the batch treated using water with VPI is shown in Table 3; the headings summarize the method.

What is calculated is the probability that an object will "survive" (be stable) at any given year after it was treated or stored. The probabilities for different batches of objects can then be compared (Fig. 4, Table 4). These survival probabilities are, in fact, familiar figures, often quoted in press reports for treatments. A probability of 0.33 can be expressed as "one in three will survive beyond *x* years," or as "a 30% chance of survival" (Fig. 5).

Results

The question being examined is: Does treatment confer greater stability than no treatment or than desiccated storage? This can be expressed as a null hypothesis: At a given year *T*, there is no difference between the probability of survival for an iron

TABLE 3. *Life table for the batch treated using water/VPI.*

Year	*Interval number*	*Years since treated*	*Observed unstable*	*Missing*	*Number entering interval*	*Number at risk* (li − ½ wi)	*Prob. of instability at interval* i	*Prob. of surviving*	*Prob. of surviving to Year T*
	i	*T*	*di*	*wi*	*li*	*ni*	*qi=di/ni*	(1−*qi*)	*PT*=(1−*qi*) (1−*qi*) (percent)
1983	0	0	0	0	31	31.00	0.00	1.00	100
1984	1	1	0	0	31	31.00	0.00	1.00	100
1985	2	2	2	7	31	27.50	0.07	0.93	93
1986	3	3	0	0	25	25.00	0.00	1.00	93
1987	4	4	0	0	25	25.00	0.00	1.00	93
1988	5	5	7	4	25	23.00	0.30	0.70	65
1989	6	6	0	0	19	19.00	0.00	1.00	65
1990	7	7	0	0	19	19.00	0.00	1.00	65
1991	8	8	3	3	19	17.50	0.17	0.83	53

FIGURE 4. *Survival probabilities compared for batches of objects treated in different ways, and batches stored untreated. Lines for the stored batches are straight because they are calculated from one assessment in year 9, while those for treatments are the result of three sequential assessments (see Table 4).*

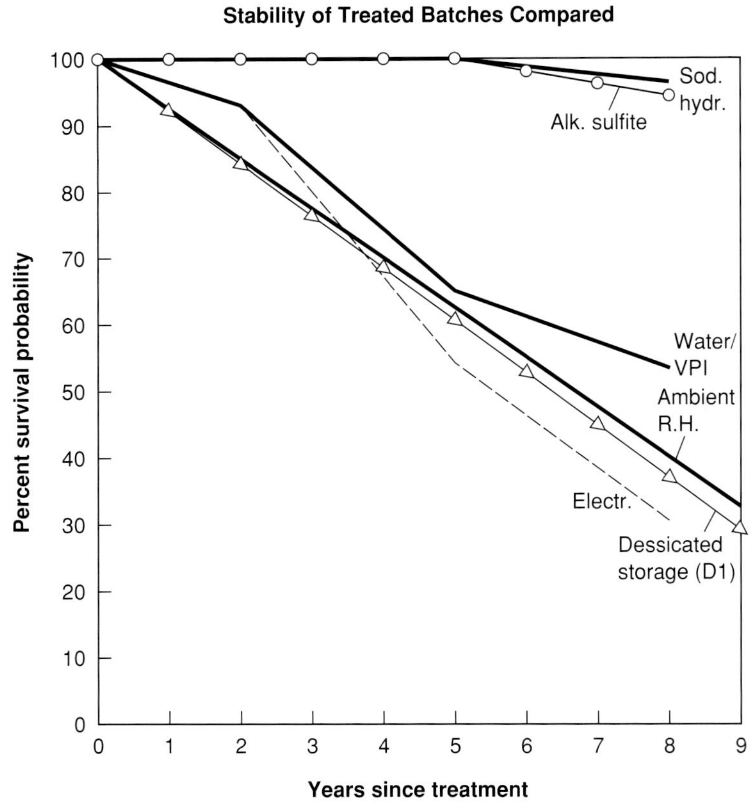

TABLE 4. *Percent survival probabilities at Year* T *for all the assessed batches.*

Year since treatment/ storage (Year T)	*Treated iron: Swan Lane, SWA 81*				*Stored at ambient RH*				*Stored at 15% RH*								
	Sodium hydroxide	*Alkaline sulfite*	*Water/ VPI*	*Electrol./ soak*	*Site codes:* MOG 86	BOY 86	SWA 81	OPT 81	*Site codes:* TF 89	TF 88	MOG 86	BOY 86	BA 85	BWB 83	SWA 81	OPT 81	WA/AM
0	100	100	100	100	100	100	100	100	100	100	100	100	100	100	100	100	100
1	100	100	100	100	—	—	—	—	—	—	—	—	—	—	—	—	—
2	100	100	83	94	—	—	—	—	77	—	—	—	—	—	—	—	—
3	100	100	83	94	—	—	—	—	—	31	—	—	—	—	—	—	—
4	100	100	83	94	—	—	—	—			—	—	—	—	—	—	—
5	100	100	65	52	13	15	—	—			13	40	—	—	—	—	—
6	100	100	65	52			—	—					21	—	—	—	—
7	100	100	65	52			—	—						53	—	—	—
8	96	94	61	26			—	—							—	—	—
9							32	72							30	39	—
10																	—
11																	—
12																	—
13																	5

object, whether it is treated, stored in desiccated conditions, or simply stored in the ambient environment.

The comparative survival probabilities at the different assessments of the Swan Lane batches are compared in Figure 4. Survival probabilities at the end date (year 8) are compared in Figure 6, showing a very considerable difference between the best treatments (soaking in sodium hydroxide and alkaline sulfite), which more than doubled the probability of survival over storage at ambient RH and over electrolysis

FIGURE 5. *The percentage of stable objects compared for batches of untreated stored objects.*

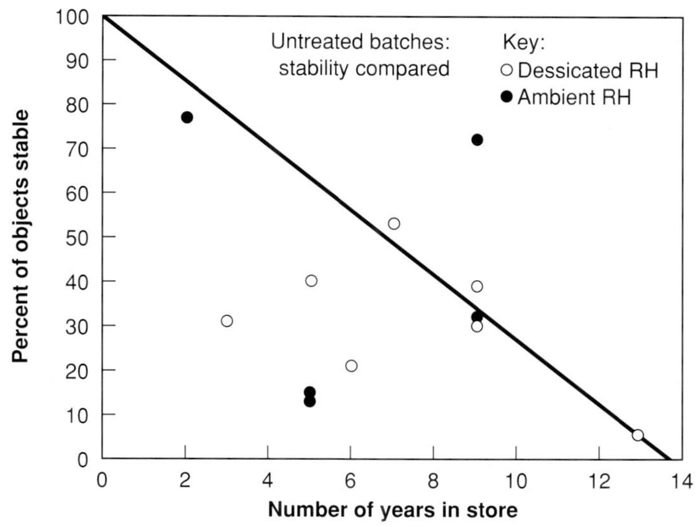

FIGURE 6. *The probability of being stable at year 8: outcome of treatments and storage compared.*

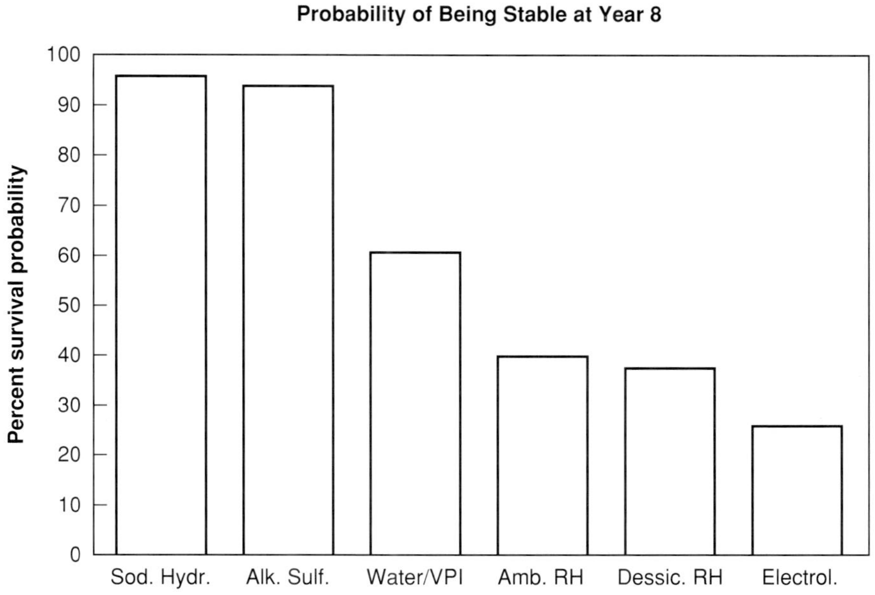

followed by soaking, which in this study gave the worst survival rate of all. Soaking in deionized water with VPI doubled the survival chance over that of untreated iron. The interpretation of these results is discussed later.

But are these differences only apparent? The chi-squared test of significance was applied to a table of numbers of objects, stable and unstable, for the treatment batches compared to the untreated iron (all for the Swan Lane site, to reduce variables). The results were significant at the 0.5% level, and so the null hypothesis can be rejected; there is a real difference between active treatment and storage alone. Further tests of significance (binomial tests) show that electrolysis is significantly worse than the average of the other three treatments, and that water with VPI is probably worse than sodium hydroxide and alkaline sulfate.

The results for the batches that were not treated but stored in different ways are equally interesting. In order to compare results with those for the treated batches, the simple percentages of objects stable at the year of inspection were calculated (Fig. 6). The result for the batch longest in storage (also the largest batch by far), from which the percentage of objects surviving is very small, has been used to plot the line in Figure 6, which gives a notional average. If objects in desiccated storage survived longer than objects stored at ambient RH, then the markers for these batches would be located in the upper right-hand part of Figure 6. But they are, in fact, randomly distributed, as are the markers for batches stored at ambient RH.

Because the batches were assessed at different numbers of years after storage, it was not possible to carry out a chi-squared test of significance for the untreated objects. Other possible tests do exist, but they are complex, and it is likely that the differences would not be statistically significant.

Interpretation

Instability in iron has a number of contributing factors, as Turgoose and others have shown (Turgoose 1982a, 1982b, 1985). Given these factors—the chemical nature of the iron's corrosion, particularly perhaps the chloride content; the physical nature (thickness and impermeability) of the mineral layers; and the relative humidity and oxygen content of the storage environment—it is difficult to evaluate the success of treatments as well as explain the causes of failure.

The nature of the object is one consideration. Watkinson (1983) has shown that if an object is completely mineralized, then it is not likely to recorrode. The Museum of London anaerobic iron does not fall into this category; the objects each have a relatively thin corrosion layer and a metal core. It seems that little of the dry-site iron from the Passmore Edwards Museum is protected by this factor either, but time will show whether, for example, the few remaining objects from the batch longest in storage (thirteen years) will eventually break down.

Iron from the Museum of London was from wet, anaerobic sites. Results for these batches are no different from those for the Passmore Edwards dry-site iron (Tables 1, 4), although the mineral layers on the objects look very different. In this case, the burial context of the objects has not had a strong influence on their survival.

The concentration of chloride ions is another factor. There is a correlation between the amount of chloride ion removed during treatment and the probability of stability, as illustrated in Figure 7. The comparative chloride-extraction rates for the different treatments have been confirmed in measurements made by other conservators at the Museum of London in the course of treating other batches of objects, as well as by Rinuy and Schweizer (1981), and North and Pearson (1975, 1978). In iron conservation, one questions the basis for any assumption, but it is widely assumed that if sufficient chloride can be removed from an archaeological object, then the object will be stable. An empirical test by Rinuy and Schweizer (1981) did indicate that iron does not corrode in the complete absence of chloride, even at elevated relative humidities.

It is quite surprising that sodium hydroxide alone produced such a good result. The treatment was tried as a result of North and Pearson's article discussing its use on marine iron (1978). The ill effects subsequently predicted by Turgoose seem not to have occurred (1985a:14). It was tried because it prevented obvious corrosion and was less disruptive than alkaline sulfite to mineral layers.

FIGURE 7. *The probability of being stable at year 8 plotted against the amount of chloride extracted during treatment.*

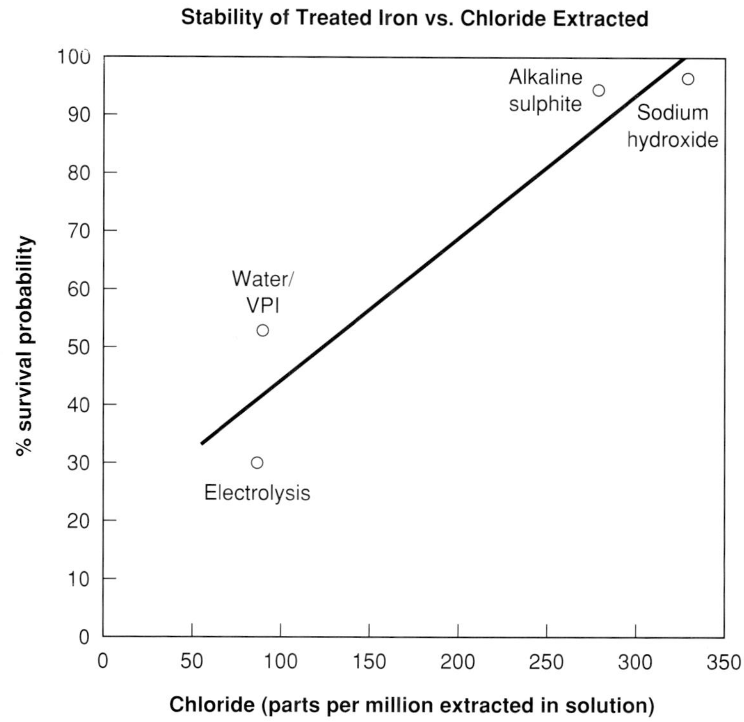

Pretreatment using electrolysis had a markedly adverse effect on the stability of the objects. This procedure was used to make the corrosion layers separate more easily for cleaning delicate, decorated surfaces; and it accomplished this purpose very effectively. It was postulated that electrolysis would make the mineral layers more porous and allow soluble chloride to escape more easily during subsequent soaking. Perhaps this pretreatment also allows oxygen and humidity to penetrate more easily during subsequent storage. If this is the course of its lack of success, then it may be possible to prevent this effect by consolidating the corrosion layers.

The results from this study seem to indicate that the storage environment has much less influence on the stability of iron than had been supposed—a great deal less than do most treatments. There may be an explanation for this. Even if a desiccated environment is maintained, as it was for the Passmore Edwards material, the phenomenon of interstitial condensation may mean that pockets of high humidity are formed in the plentiful pores and cracks in the mineral coatings of the objects (Schreir 1976). Any free water will leach out soluble salts, such as chlorides, and concentrate them in these small areas.

Should it be confirmed that desiccated storage has no real advantage over storage in ambient conditions, this information will have far-reaching consequences for the techniques used to store archaeological iron. Many more samples need to be assessed, however.

Conclusions

These results reinforce the tentative conclusions from the first study that, in real-time assessments after long periods in storage, active treatment improves the survival chances of iron. Treatment appears to have a much greater effect than does controlled storage at low relative humidity. The most effective treatments are those that remove the greatest quantity of chloride, but this study cannot demonstrate cause and effect here.

Following the first study and the encouraging early results of the second, the Museum of London has continued to treat its archaeological iron, using the procedures outlined in Appendix 2. These include soaking in the corrosion inhibitor triethanolamine, as suggested by Argo (1985:31). This method has not yet been formally assessed, although it has been observed that quite a large amount of chloride is extracted in this way. The museum will revise its standard treatments again to take account of present findings, and and will also consider whether to use sodium hydroxide rather than alkaline sulfite, should it be less disruptive to the mineral layers.

The effects of treatment on the objects' surfaces are also being studied by Dana Goodburn Brown, using the scanning electron microscope to record surface detail before and after treatment. At present, selected objects are being carefully cleaned, and silicone-rubber molds (Dow Corning 9161) are being taken of the surfaces before and after treatment. The four different treatments described in Appendix 2 are being assessed; and, in addition, a treatment is being developed at the Institute of Archaeology at University College, London. The same area of each object will be examined in order to evaluate the effect each treatment has on the surface.

Further work is suggested. Turgoose is currently engaged in a collaborative research project under the European STEP initiative at the University of Manchester Institute of Science and Technology, assessing the corrosion potential of archaeological iron. In parallel with this work, studies based on the real-life assessment of actual objects are clearly useful. Accelerated aging tests in high relative humidity for mineralized iron are likely to introduce other complications arising from the nature of the mineral layers and will not give convincing results. Some of the protocols set out for clinical trials in medicine, described by Mould (1981), might be adopted.

The development of more effective treatments—including less aggressive ones, perhaps not involving chloride removal—should also be encouraged. It is tantalizing to read abstracts of published work on the effects of silicates and corrosion inhibitors (Art and Archaeology Technical Abstracts 27-815, 27-907), and discussions of the use of corrosion inhibitors (Turgoose 1985b). It is encouraging to report that Ahmed al Zaid, at the Institute of Archaeology at University College, London, is researching the use of silane products on iron.

Appendix 1. Treatment Procedures Used in the Trial

Mechanical Removal

Mineral layers overlying the original surface were first removed using mechanical means: a scalpel or power pen. For ease of comparison of treatments, each batch consisted of iron objects totaling approximately the same weight, and all soak baths contained 1 L of solution. The temperature of all the treatment baths was maintained at 50 °C in a laboratory oven. Sealed polythene boxes, completely filled with liquid, were used to keep oxygenation to a minimum.

Alkaline Sulfite

For general discussion of this method, see North and Pearson (1975). Only relatively robust objects with no tinning were selected for this treatment, since tin will dissolve in alkaline solutions. The procedure was as follows:

1. Chloride removal: Objects were immersed in four successive baths of 0.5 M alkaline sulfite (Na_2SO_3 + NaOH). The duration of each bath was one week.
2. Removal of alkaline solution: Objects were immersed in changes of deionized water and Dichan VPI until no more chloride could be detected. There were fifteen solution changes. The total time of treatment was twenty weeks.[1]

Sodium Hydroxide

For a discussion of the benefits of using washing solutions at raised pH, and especially sodium hydroxide, see North and Pearson (1978). Only relatively robust objects with no tinning were selected for this treatment, since tin will dissolve in alkaline solutions. The procedure was as follows:

1. Chloride removal: Objects were immersed in five successive baths of 0.5 M sodium hydroxide (NaOH).
2. Removal of alkaline solution: Objects were immersed in changes of deionized water with 0.5% w/v Dichan VPI until no further chloride could be detected. There were thirteen solution changes. The total time of treatment was nineteen weeks.

Electrolysis Followed By Deionized Water

The procedure was as follows:

1. Corrosion softening: The electrolyte was sodium carbonate. A low density current was passed until the corrosion layers had softened and separated sufficiently for easy removal.
2. Chloride removal: Objects were soaked as a batch in changes of deionized water with 0.5% w/v Dichan VPI until no more chloride could be detected. There were fifteen changes of solution. The total time of treatment was twenty weeks.

Deionized Water and VPI Soaking

The procedure was as follows:

Chloride removal: Objects were soaked in changes of deionized water with 0.5% w/v Dichan VPI until no more chloride could be detected. There were fifteen changes of solution. The total time of treatment was twenty weeks.

Drying and Protection

After treatment, objects were dewatered through two changes of industrial methylated spirits and coated with two coats of Incralac lacquer.

Appendix 2. Desalination Treatments in Current Use

Desalination treatments for iron currently used as standard at the Museum of London involve a choice of method as follows:

Alkaline Sulfite

This method of chemical reduction makes corrosion more porous and allows $Cl-$ to diffuse out. It also extracts the most chloride and appears not to damage silver or copper (black deposit on the latter can be scraped off).

Contraindications: Alkaline sulfite dissolves tin, weakens adhesives, and can disintegrate fragile, heavily mineralized objects. It may also damage original surfaces.

Soaking In Triethanolamine (TEA) Solution

TEA is thought to react with iron oxychlorides, freeing chloride ions. It has the next best chloride extraction rate after alkaline sulfite. TEA is a corrosion inhibitor, so it inhibits flash rusting. It seems not to damage silver or copper (copper may stain) and does not damage fragile objects. It is possibly suitable for organics such as bone.

Contraindications: TEA damages tin. It is less damaging to adhesives than is alkaline sulfite.

Soaking In Deionized Water with Corrosion Inhibitor (Dichan VPI)

The corrosion inhibitor prevents rusting during the soak; this may enhance chloride removal. It is possible that traces remaining in the object will confer some corrosion protection. Deionized water with VPI does not damage nonferrous metals, adhesives, or leather.

Contraindications: Do not use this method with wood or iron composites, as the VPI soaks into the wood and produces quantities of crystals as it dries.

Tannic Acid

This is used as a surface coating for "maintenance" of unstable objects.

Acknowledgments

The author thanks Ruth Waller, the research assistant for this project. Reg Davis, of the Royal Marsden Hospital, kindly drew the author's attention to the use of medical statistics for monitoring survival rates. Clive Orton, the coauthor of the earlier study, has again provided invaluable advice and support with the statistical assessment. The conservators at the Museum of London have encouraged this renewed assessment of treatments, and particular thanks are extended to Rose Johnson and Jill Barnard, who organized a stimulating colloquium on the subject in early 1992.

Note

1. North and Pearson recommend neutralizing the sodium sulfite using barium hydroxide. Simply soaking to remove it was judged preferable, since barium hydroxide can leave unsightly white deposits.

References

GILBERG, M. R., AND N. J. SEELEY

1982 The alkaline sodium sulfite reduction process for archaeological iron: A closer look. *Studies in Conservation* 27:180–84.

KEENE, S., AND C. ORTON

1985 Stability of treated archaeological iron: An assessment. *Studies in Conservation* 30:136–42.

MOULD, R. F.

1981 Introductory Medical Statistics. London: Pitman Medical.

NORTH, N., AND C. PEARSON

1975 Alkaline sulfite reduction treatment of marine iron. ICOM Committee for Conservation, 4th Triennial Meeting, Venice, vol 2.

1978 Washing methods for chloride removal from marine iron artifacts. *Studies in Conservation* 23:174–86.

RINUY, A., AND F. SCHWEIZER

1981 Méthodes de conservation d'objets de fouilles en fer: Étude quantitative comparée de l'élimination des clorures. *Studies in Conservation* 26:24–28.

SCHRIER, L. L.

1976 *Corrosion Handbook.* London: Newnes-Butterworth.

TURGOOSE, S.

1982a The nature of surviving iron objects: Conservation of iron. Maritime monographs and reports. London: National Maritime Museum.

1982b Post-excavation changes in iron antiquities. *Studies in Conservation* 27:97–101.

1985a The corrosion of archaeological iron during burial and treatment. *Studies in Conservation* 30:13–18.

1985b Corrosion inhibitors for conservation. Occasional Papers No. 4. London: UKIC.

TYLECOTE, R. F., AND J. BLACK

1980 The effect of hydrogen reduction on the properties of ferrous materials. *Studies in Conservation* 25:87–96.

WATKINSON, D.

1983 Degree of mineralization: Its significance for the stability and treatment of excavated ironwork. *Studies in Conservation* 28:85–90.

Biography

Suzanne Keene qualified in conservation at the Institute of Archaeology in London. She worked in the Winchester Research Unit on the material from the well-known excavations there. After a period in private practice, she became head of Archaeological Conservation at the Museum of London, then head of the Department of Conservation there. She now works as head of Collections Management at the Science Museum of the National Museum of Science and Industry in London. Her research interests include conservation information and the stabilization of archaeological iron.

Conservation of Corroded Metals: A Study of Ships' Fastenings from the Wreck of HMS *Sirius* (1790)

IAN DONALD MACLEOD

The study of corroded metal objects recovered from historic shipwrecks provides conservators with a unique opportunity to examine the effects of salts, water movement, depth of burial, and the level of dissolved oxygen on the degradation of a wide range of artifacts. Normally, the chronology of shipwrecks is well defined, and this helps quantify the rates of deterioration that occur. Previous work in the Materials Conservation Laboratory of the Western Australian Maritime Museum at Fremantle has concentrated on the identification of corrosion processes and how, by studying the deterioration of these objects, conservators can improve their techniques for stabilizing such artifacts (North 1976:253–58; MacLeod 1987a:25–40).

In the field seasons of 1985, 1987, and 1988, a team of maritime archaeologists led by Graeme Henderson of the Western Australian Museum's Department of Maritime Archaeology recovered a variety of metal artifacts from the wreck of HMS *Sirius*. The wreck is located on a coralline reef off Norfolk Island, which lies at 29.48° south latitude and 167.59° east longitude, about 1,500 km from Sydney and 1,000 km from Auckland, in the south Pacific Ocean. The copper, brass, and bronze objects recovered—including nails, bolts, sheathing, and other fittings (Figs. 1–3)—were all worn and had apparently been used in the construction of the *Sirius*.

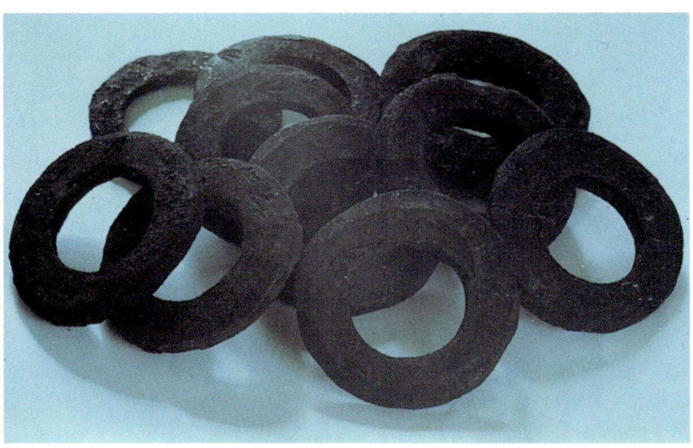

FIGURE 1. *Clinch rings from the HMS* Sirius.

FIGURE 2. *Bronze sheathing tacks from the HMS* Sirius.

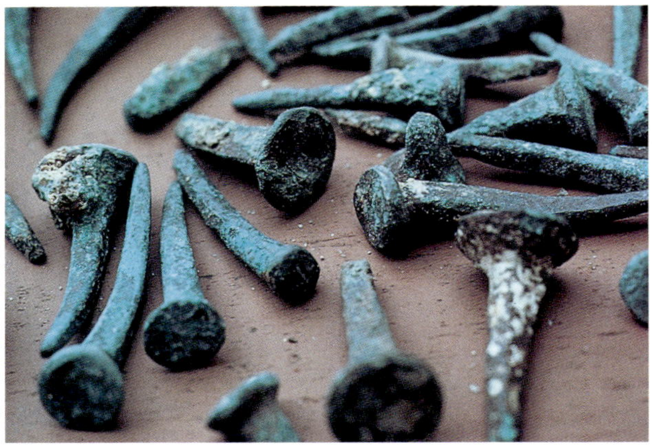

FIGURE 3. *Brass belt buckles from the HMS* Sirius.

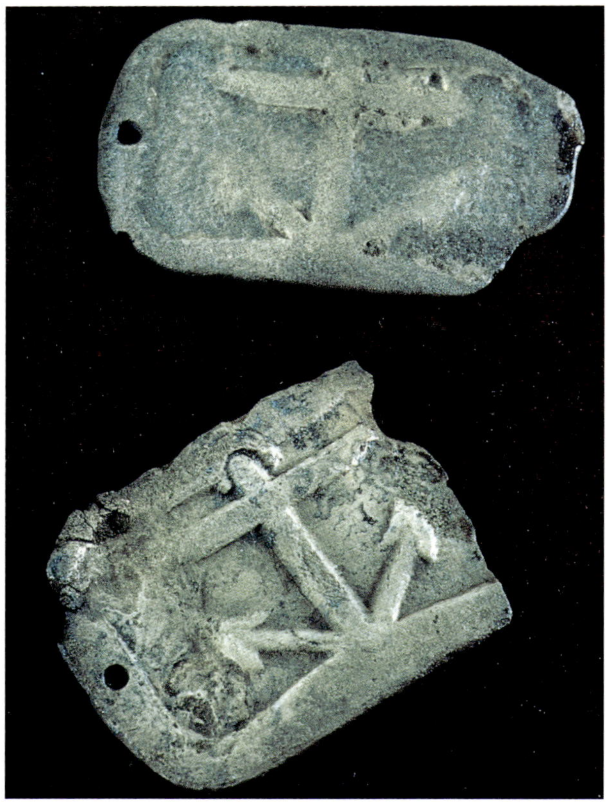

Historical Background

HMS *Sirius* was originally named the *Berwick*. It was built in 1780–81 as a Baltic trading vessel, but in 1781, while the ship was still in dry dock at Christopher Watson's shipyard at Rotherhithe on the Thames in England, it was purchased by the British Admiralty. The Admiralty intended to make the *Berwick* the flagship of the First Fleet, to serve in establishing settlement in the newly founded colony of Australia. Initially, however, the *Berwick* was fitted out as an armed storeship for service to North America and the West Indies. After being coppered in early 1782, it sailed for Nova Scotia and New York. In 1783 the *Berwick* returned to England, where it was docked for repairs before sailing to the West Indies. Finally, in 1786, after undergoing a comprehensive refit, the ship was renamed HMS *Sirius* and set sail on its First Fleet voyage to Australia.

The *Sirius* was wrecked at Kingston on Norfolk Island at noon on March 19, 1790 (Fig. 4). Due to a change in wind direction, the ship was driven against a reef while unloading cargo and passengers. Other ships that went down in the rough surf off Norfolk Island include the *Mary Hamilton*, sunk to the west of the *Sirius's* principal stranding site in 1873; and the *Renaki,* a three-masted auxiliary schooner thrown up on the reef in 1943. The site of the *Rapid,* which sank in 1811 near the Northwest Cape, western Australia, is more sheltered than that of the *Sirius,* and objects recovered from the former were used for corrosion comparisons. The *Mary Hamilton* went down west of the principal stranding site of the *Sirius*, causing some contamination of a portion of the site, although testing results reveal differences between the artifacts of the two vessels. The *Renaki* was wrecked in an area so distinct from that of the *Sirius* that site contamination did not pose a problem; the few remains of the schooner that were not salvaged are clearly identifiable.

Because of where the *Sirius* was stranded, most of the ship's supplies were salvaged in 1790, and the bulk of the shipwreck remained intact for almost two years thereafter. This fact clearly attests to the skill and strength of the ship's construction, considering the area's rough-breaking seas. To this day, large ships cannot tie up at wharf on the island because of the treacherous waters, and all cargo must be unloaded in deeper water and brought ashore with small boats.

The artifacts from HMS *Sirius* are scattered around the main site, which is located some 70 m offshore. The undersea terrain in this area is a very gently sloping plane of hard calcareous rock in water varying in depth from 1.5 to 4.5 m. The wreck site itself consists of flat beds of overlying coralline reef materials with a series of gullies and depressions in the seafloor. The configuration of the island and the particular location of the wreck site means that submerged objects are subjected to the effects of continuously breaking surf. During a storm, the wave action is sufficient to shift a 1.5 ton concrete plinth 30 m across the seabed. The ship's submerged remains primarily consist of about two hundred concreted cast-iron ballast blocks, a few

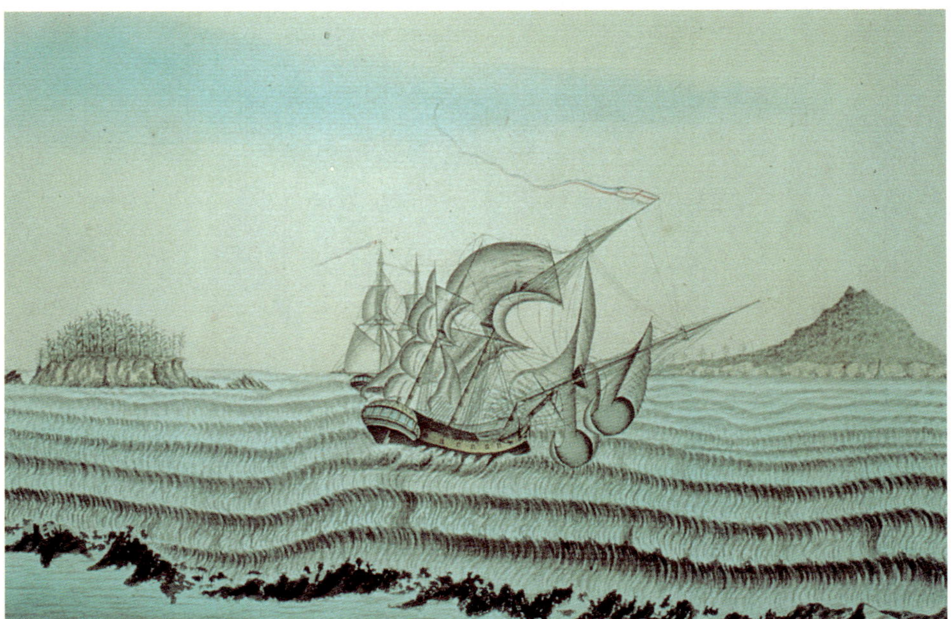

FIGURE 4. *"The Melancholy Loss of His Majesty's Ship Sirius, wrecked on Norfolk Island, on Friday Noon March 19th 1789 taken from the Flag Staff on the Beach." Watercolor by George Raper, 1790.*

anchors, and one remaining carronade. Smaller artifacts are either trapped within the coralline corrosion matrices or under the ballast. The warm seawater (20.80 ± 1.96 °C), a mean annual salinity of 35.77 ± 0.04 ppt, and dissolved-oxygen levels of 7.7 ± 0.3 ppm constitute a highly corrosive environment for these objects.

Analysis of the corrosion properties and materials performance of the various alloys of the recovered artifacts was undertaken for two purposes, one of history and one of materials science. The primary historical purpose was to establish the quality of the materials used in the construction of the *Sirius;* several accounts indicate that it was poorly built, whereas the archaeological evidence supports the contrary. The second purpose of this research was to learn how a very aggressive shipwreck site degrades different materials, and what conservation problems this presents to the conservator (MacLeod 1989b:227–29).

Experimental Method

Eight samples of copper sheathing, a brass bolt, and bronze sheathing nails were examined metallographically and electrochemically after they had been sectioned and embedded in Araldite D (Ciba-Geigy's two-part epoxy resin, which is cured with an amine catalyst). Surfaces were prepared by grinding with wet and dry carborundum paper to 1,200 grit and polishing with diamond paste to 0.25 µm; the etchant was 2 wt % ferric chloride in ethanol. Some objects were sectioned in several places with a number of longitudinal sections (LS) and transverse sections (TS) representing the tail (shank), body, and head regions of the fittings.

Drilled core samples (360 ± 100 mg) were analyzed at a commercial laboratory (ANALABS) using atomic absorption techniques. Core samples were taken in an attempt to overcome the problems associated with analysis of archaeological metals (Caley 1964:1–15). Results are listed in Table 1.

TABLE 1. *Composition of ships' fastenings and sheathing from HMS* Sirius *(1790).*

Type	*Copper sheathing*	*Copper sheathing*	*Copper sheathing*	*Brass bolt*	*Sheathing tack*	*Clinch ring*	*Eroded tack*	*Sheathing nail*
Sample	*SI 357*	*SI 465A*	*SI 491*	*SI 15*	*SI 465*	*SI 254*	*SI 228A*	*SI 228B*
Cu	99.4%	99.4%	98.9%	67.7%	91.1%	99.1%	91.4%	90.0%
Sn	25.6	15.3	75.9	402	7.69%	1010	8.16%	8.00%
Pb	1750	1710	99	6670	520	1000	2720	4680
Zn	48	16	28	31.5%	108	120	257	7680
As	2410	2550	8260	460	6390	4380	<200	<200
Sb	300	302	185	92.7	216	183	52.9	49.3
Bi	500	525	1130	173	2240	935	265	1020
Ag	177	665	49	161	680	394	400	530
Ni	190	185	350	275	850	300	180	185
Fe	44	76	104	370	170	220	460	5130

Copper values are calculated by difference. Results are in ppm unless otherwise stated. Due to the small size of some samples it was not possible to accurately determine the amount of arsenic. The symbol < denotes values less than the detection limit.

Vickers microhardness measurements on the polished sections were made using a Tukon model 300 operating on a 600 g load with a x20 objective. All the sections were examined under the scanning electron microscope (SEM) at the Commonwealth Scientific and Industrial Research Organisation's Division of Mineralogy, using the back-scattered electron/low-vacuum mode before and after the corrosion experiments. Semiquantitative elemental analyses were made using energy dispersive analysis by X rays (EDAX) attached to the JEOL scanning electron microscope (JSM2) after calibrating the instrument with standard bronze, brass, and copper alloys (Robinson and Nickel 1979:1322–28).

Electrical connections were soldered to the metal sections prior to encapsulation in resin. The corrosion potentials were measured using a Titron silver-silver chloride reference electrode (saturated with KCl), which was calibrated against a platinum electrode in a quinhydrone solution at pH 4.0. (Voltages are all relative to the Ag/AgCl unless otherwise stated.) The seawater temperature in the laboratory studies was 24 ± 1 °C. Dissolved-oxygen measurements were taken using an Imperial Chemical Industries (ICI) oxygen meter (411) after correction for the salinity of the seawater (35.67 ppt). The seawater was filtered after collection from the ocean, but no preservatives were added. The current voltage curves for polarization-resistance measurements were recorded using a Princeton Applied Research (PAR) potentiostat with a platinum auxiliary electrode and an XY recorder; the data were collected over a range of ± 25 mV around the corrosion potential (E_{corr}) with the voltage scanned at ± 1 mV per second. The effects of dissolved oxygen and stirring on E_{corr} were determined using a combination of E_{corr} and polarization measurements. Tafel slopes were determined on the same samples used in the polarization studies. The measurements were made in a specially designed polarographic cell that allowed side mounting of the metal sections.

Corrosion Potentials and Polarization Resistance

When copper and its alloys are immersed in oxygenated seawater, they will corrode at a rate that is dependent on their chemical composition and microstructure and on the amount of dissolved oxygen in the water (Gilbert 1982:47–53; Bjorndahl and Nobe 1984:82–87). In aerated solutions, the E_{corr} is the voltage of a corrosion cell consisting of the anodic (oxidation of metal) and cathodic (oxygen reduction) half-cells. The way in which the corrosion rate changes with the voltage of a corrosion cell is very complex, but it can be simplified in two extreme sets of conditions. At voltages close to the corrosion potential, the metal obeys Ohm's law, since there is a direct relationship between the voltage and the current, according to the formula $E = iR$, where E is the voltage, i is the current, and R is the resistance. In other words, there is a linear relationship between ohmic resistance and the voltage. At applied voltages several hundred millivolts removed from the E_{corr}, the increase in corrosion rate with applied voltage has a logarithmic rather than a linear response. The characteristic slope is called the Tafel slope. Corrosion in the Tafel region is often characterized by the presence of a passivating film on the surface of the metal.

The corrosion rate (current) at the E_{corr} can be calculated from the rate at which the current varies with small variations in the potential. For voltages within 30 mV of the E_{corr}, the polarization resistance (R_p) is calculated from the slope of the current voltage plot via the expression:

$$\left(\frac{\delta E}{\delta i}\right)_{\delta E \to O} = R_p$$

The corrosion rate (current) is inversely related to the polarization resistance by the expression:

$$R_p = B/i_{corr}$$

The constant B can be calculated via Faraday's laws relating to weight-loss data, or it can be determined from analysis of electrochemical polarization data. In physical terms, the value of B can be obtained from the Stern-Geary (1957) relationship,

$$B = \frac{b_a \cdot b_c}{2.303\ (b_a + b_c)}$$

where b_a and b_c are the anodic and cathodic decadic Tafel slopes, as described above. The values of B depend on the composition of the objects and the types of corrosion products on the surfaces.

The Tafel slopes of the various fittings tested fell into two groups, one of which had characteristic bc values of 52 ± 5 mV, while the other had values of 39 ± 2 mV. The former is typical of cathodic Tafel slope values for concreted bronzes, and the latter is characteristic of machined samples. The individually calculated values of B were used to determine the corrosion currents, shown in Table 2, using the above relationships. The anodic Tafel slope (b_a) of the samples varied from 42 to 85 mV. These calculated corrosion currents are based on measurements of R_p after twenty-one days of exposure of the samples to seawater. As such, they are approximately one order of magnitude too high for the bronzes and should be regarded as initial

TABLE 2. *Corrosion rates of copper alloys from the wreck of HMS* Sirius.

Registration	*Type*	*Corrosion current $\mu A\ cm^{-2}$*	*n^**
SI 465	Bronze tack	3.3 ± 0.3	1.04
SI 228A–1	Bronze tack	3.6 ± 0.6	1.05
SI 228A–2	Bronze tack	8.7 ± 1.1	1.05
SI 357	Copper sheathing	2.0 ± 0.4	1.00
SI 254	Copper ring	4.3 ± 0.1	1.00
SI 15	Brass bolt	19.2 ± 4.1	1.31
SI 228B–3	Bronze tack	9.2 ± 0.2	1.05
SI 228B–1	Bronze tack	10.8 ± 0.6	1.07
SI 228B–4	Bronze tack	13.0 ± 0.6	1.08
SI 228B–2	Bronze tack	16.6 ± 1.4	1.10

*n is the number of electrons in the rate-determining step.

corrosion rates rather than average long-term values (Taylor and MacLeod. 1985:100–4). In contrast, the copper corrosion currents fall within the normal range of long-term corrosion rates (Lush and Carr 1979:1079–88).

Results and Discussion: Corrosion Potentials

The sectioned copper alloys were immersed in fresh seawater (pH 8.18), and the corrosion potential was monitored for a period of three weeks. All the samples showed significant changes in the corrosion potential as the polished surfaces reacted with the seawater during the first ten days. The bronze sheathing tack (228B) shifted by more than 200 mV, whereas the copper sheathing samples and the clinch ring showed anodic shifts of only 20 mV over the same period. During the monitoring of E_{corr}, the level of dissolved-oxygen seawater was varied to see how sensitive the metals were, since the microenvironment at the site may be significantly different from the macroenvironment, which is fully oxygenated. The solutions were saturated with pure oxygen (14 ppm) or with air (8.2 ppm) or degassed with instrument-grade nitrogen to give a reading of 0 ppm. Three distinct types of response were observed. The E_{corr} of the sections of the bronze sheathing tack moved anodically by 9.0 ± 0.4 mV/ppm, and the brass bolt moved by 4.8 mV/ppm O_2, while all the copper fittings and the copper sheathing tacks (465 and 228A) moved by 2.4 ± 0.5 mV/ppm O_2. The latter samples also showed no increase in corrosion potential above the air-saturated, dissolved-oxygen level. These results indicate that, above this concentration of oxygen, the rate of corrosion is controlled by the anodic oxidation of copper metal. For the copper samples, the anodic shift of 24 mV per tenfold increase in oxygen concentration corresponds closely to the theoretical response of the oxygen half-cell of 29 mV. The corresponding values for the brass and bronze fittings are much higher than the theory would suggest, but they are consistent with the observations of others who saw shifts as great as 160 mV per decade for pure tin (Butler et al. 1969: 715–38). The ratios of the E_{corr} sensitivity of the alloys to oxygen are 4:2:1 for the bronze, brass, and copper fittings, respectively. These ratios are the same as the number of electrons commonly involved in the oxidation of tin, zinc, and copper.

Although the wet-chemical analyses listed in Table 1 show the overall composition of the alloys, it is readily apparent from the SEM EDAX analysis of the corroded surfaces that wide variations exist between the composition of the surface layers and the inner sections of some of the cast-bronze sheathing tacks. Apart from the brass bolt (SI 15), all the other materials from the *Sirius* site showed a similar dependence of E_{corr} and the size of the voltage shift when the seawater was deoxygenated. The response of the fittings to the change in levels of dissolved oxygen is given by the relationship:

$$^{O_2}E_{corr} = 0.236 \cdot {}^{N_2}E_{corr} - 99 \text{ mV}$$

The apparently anomalous behavior of the brass bolt is not really surprising, since it came from a part of the wreck site that had been contaminated with material from another ship. Inspection of the analysis in Table 1 shows that this anomalous bolt has a composition essentially identical to a modern, yellow, low-leaded brass (Miner and Seastone 1955:2–325). When compared with artifacts recovered from vessels of similar age to the *Sirius*, the low level of impurities in this fitting adds further weight to the conclusion that the bolt did not originate from the flagship of the First Fleet (MacLeod 1987b:280–91). The bolt most probably comes from the wreck of the *Mary Hamilton* (1873).

Composition and Corrosion Rates

Bronzes

Several fittings were sectioned in sequentially to gain an understanding of how differences in microstructure and composition affect the rate at which different parts of the same object corrode. It has always been a major problem in stabilizing corroded artifacts to reliably determine what sort of variation is possible within one object. Results from the corrosion of the square-headed bronze sheathing nail (228B) illustrate the point. The corrosion current increases linearly with increases in the surface concentration of tin (Fig. 5), according to the equation:

$$i_{corr} = 0.886x + 1.44$$

where x is the wt % tin in the sample and the corrosion current is expressed in $\mu A/cm^2$. The effect of the increased tin concentration is also seen in the corrosion potentials of the fastenings in oxygenated seawater: the lower the tin content, the more negative the E_{corr}, and the lower the corrosion rate.

It should be noted that the corrosion data above relate to the fully oxygenated site of HMS *Sirius*. The sensitivity of the bronzes to dissolved oxygen has previously been noted (MacLeod 1985:10–13). A difference appears to exist between these studies of the *Sirius* fittings and the data from studies of long-term corrosion of artifacts

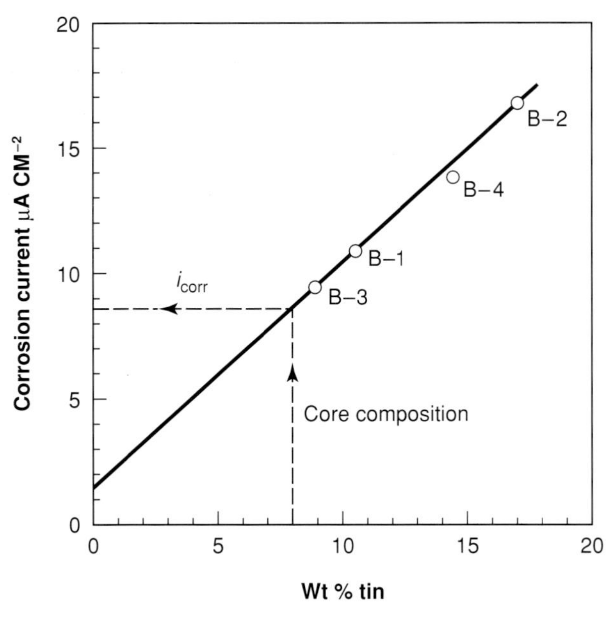

FIGURE 5. *Diagram showing the effect of tin on the corrosion rate of a bronze sheathing tack (SI 228B).*

recovered from other less aggressive sites. Interestingly, the corrosion of archaeological bronzes on land sites is similar to that of bronze objects found in marine sites with a low level of oxygenation.

From the SEM EDAX analysis of the surfaces of the corroded sections of the 228B bronze sheathing tack, it is clear that significant amounts of antimony are present in the tin-rich phases despite the total content amounting to only 0.00493 wt %. Tin and antimony readily form intermetallic compounds that are less reactive than the parent, tin-rich areas (North and MacLeod 1987:90–91). Increased amounts of antimony in the $(\alpha + \delta)$ phase of the bronze effectively promote interdendritic corrosion of the sections that have higher tin concentrations, as seen in Figure 6. In addition, increased amounts of antimony in the bronze fittings decrease the sensitivity of the corrosion potentials to changes in the value of E_{corr} when the seawater is deoxygenated. These observations confirm the theory that very small amounts of impurities, such as antimony, can have a major effect on corrosion performance.

The sensitivity of bronzes to impurities such as iron can be seen by comparing the corrosion currents, shown in Table 2, for the SI 228A and SI 228B sheathing tacks. Inspection of the analytical data in Table 1 shows that 228B has eleven times as much iron as 228A. The effect of this greater concentration of iron is seen in a doubling of the mean corrosion rate of SI 228B compared with SI 228A, which otherwise has a very similar composition. For the sections of SI 228B that corroded the fastest, SEM analysis showed that significant amounts of iron were found in the corroded, tin-rich surfaces.

The variability of the corrosion rates in the bronze fittings is also seen in the microhardness measurements, which show a direct dependence on the composition of the alloys. The bronze sheathing tack (SI 465) showed a linear dependence of microhardness on the amount of tin over the range of 4–10 wt % according to the following relationship:

$$HV_{465} = 41.3 + 10.2x$$

where x is the amount of tin at the point of measurement. The maximum value of 145 is essentially the same as that measured by Samuels (1983) on an identical bronze tack from the *Sirius*. Apart from localized hardness, possibly due to so-called tin sweat and coring segregation, the other cause of the decrease in microhardness from the heads of the nails to the tips is that the nails had all been hammered in use; in other words, the fittings had not been cargo but part of the vessel, and thus they had been work-hardened.

Copper

Corrosion performance of the copper hull fastenings and the copper sheathing was an essential factor in ensuring that the *Sirius* remained a seaworthy vessel. Physical examination of two samples of copper sheathing, SI 357 and SI 491, indicates that they are similar, whereas another sample, SI 465A, is much thicker and appears to be tougher. Analysis of the samples shows that the chemical composition of SI 357 and

SI 465A is identical and that the apparently similar material clearly has a different composition from that of the other samples, the latter sheathing being typical of most of the recovered materials.

The main impurities in the copper sheathing are arsenic and lead. One major effect of arsenic in copper is that it causes an increase in the microhardness of the metal during cold-working. Using the average values for each of the sheathing samples and the clinch ring, the following relationship between microhardness and the amount of arsenic was established:

$$HV_{Cu} = 26y + 74$$

where y is the wt % of arsenic in the material. It should be noted that the maximum amount of arsenic in these alloys is 0.826%, which is at the lower end of the range for ancient alloys (Scott, personal communication, 1991).

Since the maximum amount of arsenic occurs in the bulk-sheathing material, it is reasonable to assume that the vessel was covered with good-quality copper sheathing. The thicker sheathing (465A) showed negligible signs of intergranular corrosion and had been extensively worked, as indicated by the presence of annealing twins and the oblate Cu_2O inclusions (approximately 30 μm across). The grain size ranged from 50 to 60 μm. Erosion appears to have been the main corrosion mechanism for this material, which was subjected to the most turbulent waters of the wreck site. Localized current flow during the descent of a typical 4 m wave breaking on the site at a water depth of 3 m produces a velocity of 12 ms^{-1} at an angle somewhat steeper than 45° (Cresswell 1989:46–70). Such flow rates are typical when erosion is an important factor in the rate of copper degradation (Lush and Carr 1979:1079–88).

The copper clinch ring (SI 254) shows twice the corrosion rate of the copper sheathing (SI 357), as listed in Table 2. The reason for this is readily apparent from the differences in their microstructures. The sheathing has a very fine recrystallized structure as a result of extensive working, with typical grain sizes of 35–45 μm. Apparently, the large amount of nonmetallic inclusions does not have a deleterious affect on corrosion performance. In contrast, the clinch ring displays severe intergranular corrosion. The small amounts of tin and lead, approximately 0.1 wt %, were apparently concentrated at the grain boundaries and promoted intergranular corrosion (Fig. 7). The iron impurities were also concentrated at the grain boundaries and also appear to be involved in the increased corrosion rate. The normally beneficial effect of arsenic as an inhibitor of intergranular corrosion may have been lost by the even distribution of the arsenic in solid solution.

The surface hardness of the clinch ring was 107 HV, but this rapidly falls to a typical annealed value of 56 ± 12 HV within 1.3 mm of the surface. The mechanical stress of the working of the surface probably increases the corrosion rate (MacLeod and Pennec 1990:732–38). The composition of the clinch ring is the same as the large copper bolt examined by Samuels (1983) and was probably made from the same batch of metal produced by the Swansea copper refinery. The properties of the different alloys provide evidence for the two periods of construction of the vessel, which confirms other archaeological evidence. The two distinct batches of copper sheathing reported in Table 1 probably relate to the ship's initial fitting out (SI 491)

FIGURE 6. *A scanning electron micrograph of the interdendritic corrosion at the seaward edge of an eroded bronze sheathing tack (SI 228A). Full width of the image is 306 μm.*

FIGURE 7. *A scanning electron micrograph of the intergranular corrosion on a copper clinch ring (SI 254). Full width of the image is 306 μm.*

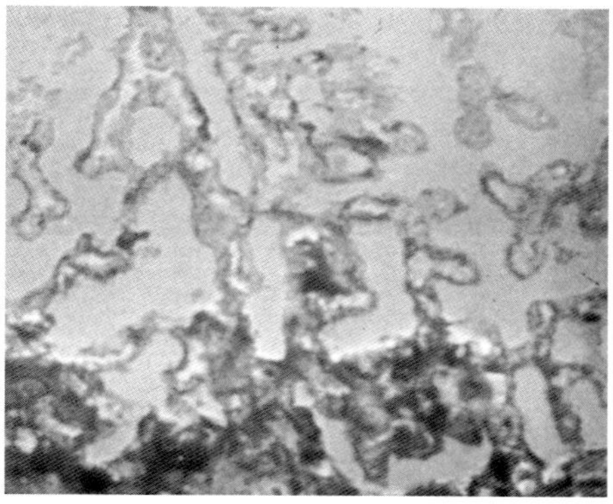

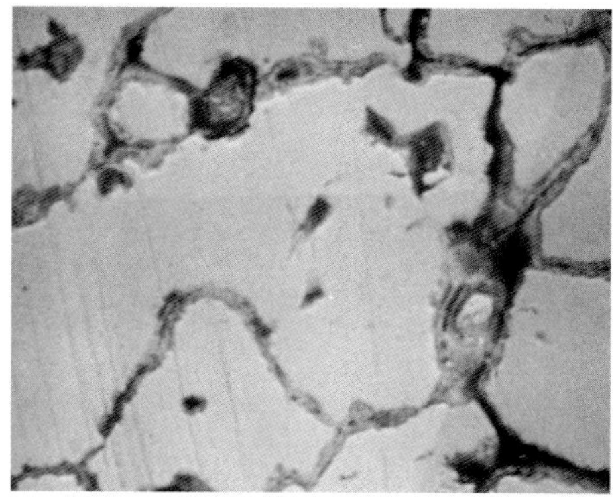

in 1781 and to its subsequent refurbishing (SI 357 and SI 465A) in 1786 (Henderson and Stanbury 1988:36–53).

Brass

Although the brass bolt (SI 15) comes from a different shipwreck, some of the corrosion properties of this important class of copper alloy are worth noting. The microstructure of the bolt indicates that it was either hot-worked or cold-worked and then annealed. Extensive intergranular corrosion is present, with preferential corrosion of the zinc-rich beta phase at the grain boundaries. Microhardness measurements of the head of the bolt show work-hardening of the outer surface to yield maximum values of 149 HV, which fall to 105 ± 13 HV within 2 mm of the surface. The shank is fairly uniform in section with an average value of 80 ± 12 HV. SEM micrographs of the surfaces reveal an uneven distribution of lead in the structure of the alloy. The high corrosion current of 19.2 ± 4.1 μA in the fully oxygenated environment reflects a maximum corrosion rate; if the material had continued to corrode at that rate, it would have ceased to exist in the form in which it was recovered. The extent of corrosion is consistent with a longer-term passivation of the corrosion process.

Conservation of the Metals

Having established the corrosion mechanisms for the various fittings from the shipwreck of HMS *Sirius,* one must look at the practical consequences of these observations. The largest difference in corrosion is found in iron artifacts, since they are inherently more reactive than copper and its alloys. A cast-iron carronade from the *Sirius* site, for example, released a total of 26.09 kg (5.36 wt %) of chloride ions during conservation treatment. In comparison, a cannon of similar weight from the more sheltered site of the *Rapid* (1811) released a total of only 4 kg (1.12 wt %) of chloride ions during the same period (MacLeod 1989a:7–16). When the amount of corrosion on the more sheltered site is normalized to account for the shorter period of immersion, the chloride content only increases to 4.6 kg (1.29 wt %). Even after this correction, results show the cannon from the *Sirius* to be four times more corroded than the similar object from the *Rapid* site.

Likewise, the copper, brass, and bronze alloys recovered from the *Sirius* wreck site show higher chloride contents than similar materials from a number of wreck sites off the Western Australian coast. Conservation treatment of the bulk of the copper sheathing on Norfolk Island involved the extended washing for two years in a series of 2.5 wt % sodium sesquicarbonate solutions. During this time, the wash solutions were analyzed for chloride ions; the total amount released was 0.07 wt % chloride, which is almost three times the average amount extracted from other wreck sites. The effect on brass and bronze artifacts was even more marked; the average amount of chloride removed under similar conditions was 0.52 wt %, a factor of more than eight times that of average sites (MacLeod 1987c:1079–85).

Measurements of the corrosion potential of artifacts on the site prior to recovery can now be used as a guide to the level of conservation that is probably needed. The in situ E_{corr} value of a copper bolt from the gun carriage of the remaining carronade was -121 mV. This voltage (an inherent indicator of the corrosion rate) is the same as the laboratory measurement of the clinch ring in oxygen-saturated seawater. The consistency of these results may indicate that similar corrosion problems will be found among other artifacts from the *Sirius*. Further studies are needed to confirm the reliability of such indicators.

Conclusion

A study of the corrosion characteristics of a series of copper, brass, and bronze fittings from the wreck of HMS *Sirius* shows that bronzes are very sensitive to the amount of dissolved oxygen in seawater. The electrochemical data reveal how differences in the distribution of alloy elements in the same object can lead to major variations in the materials performance of that artifact. The sensitivity of bronzes to the amount of dissolved oxygen indicates that artifacts recovered from low-oxygen microenvironments, such as those found under the concretions associated with iron objects, should be very well preserved without the need of direct galvanic protection. There is, however, a limited range of conditions that provide low-oxygen microenvironments without stimulating the activity of anaerobic bacteria. In addition, copper alloys recovered from sites with a high level of microbiological activity can be extensively corroded as a result of the high concentration of sulfides.

Minor amounts of impurities, such as antimony and iron, have a major influence on the corrosion mechanism of archaeological bronzes. Differences in performance of copper fastenings and sheathing can be rationalized in terms of the influence of arsenic and the metallurgical microstructure. The turbulent and highly oxygenated nature of the HMS *Sirius* wreck site has a direct influence on the extent of corrosion of the artifacts found there. In light of these findings, the conservator should be alert to the special care that is needed to treat such materials.

Acknowledgments

The author gratefully acknowledges the assistance of the CSIRO Division of Mineralogy, Floreat, for access to the SEM; Curtin University of Technology for use of the microhardness instruments; Murdoch University, Perth, for access to their electrochemical instruments; and Myra Stanbury, curator of the Western Australian Maritime Museum's Department of

Maritime Archaeology, for providing historical details. The assistance of laboratory trainees Mario Zaza and Paul Smart was helpful. The permission of the government of Norfolk Island to carry out this work, and the support of John Bannister and Graeme Henderson, is gratefully noted. Helpful comments from the editors are duly acknowledged with thanks.

References

BJORNDAHL, W. S., AND K. NOBE

1984 Copper corrosion in chloride media: Effects of oxygen. *Corrosion* 40(2):82–87.

BUTLER, G., P. E. FRANCIS, AND A. S. MCKIE

1969 The corrosion potentials of some common metals in oxygenated solutions at 25 ° and 150 °C. *Corrosion Science* 9:715–38.

CALEY, E. R.

1964 The heterogeneity of ancient metals and the sampling problem. In *Analysis of Ancient Metals,* 1–15. London: Pergamon Press.

CRESSWELL, G.

1989 The oceanography of the Norfolk Island vicinity. In *1988 Expedition Report on the Wreck of HMS Sirius (1790)*, 46–70. Report compiled by Graeme Henderson. Norfolk Island Government Project.

GILBERT, P. T.

1982 A review of recent work on corrosion behavior of copper alloys in seawater. *Materials Performance* 21:47–53.

HENDERSON, G., AND M. STANBURY

1988 *The* Sirius *Past and Present*. Sydney: Collins.

LUSH, P. A., AND M. J. CARR

1979 Copper dissolution in sea water under mixed activation and diffusion control. *Corrosion Science* 19:1079–88.

MACLEOD, I. D.

1985 The effects of concretion on the corrosion of nonferrous metals. *Corrosion Australasia* 10:10–13.

1987a Conservation of corroded copper alloys: a comparison on new and traditional methods for removing chloride ions. *Studies in Conservation* 32:25–40.

1987b Secondhand metal: Conservation and industrial archaeology of shipwreck artefacts. In *Archaeometry: Further Australasian Studies,* 280–91. W. R. Ambrose and J. M. J. Mummery, eds. Canberra: Australian National University Productions.

1987c Stabilisation of corroded copper alloys: A study of corrosion and desalination mechanisms. In *ICOM Committee for Conservation Preprints, 8th Triennial Meeting, Sydney*. Vol. 3: 1079–85.

1989a The application of corrosion science to the management of maritime archaeological sites. *The Bulletin of the Australian Institute for Maritime Archaeology* 13(2):7–16.

1989b The electrochemistry and conservation of iron in sea water. *Chemistry in Australia* 56(7):227–29.

MACLEOD, I. D., AND S. PENNEC

1990 The effects of composition and microstructure on the corrosivity of copper alloys in chloride media. In *ICOM Committee for Conservation Preprints, 9th Triennial Meeting, Dresden.* Vol. 2, 732–38.

MINER, D. F., AND J. B. SEASTONE

1955 *Handbook of Engineering Materials.* London: Wiley.

NORTH, N. A.

1976 Formation of coral concretions on marine iron. *International Journal of Nautical Archaeology and Underwater Exploration* 5(3):253–58.

NORTH, N. A., AND I. D. MACLEOD

1987 Corrosion of metals. In *Conservation of Marine Archaeological Objects,* 90–91. Colin Pearson, ed. London: Butterworths.

ROBINSON, B. W., AND E. H. NICKEL

1979 A useful new technique for mineralogy: The back scattered electron/low vacuum mode of SEM operation. *American Mineralogist* 64:1322–28.

SAMUELS, L. E.

1983 The metallography of some copper–alloy relics from HMS *Sirius. Metallography* 16:69–79.

STERN, M., AND A. L. GEARY

1957 Electrochemical polarization 1: A theoretical analysis of the shape of polarization curves. *Journal of the Electrochemical Society* 104:56–63.

TAYLOR, R. J., AND I. D. MACLEOD

1985 Corrosion of bronzes on shipwrecks: A comparison of corrosion rates deduced from shipwreck material and from electrochemical methods. *Corrosion* 41(2):100–4.

Biography

Ian Donald MacLeod received his doctorate from the University of Melbourne, where he studied the electrochemistry of metal fluorides dissolved in anhydrous hydrogen fluoride. During a two-year fellowship at the University of Glasgow, he studied organometallic chemistry. He spent three years at Murdoch University in Perth, where he concentrated on copper chemistry in association with mineral processing. For the past thirteen years, Dr. MacLeod has worked in the Materials Conservation Laboratory of the Western Australian Maritime Museum at Fremantle and is currently head of the Department of Materials Conservation. His research has focused on the problems of conservation and corrosion of metals recovered from historic shipwrecks. He has learned to scuba dive and perform in situ conservation experiments on iron artifacts on the seabed. In addition to his work with shipwrecks, he has an active research interest in the conservation of aboriginal rock art.

The Conservation of Outdoor Zinc Sculpture

CAROL A. GRISSOM

Zinc sculpture of the nineteenth century was principally an imitative art form and, probably for this reason, little has been written about its history. In Europe a few monumental zinc sculptures were widely discussed; these include the *Amazon on Horseback Attacked by a Tiger* by August Kiss and a statue of Queen Victoria cast by Paillard, which were prominently displayed in 1851 at the Crystal Palace Exhibition in London (*The Crystal Palace Exhibition Illustrated Catalogue* [1851]1970:xvii, 37; Tomlinson 1852:vol. 1, xcvi; vol. 2, 1047). In America, a few monumental pieces were made by known sculptors. For example, the Bohemian-born Caspar Buberl, best known for the terra-cotta frieze on the Pension Building in Washington, D.C., was commissioned to make the grouping *Columbia Protecting Science and Industry* (1880), which decorates the north entrance to the Arts and Industries Building of the Smithsonian Institution. Commissioned sculptures were the exception, however, and by far the majority of zinc sculptures were modeled by anonymous artists.

While the influential architect Karl Friedrich Schinkel did promote zinc for architectural use, citing its use for a cornice during restoration at "our university" in Berlin (*The Builder* 1849; Snodin 1991); this occurred relatively early, when the use of zinc for artistic purposes was still a novelty. Most sculptors seem to have held zinc in low esteem, including the prominent American sculptors John Quincy Adams Ward and Daniel Chester French, who rejected the use of zinc for sculpture (Shapiro 1985:79–80). Although in his youth Lorado Taft made models for a zinc company,[1] he does not mention zinc sculpture even once in his chronicle of American sculpture first published in 1903 (Taft [1903, 1924] 1969), nor is zinc sculpture mentioned in the most comprehensive present-day book on American sculpture (Craven 1968).

Written information is limited principally to accounts of nineteenth-century industrial expositions and ephemera such as advertising pamphlets and the trade catalogues through which most sculptures were sold. One exceptional book, bearing the unlikely title of *Artists in Wood*, discusses early American zinc foundries and models (Fried 1970). Jacques de Caso's discussion of nineteenth-century serial sculpture in France also has much pertinence to zinc sculpture in America (1975).

In the summer of 1978, when Phoebe Dent Weil and the author saw the parts of a disassembled monument that had been identified as being made of White Bronze, it was not known for certain that the material was zinc because the use of zinc for sculpture was unfamiliar at the time. However, the Fourth Ohio Infantry Monument at Gettysburg National Military Park, Pennsylvania, had been leaning dangerously prior to disassembly, and the Center for Archaeometry received a small contract to examine the monument and recommend treatment. Preliminary research quickly revealed that the conservation literature contained no information on the subject. A National Museum Act Grant was obtained which allowed in-depth study of written information, and coincident requests for the treatment of a range of zinc sculptures provided experience in treatment (Grissom 1982). Since then, a number of conservators have studied and treated zinc sculptures, and now a small body of literature on the subject can be found (Gayle et al. 1980; Keck et al. 1983; Nosek 1987; Scott 1987; Weil 1987; Klingelhofer 1990; Wypiski 1990).

Several reasons may be proposed for this recent interest in zinc sculpture. In the past, zinc sculptures were likely to have been repaired by local metalworkers or concerned citizens; many are located in small towns where knowledge about conservation would have been minimal (Knapp 1980). Awareness of the deterioration of outdoor sculpture of all kinds has increased dramatically during the last twenty years, however, and now a number of outdoor-sculpture conservators work throughout the country. Perhaps because taste has changed or because zinc sculptures now qualify as antiques, these sculptures have increased in value, warranting the expense of professional care. Recently a pair of zinc Newfoundland dogs was sold for $18,500 at the Winter Antiques Show in New York (Reif 1987). Finally, the likelihood of damage has increased because zinc artwork has been weakened by a century outdoors. A sculpture such as the larger-than-twice-life-sized *Goddess of Liberty* (1888), which crowned the Texas State Capitol until 1985, cannot be ignored when its security becomes questionable (Weil 1987:168).

While the following discussion is intended to summarize all aspects of the conservation of zinc sculpture, an effort has been made to present material that has not been previously published. In particular, the emphasis is on American material.

History

Zinc sculpture was popular from about 1850 to the beginning of the twentieth century, although a reference to a zinc sculpture made as early as 1826 has been found (Servant 1880:7). The use of zinc for sculpture began after large-scale refinement of zinc in the West made the metal available at relatively low prices.[2] The invention of electroplating (1838) and the introduction of slush casting (1845) contributed to the use of zinc, particularly in France, because it allowed inexpensive reproductions to be made that looked like bronze but cost about one-sixth to one-eighth the amount (Tomlinson 1852, vol.1:xcvi). Later, the introduction of stamping allowed for even less expensive items to be made. A stamped zinc soldier cost $300 in 1891, while the same-sized sand-cast statue cost $450 (Mullins 1891; Monumental Bronze Company 1882).[3]

Zinc was used to depict a variety of sculptural subjects, and the sculptures themselves were displayed in a range of situations. Zinc copies of antique sculptures and works by contemporary masters such as Canova, Carrier-Belleuse, and Carpeaux were available on both sides of the Atlantic (Mott 1875; Poiré 1880:624). However, while small-scale copies of such sculptures were made in large quantities in Paris for domestic display, they do not seem to have been much produced in the United States.

Many zinc sculptures were publicly displayed. In Europe they were used to decorate the Cathedral in Helsinki, Finland; the Royal Opera House in Berlin; and a post office in Hanover, Germany (Deveranne 1845, 1847; *Scientific American* 1882). During this period in the United States, the rapid growth of cities created a demand for sculpture, and zinc sculptures were purchased throughout the country to decorate new county courthouses, parks, and cemeteries. Statues of Justice, fountains, and Civil War memorials were typical. The Civil War memorials, together with tombstones produced by the same companies, are probably the most numerous zinc items in the United States, and they are largely an American phenomenon.

Another type of zinc sculpture that seems peculiar to the United States is the show figure. Often used for advertising, show figures were frequently humorous. For example, a sculpture of a baby elephant that had been born to the Barnum, Bailey and Hutchinson Circus was available holding either a clock or cigars in its trunk. It was to be used as a sign for a watchmaker or tobacco shop, or as a fountain with water spewing from its trunk (Monumental Bronze Company 1882). A rotund King Gambrinus, now exhibited at the Arts and Industries Building of the Smithsonian Institution, was once mounted on the American Brewery in Baltimore, Maryland, as its symbol. A Rampant Colt decorated the dome of the Colt gun factory in Hartford, Connecticut, and two statues of Puck, the symbol of *Puck Magazine*, still decorate the Puck Building in New York City. Sculptures of stereotypical American Indians, originally introduced as substitutes for wooden cigar-store Indians, were popular items for parks and municipalities (Fried 1970; Knapp 1980).

In addition, a variety of utilitarian materials and objects, including architectural decoration, roofing, weather vanes, toys, coins, and household items such as buckets, were made with zinc.

Surface Treatment

The surfaces of zinc sculpture were invariably treated to imitate other materials, including bronze, gold, marble, brownstone, granite, polychromed wood, and plaster. Surface appearance was of primary importance to the original owners of these sculptures; hence, it remains a principal concern today. The introduction to an American zinc catalogue of the period contains a description of surface treatment that is worth quoting at length because it is both explicit and rare (Mott 1875).

> All the Statuary represented in this Catalogue is cast in Zinc, which is superior to Iron for this purpose from the fact that it can be cast in pieces, and when carefully fitted by skilled workmen, retains nearly all the delicate finish and beauty of line to be found in the original. It is also more enduring, as it cannot be affected by the weather like stone or marble; and again, should any part of the figure by accident be broken, it can easily, and at a mere nominal expense, be repaired; whereas, with Iron or Marble such damage is irreparable.
>
> Although the figures can be painted or bronzed in any style to suit the taste of the purchaser, we however, take this opportunity of giving our opinion as to the style and manner of painting or bronzing, which we think most suitable for the various figures.
>
> Animals, such as Game Dogs, Deer, &c. may be painted to imitate their natural color with very good effect. Mythological subjects, such as the Sphinx, Griffins, Dragons, large Dogs such as the Antique and St. Bernard, also Lions and Horses, should either be painted a stone color, or bronzed; large emblematic figures such as Industry, Justice, Commerce, and other figures used chiefly for buildings, look well bronzed, but can be painted white, or sanded in imitation of brown stone, whichever is more suitable for the building on which they may be placed.
>
> Our Figures, representing the best works of art, ancient and modern, can either be bronzed or painted white to suit the taste of the purchaser. Such figures as Kiss' *Amazon*, for instance, should always be bronzed, in fact nearly all the Statuary herein represented look well bronzed, although we should prefer such Figures as are used for Cemetery purposes painted white as being more appropriate, also Figures which show but little drapery can be painted white with good effect.

In practice, identification of the original paint on zinc sculptures can be difficult. Paint adheres poorly when exposed outdoors for as long as one hundred years, and original paint is either substantially lost or overpainted. Moreover, because the choice of a particular painted coating was often left to the purchaser, the type of paint chosen is not always predictable. Analyses of cross sections along with historical research can prove successful in determining the original coating, however. Although many sculptures were probably sold unpainted, some American trade catalogues do specify the availability of sculptures with different paint coatings at different prices. These include "painted one coat," "bronzed," "gilded," "painted natural," and "paint in colors" (Mott 1890). A listing of such catalogues and their library locations is found in Grissom (1982), and a fuller version that includes itemization of sculptures in trade catalogues is planned for future publication.

A discussion of what was meant by *bronzed* in American sculpture catalogues merits further discussion because it is not clear if the term denotes copperplating or a metal-flake paint (usually copper, but it could be brass or another alloy). Copperplating of zinc is almost unfailingly specified in European discussions of zinc sculpture, and many copperplated examples of European origin are known, including the famous statues of Amazon, Queen Victoria, and Liberty Enlightening the World. However, one may speculate that when American sculptures were bronzed, this generally meant that a copper-flake paint was applied. Although a translation of an influential French book on electroplating and electrotyping was published in Philadelphia (Roseleur 1872), these methods seem to have been less well known in the United States (Shapiro 1985:98–100). No specification of the use of electroplating of zinc sculptures is found in American literature until well into the twentieth century (Stone 1914; Rigg and Morse 1915, 1916). Indeed, it seems implausible that catalogues such as Mott's would not mention copperplating if it had been used, since these catalogues describe materials with some degree of specificity, and the use of plating would have been a feature to boast about.

Only one very late American sculpture is known which was originally copperplated.[4] However, evidence of the application of copper-flake paint has been found on a number of zinc sculptures. Bronzing is specified in a 1893 letter from the J. W. Fiske Iron Works regarding the purchase of a fountain at Sailors' Snug Harbor: "I will forward to you my #209 Neptune fountain in one coat of paint and when delivered at the grounds and set in position I will send one of my best men to bronze it in a first class manner giving it a final coat of spar varnish" (Fiske 1893). Analyses of cross sections of paint from the fountain show that a white-lead paint was applied first, apparently as the delivery paint; that thereafter a reddish-brown paint was applied, presumably as a background layer for the copper-flake paint; and, finally, that the copper-flake paint and two coats of varnish were applied (Fig. 1).[5] A bronze paint might be confused with plating on cursory examination, but when exposed to moisture during outdoor display, the surfaces of copperplated statues acquire an unmistakably pimpled appearance caused by pinhole corrosion (Fig. 2). Paint can also be readily distinguished from plating when examined in cross section with a microscope, although it would not be distinguished by elemental analysis.

Traces of gilding have been found on a number of zinc sculptures, including five larger-than-life-sized sheet-metal sculptures made by the Bakewell and Mullins Company for the San Diego County Courthouse, now in the collection of the San Diego Historical Society; and a statue of Puck on the Puck Building in New York City. On the courthouse sculptures, an oil size appears to have been applied to the sheet metal and then to the metal leaf, which proved to be nearly pure gold when

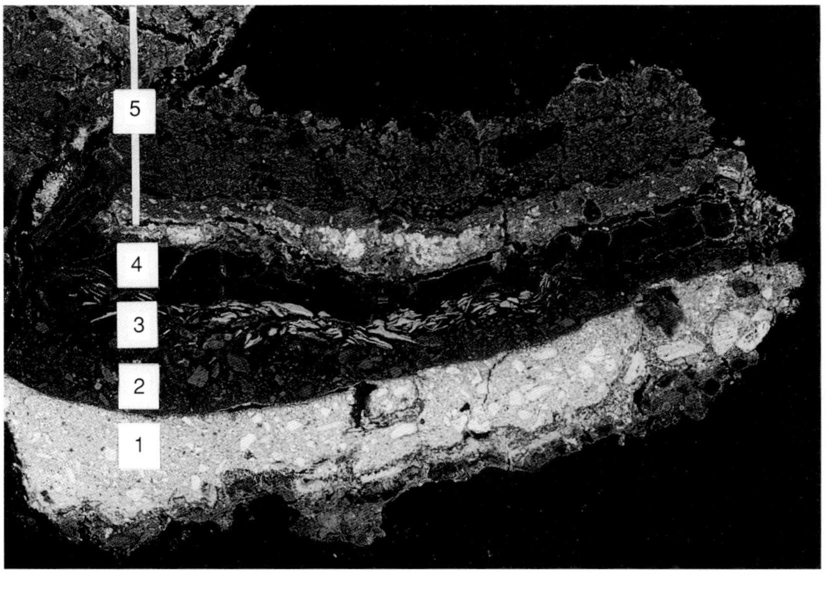

FIGURE 1. *Scanning electron photomicrograph (backscatter mode) of a cross section of paint from the Neptune Fountain, Snug Harbor, N.Y., sand-cast by the J. W. Fiske Iron Works, 1893. Numbered layers are as follows: (1) white-lead paint, (2) reddish-brown paint, (3) copper-flake paint, (4) varnish, and (5) overpaint. Magnification 56×.*

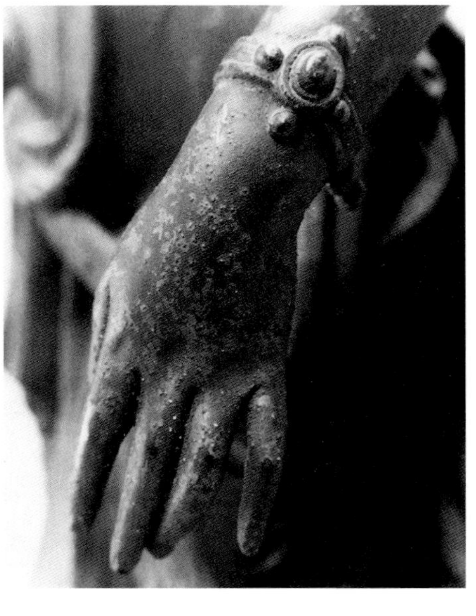

FIGURE 2. *Detail of pinhole corrosion on the left hand, Candelabrum, from the Old Merchant's Exchange Building, St. Louis, made of copperplated zinc.*

analyzed. On the sculpture of Puck, the leaf gold was applied over a white lead paint (Keck et al. 1983).

On some European sculptures, the zinc was patinated directly, usually black. Many patination recipes appear in the European literature, and examples of patinated European sculptures are known (Neumann 1869; Richter 1883:34; Hiorns 1892; *La coloration des métaux* n.d.; Jett 1982). However, there are no indications that this treatment was used in the United States.

The original surface treatment is fairly obvious to the informed observer when one technique was used. White Bronze was the trade name for a material produced by the Monumental Bronze Company and its affiliates.[6] The process was unusual: After models were sculpted to imitate stonework and cast in large sections in sand molds, they were sandblasted at the foundry to produce a uniform stonelike appearance (Fig. 3). White Bronze was used mainly for Civil War memorials and cemetery monuments, many of which bear the marks of one of the foundries.

Metal Fabrication

The fabrication methods used for zinc sculptures were sand casting, slush casting, and stamping.[7] Sand casting was the primary method for both iron and bronze casting in the nineteenth century, and it was the first method widely used for the fabrication of zinc sculptures. Sand casting was almost certainly used for unique sculptures and monuments, because there would have been less labor in making sand molds than in making metal molds for slush casting. It was also invariably the method used to make White Bronze monuments. Because the White Bronze monuments were sandblasted rather than coated, a minimum number of seams was desirable; this could be achieved by casting very large sections in sand. Sandcast sculptures, stiff and with relatively thick walls, were generally made without armatures unless they were very large; large White Bronze monuments, for example, have rudimentary supports in the lower sections. A sand cast can be identified by the remnants of sprues on the inside, and interior surfaces often have sandy textures.

Slush casting is said to have been first used to cast zinc in 1845. For a time thereafter in Paris it seems to have been used almost exclusively for the serial pro-

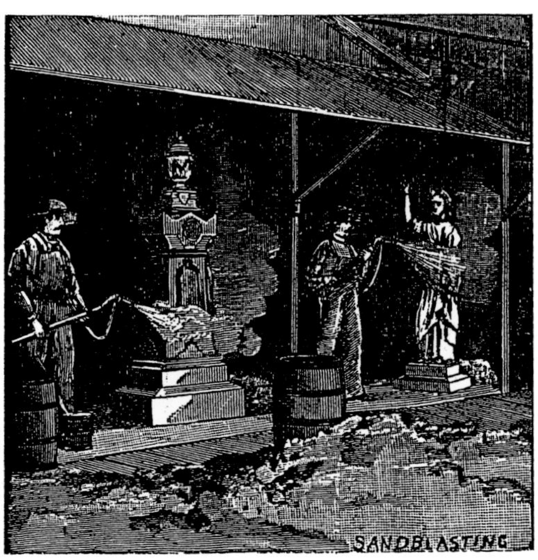

FIGURE 3. *Sandblasting white bronze (Western White Bronze 1893).*

duction of large quantities of zinc sculptures (Servant 1880:7). The molten zinc was poured into open bronze molds and the excess immediately poured back out, leaving a layer of zinc. This process was quick, and many copies could be made that retained fine details from the bronze molds, precluding the need for expensive chasing. However, the process permitted no undercutting within individual castings, and only relatively small sections were produced because the mold size was limited to that which a man could hold and pour. An 1871 advertisement for American show figures notes "a heavy outlay for Design, Moulds &c." This suggests that metal molds might have been purchased. By contrast, sand molds would have been remade for each casting (Fried 1970:34).

There is reason to question the extent to which slush casting was used in America, however. Bronze molds could have been readily supplied in Paris because of the active bronze industry there, but in the United States it would have been difficult to obtain good molds until very late in the century, given the state of sculpture founding (Shapiro 1985). An 1871 advertisement for American show figures notes "a heavy outlay for Design, Moulds & C." This suggests that metal molds might have been purchased; by contrast, sand molds would have been remade for each casting, and they could not have been purchased (Fried 1970:34). However, no American sculptures have shown evidence of slush casting, and no mention has been found of zinc slush casting in America until well into the twentieth century, and then the method was used mainly for light fixtures and small ornaments (Stone 1914; Rigg and Morse 1915, 1916).

Slush casting was generally not used for large sculptures, and sculptures cast in this way do not have armatures. Slush casts have smooth rather than sandy interior surfaces. Sometimes an artifact of pouring the molten zinc out of a mold can be found, visible as a lip marking the edge of the pour. Sprues, impressions of mold-parting lines, and undercuts within individual cast sections—all characteristics of sand casting—are not present on slush-cast sculptures.

Stamping of ornaments from sheet zinc is reported to have been discovered in France in 1849 and was in full force by 1865 (Turgan 1865:236). In America, the use of stamping is said to have begun in 1871 with importation of samples, machinery, and labor from France and Germany by a major zinc producer, the Matthiessen Zinc Company in La Salle, Illinois (*Sheet Metal Builder* 1874–75:120). The industry evidently grew rapidly, and after about 1880 most of the American trade catalogues selling zinc artwork are those of sheet-metal manufacturers. Rarely were stamped and cast items sold by the same companies. Stamped items were made primarily for architectural use, with the same designs made from rolled sheets of copper, brass, bronze, or zinc as the purchaser desired. One account describes the process in some detail: Dies were made from plaster models, the lower die of cast zinc and the upper die of lead. The metal was heated, then placed between the dies and smashed into shape with a drop hammer (Lemonte 1947:28). Sheet-metal sculptures are readily distinguished by their thinness, light weight, and smooth surfaces. They always have armatures because of their flexibility, although the armatures are often relatively flimsy.

Most zinc sculptures were made in small sections and subsequently joined with lead-tin solder. This was apparently because of the ease of joining zinc with solder, the fact that small sections were easier and cheaper to cast or stamp, and the fact that a surface coating would invariably hide the solder seams. Solder can be readily identified by its smooth, shiny surface, which often appears dark against the white corrosion products of weathered zinc. According to several catalogues, riveting was sometimes used for assembling stamped zinc sculptures.

The method for joining White Bronze monuments was altogether different. Lead-tin solder was not used except to spot-join sections prior to fusing, a process in which a large quantity of molten zinc was poured into the joint area from the back. The identification of the fusing material as zinc has been corroborated by analysis (Table 1).

Identification of the Metal

A zinc sculpture can be readily identified by simple examination techniques. Zinc has a characteristic luster, light bluish-gray color, and white corrosion products. The hardness of the metal is about the same as a fingernail (2.5 on the Mohs' hardness scale), and it is harder than lead. It does not rust like iron, nor is it magnetic. The metal to which zinc is most similar in appearance is aluminum, but zinc was used for artwork when aluminum was still relatively expensive. When broken, cast zinc reveals a highly characteristic large-grained fracture to the naked eye.

Other properties that are pertinent to conservation include the following: Zinc has a low melting point (419.58 °C), is brittle at ordinary temperatures, becomes malleable from 100 to 150 °C, and is brittle again at 200 °C. The thermal expansion of polycrystalline pure zinc is 39.7×10^{-6} cm/cm per °C, which is the highest of all common metals (Lyman 1948:1086). It recrystallizes at room temperature or below; as a result, it does not strain harden (nor can it be annealed) like copper alloys, contributing to its inability to resist continuous deformation.

Compositional analyses of zinc artwork are compiled in Tables 1 and 2. While performed by several analysts using a range of methods, the results nevertheless are remarkably consistent and show that the zinc used for statuary was relatively pure. The highest total for nonzinc elements is about 5%, but 1–2% is more typical.[8] Chiefly, lead, iron, and cadmium—which occur naturally in zinc ore—are found in zinc sculptures. However, lead may also have been added to improve fluidity, and iron may have been picked up from the crucible or ladle. The addition of 3–5% tin is reported to have been desirable in nineteenth-century German-language literature, and tin does appear in some analyses of American sculptures (Richter 1883:26; Hartmann 1863:288). Copper is present in these analyses only in low amounts. Aluminum was found in a measurable quantity in only one sculpture, in contrast to modern zinc alloys, which often contain 4% aluminum for use in die casting.

Although the number of comparisons is limited, higher-quality workmanship seems to correlate with higher percentages of zinc. Monuments produced by the Monumental Bronze Company are well made and carefully finished; several analyses show high purity, equivalent to the refined Missouri zinc from which they would have likely been made. This also confirms a contemporary statement that the company used "refined zinc of almost absolute purity" (Saw 1910:191).

TABLE 1a. *Elemental compositions of Civil War monuments made by the Monumental Bronze Company, Bridgeport, Connecticut.*

	Percentages											
	Zn	*Cd*	*Fe*	*Pb*	*Al*	*Cu*	*Ni*	*Sn*	*Mg*	*Ti*	*Mn*	*Total Non-Zn Measured*
Academy Hill Monument[a] (1889)												
	99.78	.091	.085	.039	.002	.001	<.002					.218
Appomattox Monument[b] (1886)												
Soldier	96.9	.084	.12	.58	.0062	.011		.69	.0063	.0013	.0003	1.499
Shaft	99.59	.054	.076	.27	.001	.0008		.0066	.0005	.0021	.0001	.411
Fourth Ohio Infantry Monument (1887)												
Shaft[c]	95.7		.29	.27	(<.03)	.002		(<.01)			(<.01)	.562
Fusing metal[d]	97.8		.19	.85	.9*	(<.01)		.12			(<.01)	2.07*

TABLE 1b. *Elemental compositions of sculptures made by other American foundries.*

	Percentages											
	Zn	*Cd*	*Fe*	*Pb*	*Al*	*Cu*	*Ni*	*Sn*	*Mg*	*Ti*	*Mn*	*Total Non-Zn Measured*
Concordia[c]												
	90.8		.38	2.41	.32	.017		1.71			(<.01)	4.837
Columbia Protecting Science and Industry[d] (1879)												
	95.0		.58	.91	(<.03)	.13		(<.01)			(<.01)	1.62
Captain Jack[d]												
	98.0		.25	1.35	.2	.09		.1			(<.01)	1.99
Flora[d]												
	96.0		.10	1.12	(<.03)	.09		.1			(<.01)	1.41

TABLE 1c. *Elemental compositions of a sculpture and fountain made by European foundries.*

	Percentages											
	Zn	*Cd*	*Fe*	*Pb*	*Al*	*Cu*	*Ni*	*Sn*	*Mg*	*Ti*	*Mn*	*Total Non-Zn Measured*
Liberty Enlightening the World[d] (1875)												
	98.5		.09	.07	(<.01)	.5*		(<.05)			(<.01)	.66*
Szymanow Fountain[e]												
Small bowl 1	97.3	.02	.21	1.2		.1		1.15			.01	2.69
2	93.3	.04	.20	1.1		.8		2.55			.01	4.7
3	84.8*	.02	.26	6.5*		.2		8.2*			.01	15.19*
Nodus	99.2	.01	.15	.25		.07		.08			.01	.57
Conch shell	99.3	.11	.1	.2		.06		.09			.01	.57
Base	99.4	.01	.15	.25		.05		.12			.01	.59

* Possible error in sampling or analysis.
(<) Indicates the detection limit.

TABLE 1d. *Elemental compositions of contemporary refined zinc from Missouri ores (1896).*[f]

	Percentages											
	Zn	*Cd*	*Fe*	*Pb*	*Al*	*Cu*	*Ni*	*Sn*	*Mg*	*Ti*	*Mn*	*Total Non-Zn Measured*
1	99.30	.006	.010	.653			trace					.704
2	99.34	trace	.032	.630			none					.661
3	99.27	.001	.055	.673			none					.727
4	99.54	trace	.052	.411			none					.463
5	99.04	trace	.023	.872			none					.956
6	99.65	trace	.028	.306			none					.346
7	99.38	trace	.036	.586			none					.623
8	99.58	trace	.039	.377			none					.416
9	99.68	.019	.048	.251			none					.318

Notes to Table 1

[a] Analyses (method not identified) by the AMAX Zinc Company at Sauget, Illinois (Michel, pers. comm. to Donegan, 1987). Percent zinc was determined by difference. The Academy Hill Monument is located in Stratford, Connecticut.

[b] Analyses by F. R. Sauerwine, Palmerton, Pennsylvania (Technical Services Report, The New Jersey Zinc Company, Sept. 7, 1983). The percentages of zinc, lead, and tin in the soldier were determined by atomic-absorption analysis, while the remainder of the analyses were made with an inductively coupled plasma spectrometer except for the percent zinc in the shaft, which was determined by difference. The Appomattox Monument is located at Jim Thorpe, Pennsylvania.

[c] Atomic-absorption analyses by Randy Korotov of the Earth Sciences department at Washington University in St. Louis, Missouri (Grissom 1979:11). The Fourth Ohio Infantry Monument is located at Gettysburg, Pennsylvania. The statue of Concordia is inscribed with the name Henry Dibblee of Chicago and is located on the Cairo Public Library Building in Cairo, Illinois.

[d] Atomic-absorption analyses by Harold Westley made at the Conservation Analytical Laboratory, Smithsonian Institution, in 1986. The statue of *Columbia Protecting Science and Industry* was sculpted by Caspar Buberl and is located over the north entrance to the Arts and Industries Building of the Smithsonian Institution. The statue of Captain Jack appears to be identical to a sculpture sold by Charles M. Demuth of New York and is located at Bear Brook State Park in Allentown, New Hampshire (Fried 1970:50, 61). These two sculptures may have been cast by M. J. Seelig, who made the monumental sculpture of Fulton for Buberl and who is said to have made the show figures for Demuth (Fried 1970:62–64). The statue of Flora, shown in a J. W. Fiske Iron Works catalogue, was located on the Waterworks of Louisville, Kentucky, from about 1895 until it was broken during a storm around 1977 (Starr, personal communication, 1991). The statue of *Liberty Enlightening the World* is inscribed with "Avoiron, Paris"; its present location is unknown. Copper found in the analysis of the Liberty statue may be contamination from copperplating on the sculpture.

[e] Atomic-absorption analyses (Nosek 1987:255–256). The small bowl is made of cast zinc. Variation in its tin and lead content is said to reflect variations in the melt, but it seems more likely that the third sample was contaminated by solder. The nodus, conch shell, and base are made of sheet zinc. The fountain is identified as eighteenth century, although this seems improbable because of the lack of availability of sheet zinc until rolling was perfected at the beginning of the nineteenth century.

[f] Analyses (method not identified) were done under the supervision of W. H. Seamon on ores from Missouri obtained from "nearly every smelter working" (Seamon 1896:462). Analyses were also done for arsenic, sulfur, bismuth, and antimony. With the exception of one sample that showed the presence of 0.06% sulfur and another that showed 0.035% arsenic, these elements were identified as either "trace" or "none." Percent zinc was determined by difference. As early as 1875 and until at least 1908, the Joplin, Missouri, district furnished ore from which 75% of the U.S. spelter output was derived (Ingalls 1908:289).

The finely finished Parisian sculpture of Liberty also shows a high degree of purity, particularly if the copper present is due to contamination from plating. The sculpture was most likely made of zinc from the Vieille-Montagne Company, not

TABLE 2a. *Percentages of six metals measured in samples from zinc sculptures made by the Monumental Bronze Company.*

	Total	*Pb*	*Fe*	*Cd*	*Sn*	*Cu*	*Al*
Academy Hill Monument (1889)	.22	.04	.09	.09	—	.001	.002
Appomattox Monument (1886)							
Soldier	1.49	.58	.12	.08	.69	.011	.006
Shaft	.41	.27	.08	.05	.01	.001	.001
Fourth Ohio Infantry Monument (1887)	.56	.27	.29	—	<.01	.002	<.03

TABLE 2b. *Percentages of six metals measured in samples from zinc sculptures made by other American foundries.*

	Total	*Pb*	*Fe*	*Cd*	*Sn*	*Cu*	*Al*
Flora	1.41	1.12	.10	—	.10	.09	<.03
Columbia Protecting Science and Industry (1879)	1.62	.91	.58	—	<.01	.13	<.03
Captain Jack	1.99	1.35	.25	—	.10	.09	.20
Concordia	4.84	2.41	.38	—	1.71	.02	.32

TABLE 2c. *Percentages of six metals measured in samples from zinc sculptures made in France.*

	Total	*Pb*	*Fe*	*Cd*	*Sn*	*Cu*	*Al*
Liberty Enlightening the World (1875)	.66	.07	.09	—	<.05	.50	<.01

(<) indicates the detection limit for atomic-absorption analysis.

only because this was the main source for the Paris trade but also because the company's product was reported to be very pure (Tomlinson 1852, vol.1:xcvii; Lami 1888:1136; Société 1900:31). At the other end of the spectrum is the statue of Concordia, a sculpture that was made casually enough to incorporate a large piece of iron into its fabric and shows the highest percentage of nonzinc elements. The sheet metal used for the Szymanow Fountain is notable for its purity and consistency, which one would expect for sheet metal.

For metallographic examination, zinc samples are most commonly etched with aqueous chromium trioxide (CrO_3) activated with sodium sulfate (Na_2SO_4; Anderson 1959:400–8). Definition of grain structure can best be obtained by using polarized light. The columnar structure of cast zinc, which crystallizes by the hexagonal close-packed system from the mold wall inward, is readily distinguished (Figs. 4, 5). Grains vary in size, but in sand-cast zinc they are often very large, as long as 3.5 mm. Mechanical twins are usually visible as light-colored bands. Iron-zinc intermetallic crystals ($FeZn_{13}$) are usually present and are very distinctive. Their size and number give an approximation of the amount of iron; a sample from the sculpture of Columbia shows many particles (Fig. 6), in keeping with its high-iron content. Lead,

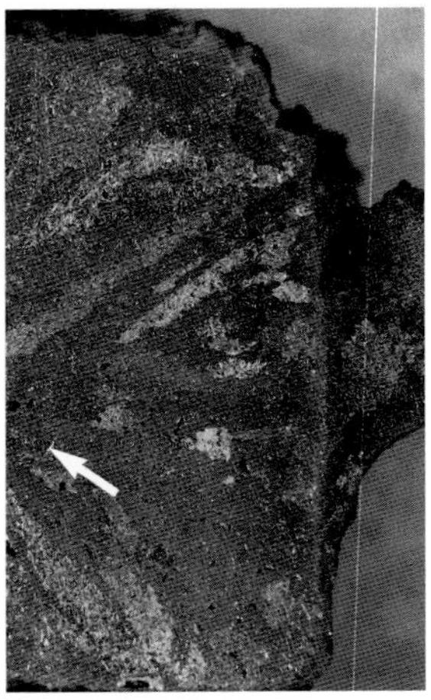

FIGURE 4. *Macrophotograph of a cross section of zinc from Columbia, sculpted by Caspar Buberl, 1890, Smithsonian Institution, Washington, D.C. Note columnar grains in the sand-cast zinc; an iron-zinc intermetallic particle is indicated with an arrow. The sample was taken from the edge of a casting; solder is visible at the right. Magnification ×15.*

which is insoluble in zinc, is often visible as dark globules segregated at grain boundaries. Copper (in amounts up to about 1%) and cadmium (in the usual low amounts) form solid solutions with zinc, and they produce no change in microstructure except for coring. Tin or aluminum in a binary system with zinc results in eutectic structures at grain boundaries, even at low concentrations.

DETERIORATION

Inherent vice is one of the principal reasons for the deterioration of zinc sculpture. The brittleness of cast zinc allows it to be easily broken, and its tendency to creep results in distortions, particularly on large monuments. Environmental factors also play a role. Painted coatings do not endure outdoors for as long as a century and leave exposed surfaces disfigured by crisscrossing lead-tin solder seams. Airborne sulfur dioxide causes serious corrosion of uncoated zinc. The presence of water as an electrolyte results in galvanic corrosion of copperplated zinc. Conservation treatments also have considerable potential for damage when inappropriate materials are used.

Breakage is probably the single most common reason why treatment for zinc sculptures is sought. Unlike bronze sculptures, from which a sword is occasionally stolen or an arm broken off, zinc sculptures tend to fragment into a large number of pieces when broken. Breakage seems to occur most frequently because of vandalism. Many zinc sculptures are situated in isolated locations in parks or cemeteries, which makes them easy targets; many are poorly maintained, which tends to attract further damage; and, above all, they are simply easy to break. Natural events sometimes cause damage, such as the tornado that resulted in the loss of zinc sculptures from the porch of the Louisville Waterworks of Louisville, Kentucky, in 1890. These were replaced with a second set of sculptures, one of which was smashed in a thunderstorm in about 1979. Damage from handling, transport, and removal from buildings is common, although usually less dramatic; solder seams open up, or protruding appendages such as arms are cracked or broken off. Breakage of stamped sculpture is much less common than for cast sculptures. Stamped sculptures are more often bent from trauma or separated at the solder seams.

Sagging and bulging as a result of the tendency of zinc to creep most often affect large zinc monuments, notably the soldiers' memorials made by the Monumental Bronze Company and its affiliates. Measuring 9 m or more in height, many of these monuments display nearly identical deterioration. The figure of the soldier shown in Figure 6, originally made with the rifle positioned vertically at the center of the body, now sinks into the base at the right heel and leans noticeably as a result, while lower sections of the monument sag and bulge on the same side.[9] In one case the soldier holds a large flag at his side and leans in that direction.[10] Although historic photographs and past treatments indicate that the sagging of these monuments has been occurring for a long time, recent concern about sagging has led to the installation of internal support systems[11] or disassembly in anticipation of treatment. In fact, so many of these monuments have either been treated or are presently disassembled that it is difficult to obtain a good photograph showing the extent of lean. Slight sagging has also been noted around points of reinforcement. Load has resulted in cracking at points of stress, as in areas adjacent to reinforced corners (Fig. 7) or where bulging has occurred. Sheet-metal sculptures are less affected by creep, but they are often distorted by physical actions. Their armatures usually require replacement because they have become bent or corroded.

Corrosion is negligible on sculptures that have been continuously coated, but many have weathered to a significant extent after coatings have been lost. In the case of White Bronze monuments, many have weathered significantly because such monuments were never coated at all. The extent of corrosion can sometimes be estimated by examining surface relief at solder seams. Assuming that the zinc and solder were originally level, loss of zinc can be measured against the solder, because the solder corrodes very slowly. Corrosion has a particularly large impact on the strength of sheet-metal sculptures because of their intrinsic thinness, made even thinner at the extremities from being stretched during stamping. Perforations are not uncommon on stamped sculpture.

Because the corrosion of outdoor bronze sculptures is familiar, it is useful to compare damage from atmospheric corrosion of zinc with the corrosion of copper and copper alloys. In cemeteries where both bronze and White Bronze sculptures of the same age are found, the zinc sculptures often appear to be in better condition. This is probably because the surfaces are of a relatively uniform white hue, while bronze sculptures are typically streaked and irregularly pitted. It is unfortunate that no data are available for cast bronze or cast zinc. Studies of long-term corrosion rates are limited to those for sheet zinc and sheet copper. Annual losses tend to be about twice as high for sheet zinc as for sheet copper, depending on the environment. It should be noted that the corrosion rates for zinc are far lower than those for steel, which is one of the reasons that zinc is used so extensively for galvanizing iron (Anderson 1956).

Relatively insoluble zinc-carbonate films form on zinc in clean air. While these films are believed to be stable and somewhat protective, data for rural State College, Pennsylvania, show a 1.12 μm per year loss of zinc sheet compared to a 0.75 μm per year loss of copper sheet over a seven-year period.[12] Never-coated White Bronze

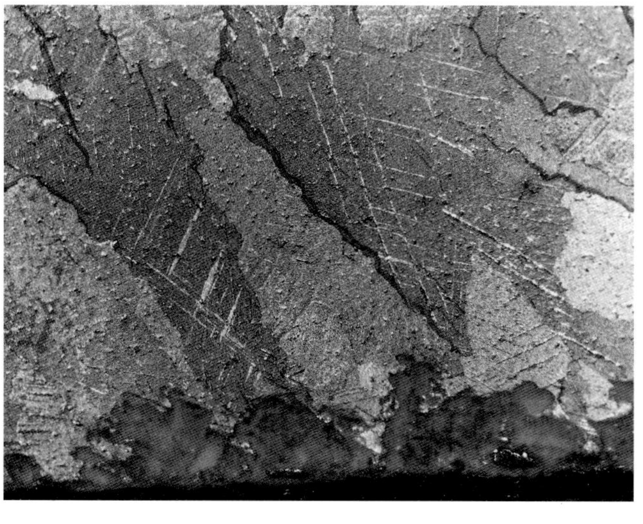

FIGURE 5. *Photomicrograph of a cross section of zinc from the Neptune Fountain. Corrosion of the surface is visible at the edge as well as between columnar grains; white bands are mechanical twins. Magnification ×12.*

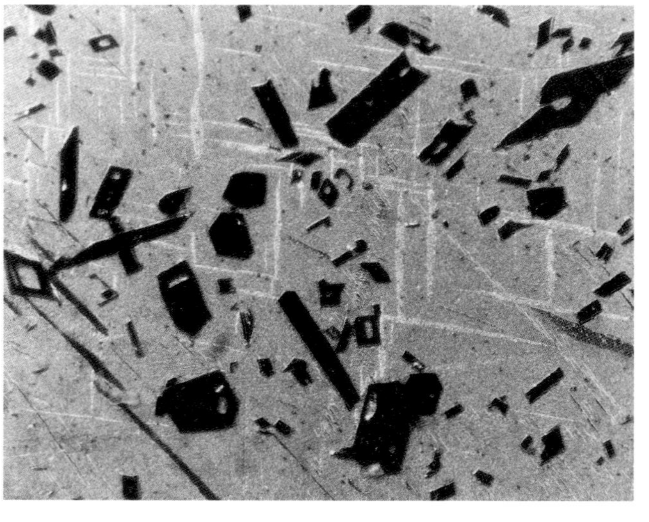

FIGURE 6. *Photomicrograph of a cross section of zinc from Columbia. The dark-colored geometric structures are iron-zinc intermetallic particles. Magnification ×35.*

FIGURE 7. *Soldiers' Monument, Winneshiek County Courthouse, Decorah, Iowa, sand-cast and sandblasted by the Western White Bronze Company, 1886.*

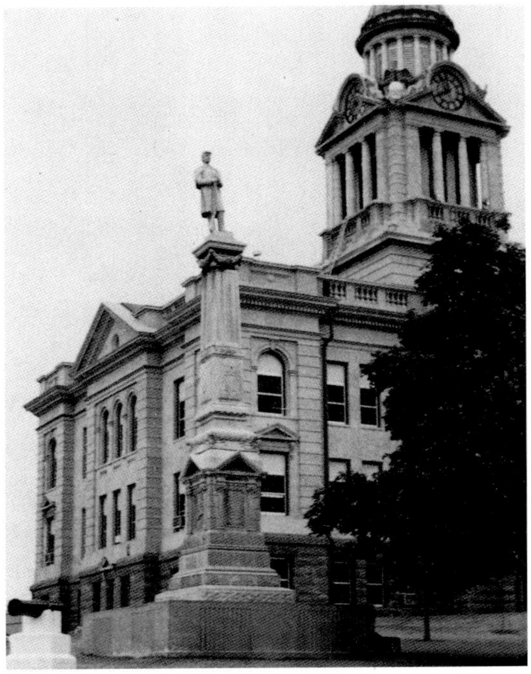

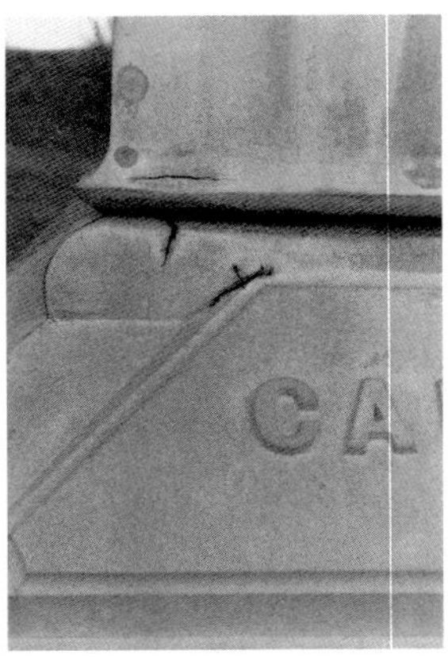

FIGURE 8. *Detail of stress cracks, Fourth Ohio Infantry Monument, Gettysburg National Military Park, Gettysburg, Pennsylvania, sand-cast and sandblasted by the Monumental Bronze Company, 1887.*

monuments located in rural environments display little apparent surface loss or pitting. However, metallographic examination of sections taken from sculptures such as the Fourth Ohio Infantry Monument at Gettysburg shows some attack at the grain boundaries.

By contrast, high levels of sulfur dioxide in the air result in dissolution of the zinc-carbonate film on the surface of zinc, and soluble zinc sulfates are formed. Data for an industrial site at Newark, New Jersey, show a 4.1 μm per year loss for zinc sheet compared to a 1.45 μm per year loss for copper sheet. The Neptune Fountain, which is located on Staten Island, New York, directly across from a large industrial area, exhibits severe corrosion (Fig. 9). Sulfur has been detected in its corrosion products, in addition to an unusual vanadium compound that might also derive from industrial pollution (Wypiski 1990). A metallographic section made from this sculpture shows fairly uniform surface dissolution on both the exterior and interior of the sample, unrelated to grain boundaries, in addition to intergranular corrosion (Fig. 5).

Chlorides are thought to play a smaller role in zinc corrosion than they do in bronze corrosion, unless sculptures are directly sprayed with seawater. While experimental samples exposed 80 feet from the sea at Kure Beach, North Carolina, showed high levels of damage similar to those for industrial areas (4.1 μm per year for zinc sheet and 1.66 μm per year for copper sheet), at 800 feet from the sea the rates of corrosion were similar to those for rural exposure. White corrosion products on a black-patinated French sculpture that was exposed outdoors have been identified by X-ray diffraction analysis as zinc hydroxychloride; however, the source of the chlorides in this case is probably the coloring solution used to patinate the sculpture (Jett 1982).

Trapped water can cause catastrophic perforation of zinc, a phenomenon known as white rust or wet-storage stain. This is believed to occur because the dense, somewhat protective zinc-carbonate film does not form when the surface has remained wet, restricting access of carbon dioxide. Many instances of the perforation of zinc roofs from water trapped on the underside are reported in the literature, especially

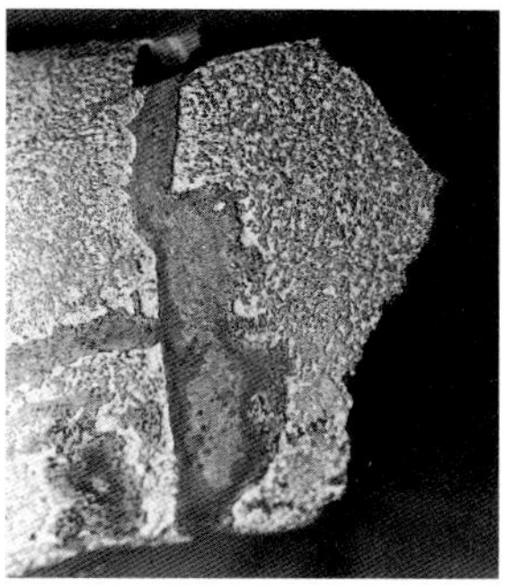

FIGURE 9. *Detail of the severely corroded right wrist, Neptune Fountain. Dark material is lead-tin solder.*

when hygroscopic roofing papers or other materials were used (Pourbaix 1948). Perforation attributable to white rust has not been observed on any zinc sculptures, probably because the sculptures are hollow and generally drain well. However, sealing between the sections of a large zinc monument, as is sometimes proposed by analogy to masonry monuments, is not desirable because moisture might be trapped inside. Furthermore, use of interior support materials—such as concrete, glass-reinforced polyester, or epoxy resins—might trap water between the zinc and the filling material if contact is imperfect or if voids occur upon weathering or aging.

Zinc is less noble than gold, silver, copper, lead, tin, or iron; in theory it should corrode in the presence of these metals. In practice the corrosion of zinc depends on the particular alloy and the atmosphere in which the two metals are found.[13] The most common alloy found in contact with zinc is the lead-tin alloy used for soldering. Solder joins often stand in relief above cast zinc, yet for the most part corrosion on these sculptures has been observed to be no more severe at the interface of solder and zinc than it is in other areas. Similar observations have been made in regard to zinc roofing (Schikorr 1965:6.1). Lead strips are sometimes used to support areas that are sagging, and no corrosion problems have been noted, although the softness and pliability of lead tend to make statues repaired in this way more vulnerable to vandalism.

Copperplating on zinc sculptures produces severe pitting when a sculpture is exposed outdoors (Fig. 2). Of two copperplated zinc exemplars of the Statue of Liberty that have been examined, only the sculpture stored outdoors shows pitting attack. Stainless steel, carbon steel, and aluminum are used as metals for auxiliary supports, and brass, carbon steel, stainless steel, galvanized iron, and aluminum bolts and washers are often used to replace the original zinc hardware used to secure sections of White Bronze monuments. Although there is considerable potential for damage from many of these metals, particularly the steel and copper alloys, in practice little damage has been observed on zinc monuments or sculptures. This is probably because the metals are usually isolated by air gaps, paint, or other materials.

Data are conflicting about the attack of zinc building materials in contact with Portland cement mortar or concrete (Jones 1957). Theoretically, concrete should attack zinc because concrete is alkaline, and damage could also be caused by calcium-sulfate salts, which are contained in Portland cement, and result in the formation of zinc sulfate.[14] Probably more serious, however, are problems that might arise because of the lower thermal coefficient of expansion for concrete, which ranges from 5 to 14 times 10^{-6} cm/cm per °C (Lea 1971:412). Stress would be created by expansion differences as the temperature rises and falls: cracking might appear in the zinc at points of weakness, or gaps might occur that could result in damage from trapped water, either from the formation of white rust or from water expansion upon freezing. Concrete is difficult to remove, and its presence inhibits repair. Finally, concrete is stronger than zinc and, should any force be applied, the latter would be damaged first. Unfortunately, concrete is a popular repair material.[15]

Zinc is attacked by strong acids and alkalis. Hence, alkaline paint strippers that contain sodium and/or potassium hydroxide (lye) should not be used to remove paint from zinc. Care should also be taken after soldering to rinse the surface of zinc immediately and thoroughly to remove all traces of flux (usually hydrochloric acid).

Treatment

Treatment of zinc sculptures that are located indoors is rarely a problem. Damage is typically minor, and these sculptures can be repaired by conventional conservation methods. Treatment of outdoor zinc sculptures is more problematic. Damage tends to be more severe, the sculptures are in worse condition, and there are few precedents for treatment. In some cases, one reluctantly comes to the conclusion that a zinc sculpture should not be placed outdoors after treatment. The Goddess of Liberty on the Texas State Capitol was replaced with an aluminum copy (Weil 1987:168).

Repair of broken zinc is probably the principal and most difficult treatment problem. Soldering is the traditional method of repair, and its past performance as the original method of joining provides evidence that it generally holds up well outdoors. It has two other desirable attributes: it tends to break instead of the zinc, and it can be removed by careful heating because of its lower melting point. Pieces can generally be soldered if the zinc is in good condition. Some conservators report difficulties with the adhesion of solder to aged zinc, however, and in some cases, the number of cracks and breaks renders soldering unfeasible. There are two other drawbacks: soldering requires aggressive surface cleaning, and heating the metal permanently alters its structure, weakening it further where it has already been weakened by breakage. While soldering may prove to be appropriate for the treatment of other broken zinc sculptures which are to be painted, it is never a suitable treatment for a damaged White Bronze monument because the solder will disfigure the uncoated monument.

Use of welding has been reported in a few instances, most particularly where it was believed that structural strength was required or where a sculpture had previously been repaired by welding. However, this method is irreversible, and when further breakage occurs, the zinc rather than the weld tends to break.

One alternative to these methods of repair is the use of glass-reinforced resins. Glass-reinforced polyester resins have been used successfully for many years for making boats, vehicle bodies, and water tanks. Glass-reinforced epoxy resins, which are more expensive but are usually considered more weather resistant, have been used for high-load-bearing and engineering applications. Moisture is considered the most serious environmental hazard for these composites, as it causes swelling of the resins. Nevertheless, survival in excess of twenty years is hypothesized (Davis and Sims 1983:266–87). While conservators may be familiar with the use of these composite resins for the treatment of museum objects, they generally have very little experience in using them for outdoor treatments. Problems could be caused by differences in thermal expansion rates between zinc and the resins, although the expansion rates of the resins can be modified by judicious use of fillers. Furthermore, removal of these resins would most likely be an unpleasant task.

In some cases, broken areas can be effectively repaired with metal armatures, straps, or plates. A metal support might be bent or cast to conform to the shape of the broken area and attached mechanically or with solder or resin.

Large losses have most often been filled with cast-zinc sections. If the new section is to be soldered in place, zinc alloys containing aluminum should be avoided.

New sheet-zinc sections can be readily shaped to fill missing areas on stamped zinc sculptures. Small losses in zinc were traditionally filled with lead-tin solder, and this method is still used (Nosek 1987). Use of epoxy and polyester putties, some metal-filled, are probably more common.

On the one hand, distortions in cast metal, whether from breakage or sagging, should be regarded as permanent, and no attempt should be made to force them back into position. On the other hand, sheet metal can tolerate some hammering. Warming to 40–50 °C ("just too hot to touch") is recommended (*Data Sheet* 1983).

Installation of an internal metal structure is generally the best solution for the large zinc monuments that are sagging and leaning. The support should be carefully designed so the zinc does not sag around points of contact and the statue at the top is supported independently of the remainder of the monument. The design should also permit adjustments of level to be made during erection, because distortions of some sections may require visual compromises to be made on site. Stainless steel is generally the material of choice, although it is expensive. A carbon-steel support could be used, but it would be difficult to maintain and would probably be more expensive over the long term. Recently a carbon-steel tower was replaced inside the Academy Hill Monument because it had been allowed to deteriorate. Points of contact between the structure and the zinc should be isolated. These monuments will likely still appear to be leaning even after great sums have been expended for support structures, because the damage is permanent; this should be emphasized to clients in advance.

A range of cleaning methods has been reported for removal of extraneous material prior to repainting or coating, but further work seems warranted on the relative effectiveness and safety of these methods. The author's microscopic examination of a corroded sculpture fragment which was abrasive-blasted with walnut shells showed that loose corrosion products were removed but more adherent corrosion products remained. The zinc did not appear to be damaged by the blasting, although at high pressures the lead-tin solder joins were slightly dimpled.[16] No studies are known on damage to zinc from blasting with glass microspheres, but their use has been reported both with an air-abrasive unit to remove thick accumulations of paint and repair putties and with standard blasting equipment to remove paint (Klingelhofer 1990; Weil 1987). Corrosion removal seemed to be controllable and safe on a copper-plated zinc sculpture that was cleaned by the author with an air-abrasive unit using glass beads; pustules of zinc corrosion products were removed without turning the surface bright copper. Sandblasting should be avoided because it damages the zinc. Successful use of high pressure washing with water has been reported for removing the localized zinc corrosion on copperplated zinc sculptures at Osborne House on the Isle of Wight.[17] High-pressure water spraying was used to remove paint from a sheet-metal cornice on the Verona apartment building in New York City, although care was required not to dimple sheet-zinc portions (Pieper 1990). Use of methylene-chloride-based paint stripper or neat organic solvents may be sufficient if removal of paint alone is required. As mentioned previously, the use of alkaline paint strippers should be avoided.

With the exception of White Bronze monuments, outdoor zinc sculpture should be coated for both durability and aesthetic reasons. An attempt should be made to ascertain the nature of the original coating, and if that cannot be determined, a suitable choice should be made based on similar sculptures. Assuming that there would be far too little of any original coating to permit simply retouching areas of loss, the appearance of the original coating should be replicated with high-quality paints. Use of a good-quality primer and compatible vinyl-acrylic copolymer paint would be suitable. If applied by a professional paint shop, use of a more durable system might be attempted, as with an inorganic zinc primer and aliphatic polyester polyurethane paint. In any case, the ephemeral nature of paint should be emphasized to clients, and a schedule for repainting put into writing.

Recopperplating a zinc sculpture that is to be displayed outdoors is not recommended. Replating requires very aggressive cleaning of a sculpture, and the weathering prospects for a copperplated sculpture are poor. A metal-flake paint of similar coloration would be the best alternative.

Regilding a sculpture requires careful selection of the metal leaf, and consideration might be given to the application of a protective coating. Ten years after regilding with 22-carat gold on a potassium chromate primer, the statue of Puck showed signs of gilding wear, as well as lifting of the paint primer in areas that tended to remain wet. Application of a transparent protective coating had originally been planned, but by the time the oil size had dried, the scaffolding had been removed (Silver, personal communication, 1991).

Application of benzotriazole to the surface of a White Bronze monument has been proposed, apparently based on the beneficial effects it can have on the corrosion resistance of copper and other metals. However, a laboratory study has shown that stimulated corrosion occurs when zinc sheets are dipped in solutions of water, sulfur dioxide, and benzotriazole, particularly when the sheets are first dipped in a benzotriazole solution and then soaked in solutions of water and sulfur dioxide. A sulfuric acid and benzotriazole solution also increased the dissolution of the zinc (Walker 1974).

Conclusion

A considerable body of knowledge is now available concerning the history and conservation of outdoor zinc sculpture. The repair of broken zinc fragments remains a major problem, and care should be given to the replication or preservation of original surface appearance. However, as always in the case of outdoor sculpture, regular maintenance is the most difficult issue.

Notes

1. Taft made models for the Western White Bronze Company from about 1886 to 1892. His letters indicate that this was primarily to pay his bills: "Papa needn't worry *yet* over my finances for I drew on the White Bronze Co., only for $50.00 to pay my rent so there is still the same am't coming from them when ever I get the critter cast" (May 13, 1886), and "I have been having so much small work in the studio of late, little medallions, advertising novelties, etc. . . . There is a prospect of making considerable money in that way even if high art seems to prove something of a failure and fame a 'goak'" (March 10, 1889).

2. European production increased from less than 5,000 English tons (2240) per annum in 1830 to 17,000 in 1840 and 70,000 in 1855. The price of zinc fell steadily from 1808 to 1820, rose somewhat from then until 1825 with the extension of sheet rolling, and then fell back to low levels. In the United States sheet zinc was imported from as early as the 1820s, but regular production of zinc did not occur until 1859. By 1870, 4,500 English tons were produced per annum and 21,000 by 1880 (Smith 1918:14–17,151–152). The price of zinc remains among the lowest of all metals.
3. These sculptures were listed as measuring 6 feet high, and they were probably copied after J. Q. A. Ward's bronze *Seventh Regiment Soldier* for Central Park in New York City. The contract for that 10-foot-high statue specified payment of $23,000, including $3,000 for installation (Seventh Regiment Memorial Association 1868). A sheet-zinc soldier of similar height (9 ft) sold for $450, which is less than one-twentieth the price of its bronze prototype (Mullins 1891).
4. The author has examined a copperplated zinc tag inscribed CHICAGO NEW YORK/ ORBRONZE/TRADE MARK/DAPRATO STATUARY CO. which was formerly attached to a lamentation group (dated to the 1920s) in the Catholic Cemetery in Fort Wayne, Indiana. The plating on the figures was described as badly abraded and erupting with "small barnacle like flowers," which led to sandblasting and painting as part of treatment in 1978 (Doyle, pers. comm. 1982). John G. Waite reports that the sculpture of an American Indian in Schenectady, New York, was originally copperplated, with copper found by instrumental analysis. The sculpture was replated after glass-bead blasting, so the original coating is no longer present. However, an identical statue located at Forest Glen in Washington, D.C., retains copper-flake paint under several coats of repaint. A second copperplated sculpture of a soldier in Berlin, New York, is reported by Waite (1991). However, no convincing evidence of plating could be found when that sculpture was examined.
5. Elemental analyses of the Neptune Fountain and the San Diego Courthouse sculptures were performed at the Conservation Analytical Laboratory of the Smithsonian Institution using a JEOL JXA-840A scanning electron microscope with energy-dispersive X-ray analyzer, with the assistance of Melanie Feather. Contradictory results for paint from the Neptune have been reported, but the sample used for those analyses has been subsequently found to consist only of overpaint (Wypiski 1990).
6. These included the Western White Bronze Company of Des Moines, Iowa, the American Bronze Company of Chicago, the Detroit Bronze Company, and the St. Thomas White Bronze Company of St. Thomas, Ontario, Canada. Further historical information on the companies is provided by Rotundo (1989).
7. Use of several other methods are mentioned in the literature, but they appear to have been rare. Gelatin mold casting is reported for small sculptures in one publication (Hartmann 1863:289–90). In the French literature, good results with lost-wax casting by the French founder Blot are noted late in the century, but the impliction is that this was unusual (Susse 1894:32–33). Examples of Danish sculptures cast by the lost-wax method have been examined by Holm (herein).
8. By contrast, considerably higher amounts are found in Indian bidri ware. Copper (2–10.1%), lead (0.5–19.9%), tin (0.1–11.4%), and iron (0.1–1.2%) were found in twenty-six atomic-absorption analyses of bidri objects (La Niece and Martin 1987).

9. The soldier on the Fourth Ohio Infantry Monument at Gettysburg National Military Park leaned backward and slightly to his proper left before being dismantled. The soldier at the top of the Appomattox Monument at Jim Thorpe, Pennsylvania, leaned backward 21 cm off-center and slightly to one side when bidding specifications were written in 1984 prior to repair. The soldier on the Western White Bronze Company's Soldiers' Monument at Decorah, Iowa, leaned backward 37 cm off-center and somewhat to its proper left prior to recent dismantling. A Western White Bronze Company monument in the Union Cemetery in Humboldt, Iowa, is an exception: it tilts forward. Two other soldiers' monuments are found in Iowa, a Western White Bronze Company monument in Eldora and a Detroit Bronze Company monument in Mason City, and they are also said to be deformed (Brown 1986).
10. The soldier at the top of the Academy Hill Monument in Stratford, Connecticut, holds a flag to his proper left. Although his right heel also sinks downward, producing a slight backward lean, most of the lean has been toward the flag at his left, measured as 20 cm off-center prior to installation of a new support structure in 1987. After reassembly, measurements show that the soldier still leans 7 cm off-center, because distortions did not permit complete correction.
11. The installation of a steel tower was proposed for the Appomattox Monument at Jim Thorpe, Pennsylvania, in 1984, and a stainless-steel tower has been installed inside the Academy Hill Monument at Stratford, Connecticut, after the failure of an earlier carbon-steel tower. Around 1986 the Hardin County Civil War Monument at Eldora, Iowa, was filled with reinforced concrete.
12. Two twenty-year exposure studies were carried out by the American Society for Testing and Materials (ASTM) at a range of sites throughout the United States, beginning in 1931 and again in 1957. While all of the more recent data for zinc has been published, only data measured at the seven-year mark have been published for copper. As a result, comparisons are made here for the seven-year data from the 1957 study, between high-grade zinc and tough-pitch copper (Dunbar and Showak 1982:534; Thompson 1968:133). Rates for tin bronze (8% Sn/92% Cu) were also determined in the 1931 study, and they are substantially higher than rates for tough-pitch copper at the most polluted and salt-ridden sites but lower than those for zinc (Copson 1956). However, in the 1931 study samples were displayed vertically, which introduces some error, contrasting with the now-standard disposition at 30° from the horizon. A recent runoff study carried out for the National Acid Precipitation Assessment Program shows lower damage rates than both ASTM studies, although neither the method of measurement nor the sites are strictly comparable (Cramer and McDonald 1990).
13. Testing has shown that in outdoor atmospheres strong effects occur when zinc is placed in contact with stainless steel, carbon steel, copper, silver, gold, and platinum. On the other hand, only small effects, except in very aggressive atmospheres, were found when zinc was placed in contact with nickel and its alloys, chromium, aluminum and its alloys, cadmium, magnesium and its alloys, lead, tin, and lead-tin alloys (Barton 1976:164).
14. On Edwardian and Victorian roofs in England, sheet-zinc soakers were widely used at the abutments between parapet walls and chimney stacks, with cement fillets in place of flashings. The destruction of the soakers has been reported at this juncture, attributed to the soluble salts in the cement (Melville and Gordon 1973:492, 524).

15. The Academy Hill Monument in Stratford, Connecticut, has been filled to a height of 28 cm, and cracking seems to be associated with the presence of the concrete. The sculpture *Columbia Protecting Science and Industry* at the Smithsonian Institution has been filled to a height of 66 cm, but lack of documentation makes it more difficult to attribute the considerable damage found on the sculpture to the concrete. The entire Hardin County Civil War Monument in Eldora, Iowa, was filled with reinforced concrete around 1986 after the interior was coated with epoxy to prevent reaction with the zinc; this monument has not been examined.
16. Blasting with English walnut shells was performed by Nicholas Veloz of the National Capitol Parks using standard equipment at the same time that a bronze sculpture was being cleaned. The walnut shells, Agroshell AD 10.5B, were purchased from MDC Industries in Philadephia, Pennsylvania. In an attempt to induce damage, a maximum pressure of 80 lbs. per square inch was tested, using a 5/32 inch nozzle held approximately 4 cm from the sample, and this pressure produced dimpling of the solder. Less pressure (35 lbs. per square inch) seemed to produce similar cleaning of the zinc without dimpling the solder and is thought to be more suitable for actual use. A fountain which was reported by a symposium participant to have been damaged by walnut-shell blasting has been subsequently identified as being made of lead.
17. The sculptures, some of which were made by the Geiss foundry in Berlin and some of which were made by French foundries, were purchased for the garden of Queen Victoria's estate (Harris, pers. comm. 1991).

References

ANDERSON, E. A.

1956 The atmospheric corrosion of rolled zinc. In *Symposium on Atmospheric Corrosion of Non-ferrous Metals*, ASTM Special Technical Publication no. 175:126–34. Philadelphia: American Society for Testing Materials (ASTM).

1959 Physical metallurgy of zinc. In *Zinc: The Science and Technology of the Metal, Its Alloys and Compounds*. C. H. Mathewson, ed. 400–22. New York: Reinhold Publishing Corp.

BARTON, K.

1976 *Protection Against Atmospheric Corrosion*. John R. Duncan, trans. London: John Wiley and Sons.

BROWN, J. L.

1986 Will last for all time. *The Bracket* 2(1):2–4.

THE BUILDER (LONDON)

1849 Cast zinc in decoration. 7:353.

DE CASO, J.

1975 Serial sculpture in nineteenth-century France. In *Metamorphoses in Nineteenth-Century Sculpture*. Jeanne L. Wasserman, ed. 1–27. Cambridge: Harvard University Press.

LA COLORATION DES MÉTAUX

n.d. (after 1914) 2d ed. Orleans, France: H. Tessier.

COPSON, H. R.

1956 Report of Subcommittee VI, of Committee B-3, on Atmospheric Corrosion. In *Symposium on Atmospheric Corrosion of Non-ferrous Metals*, ASTM Special Technical Publication no. 175,: 3–19. Philadelphia: ASTM.

CRAMER, S. D., AND L. G. MCDONALD

1990 Atmospheric factors affecting the corrosion of zinc, galvanized steel, and copper. In *Corrosion Testing and Evaluation: Silver Anniversary Volume, STP 1000*. Robert Baboian and Sheldon W. Dean, eds. 241–59. Philadelphia: ASTM.

CRAVEN, W.

1968 *Sculpture in America*. New York: Thomas Y. Crowell Co.

THE CRYSTAL PALACE EXHIBITION ILLUSTRATED CATALOGUE.

[1851] 1970 Reprint. New York: Dover Publications.

DATA SHEET 48

1983 Zinc. *Metal Construction* 15(6):345–46.

DAVIS, A., AND D. SIMS

1983 *Weathering of Polymers*. London: Applied Science Publishers.

DEVERANNE, S. P.

1845, 1847 *Zinkguss Arbeiten nach Entwuerfen und Modellen bewaehrter Baumeister, Bildhauer und Zeichner unserer Zeit; so wie nach Antiken Skulturen und aelteren und neueren Kunstwerken geformt und ausgefuehrt in der Zincgiesserei fuer Kunst und Gewerbe*, 6 vols.

DUNBAR, S. R., AND W. SHOWAK

1982 Atmospheric corrosion of zinc and its alloys. In *Atmospheric Corrosion*. W. H. Ailor, ed. 529–52. New York: John Wiley and Sons.

J. W. FISKE IRON WORKS

1893 Letter to Gov. G. D. S. Tract, Sailors' Snug Harbor. New York: Snug Harbor Archives.

FRIED, F.

1970 *Artists in Wood: American Carvers of Cigar Store Indians, Show Figures, and Circus Wagons*. New York: C. N. Potter.

GAYLE, M., D. W. LOOK, AND J. G. WAITE

1980 *Metals in America's Historic Buildings*. Washington, D.C.: Heritage Conservation and Recreation Service, Technical Preservation Services Division.

GRISSOM, C. A.

1979 *A conservation study of the Fourth Ohio Infantry Monument at the Gettysburg National Military Park, Gettysburg, Pennsylvania*. Report. Washington University No. 78021/13. St. Louis: Center for Archaeometry.

1982 *The conservation of zinc sculpture*. Final report for National Museum Act Grant FC-105379, Washington, D.C.

HARTMANN, C. F. A.

1863 *Das Kupfer und das Zinc*. Weimar: B. F. Voigt.

HIORNS, A. H.

1892 *Metal-colouring and Bronzing*. London: Macmillan and Co.

INGALLS, W. R.

1908 *Lead and Zinc in the United States*. New York: Hill Publishing Co.

JETT, P.

1982 *Conservation record, TL 19.588*. Report. Research Laboratory, Museum of Fine Arts, Boston.

JONES, F. E.

1957 Corrosion of metals in buildings. *Chemistry and Industry*. August 3:1050–63.

KECK, L., F. MATERO, AND C. S. SILVER

1983 What fools these mortals be! Typescript. *The Conservation and Restoration of a Zinc Statue of Puck*.

KLINGELHOFER, A.

1990 Conservation of a cast zinc sculpture. In *American Institute for Conservation, Abstracts of papers presented at the 18th Annual Meeting, Richmond, Virginia,* 44–45. Washington, D.C.: American Institute for Conservation.

KNAPP, O. A., JR.

1980 *Chief Kisco and His Brothers*. Mount Kisco, N.Y.: Mount Kisco Historical Committee.

LAMI, E.O.

1888 *Dictionnaire encyclopédique et biographique de l'industrie et des arts industriels*, vol. 8. Paris: Librairie des dictionnaires.

LA NIECE, S., AND G. MARTIN

1987 The technical examination of bidri ware. *Studies in Conservation* 32:97–101.

LEA, F. M.

1971 *The Chemistry of Cement and Concrete*. New York: Chemical Publishing Co.

LEMONTE, A. D., ED.

1947 *A Story of People: The Story of Mullins: The Story of Your Job*. Salem, Ohio: Mullins Manufacturing Corp.

LYMAN, T., ED.

1948 *Metals Handbook*. Cleveland: The American Society for Metals.

MELVILLE, I. A., AND I. A. GORDON

1973 *The Repair and Maintenance of Houses*. London: The Estates Gazette Limited.

MONUMENTAL BRONZE COMPANY

1882 *Illustrated Catalog of White Bronze Monuments, Statuary, Portrait Medallions, Busts, Statues and Ornamental Art Work for Cemeteries, Public and Private Grounds and Buildings.* Bridgeport, Conn.

J. L. MOTT IRONWORKS

1875 *Illustrated Catalogue of Statuary, Fountains, Vases, Settees, etc. for Parks, Gardens and Conservatories.* New York: E. D. Slater.

1890 *Illustrated Catalog and Price List of Statuary and Animals.* New York: E. D. Slater.

W. H. MULLINS COMPANY

1891 *Catalogue of Architectural Ornaments and Statuary in Sheet Zinc, Brass or Copper,* no. 6, E.

NEUMANN, P.

1869 Ueber Herstellung einer dauerhaften schwarzen Patina auf Zinc. *Dinglers Polytechnisches Journal* 192:479–86.

NOSEK, E. M.

1987 Conservation of outdoor zinc monuments. In *Conservation of Metal Statuary and Architectural Decoration in Open-air Exposure,* 253–65. Rome: ICCROM.

PIEPER, R.

1990 Restoring metal roof cornices. *Preservation Tech Notes, Metals,* no. 2. Washington, D.C.: National Park Service.

POIRÉ, P.

1880 *La France industrielle ou description des industries francaises.* 3d ed. Paris: Librairie Hachette.

POURBAIX, M.

1948 Corrosion des toitures en zinc. *Metaux et Corrosion* 23(278):215–25.

REIF, R.

1987 Old pets find new homes. *The New York Times,* February 15.

RICHTER, K.

1883 *Zinc, Zinn, und Blei.* Vienna: Hartleben Chemisch-technische Bibliothek.

RIGG, G., AND H. C. MORSE

1915 Spelter impurity in slush castings. *The Foundry* 43:494–501.

1916 The effects of the common impurities in spelter upon slush castings. *Transactions of the American Institute of Metals* 9:26–59.

ROSELEUR, A.

1872 *Galvanoplastic Manipulations. A Practical Guide for the Gold and Silver Electroplater and the Galvanoplastic Operator.* A. A. Fesquet, trans. Philadelphia: Henry Carey Baird.

ROTUNDO, B.

1989 Monumental Bronze: A representative American company. In *Cemeteries and Gravemarkers: Voices of American Culture*. Richard E. Meyer, ed. 263–91. Ann Arbor, Michigan: UMI Research Press.

SAW, A. B.

1910 Ornamental white bronze castings. *The Foundry* 35:191–95.

SCHIKORR, G.

1965 *Atmospheric Corrosion Resistance of Zinc*. New York: American Zinc Institute.

SCIENTIFIC AMERICAN SUPPLEMENT

1882 Group of statuary, representing the International Postal Union, on the new Post Office General, at Hanover, Germany. 13(329):5239.

SCOTT, J., JR.

1987 Examination and conservation of zinc sculpture. In *American Institute for Conservation*. Preprints of papers presented at the 15th annual meeting, Vancouver, British Columbia, Canada.

SEAMON, W. H.

1896 The application of sheet zinc for roofing and other purposes. *The Engineering and Mining Journal* 62:389–90, 413–14, 437–39, 461–62.

SERVANT, M. G.

1880 *Les bronzes d'art*. Paris: Imprimerie nationale.

SEVENTH REGIMENT MEMORIAL ASSOCIATION

1868 *Contract with John Quincy Adams Ward*. Ward Material, Box 36, Albany Institute of History and Art, May 11.

SHAPIRO, M. E.

1985 *Bronze Casting and American Sculpture 1850–1900*. Newark, Del.: University of Delaware Press.

SHEET-METAL BUILDER

1874–75 The use of stamped zinc. 1:120.

SMITH, E. A.

1918 *The Zinc Industry*. London: Longmans, Green and Co.

SNODIN, M., ED.

1991 *Karl Friedrich Schinkel: A Universal Man*. New Haven and London: Yale University Press.

SOCIÉTÉ ANONYME DES MINES ET FONDERIES DE ZINC DE LA VIEILLE-MONTAGNE

1900 *L'industrie du zinc*. Paris: Exposition universelle.

STONE, G. C.

1914 Spelter-manufacture and properties. *Transactions of the American Institute of Metals* 8:174–95.

TAFT, L.

1886–92 Various letters to family and to and from the Western White Bronze Company in Des Moines, Iowa. Taft Papers. Urbana, Ill.: University of Illinois Library.

[1903, 1924] 1969 *The History of American Sculpture.* Reprint. New York: The Macmillan Co.

THOMPSON, D.H.

1968 Atmospheric corrosion of copper alloys. In *Metal Corrosion in the Atmosphere, STP 435*, 129–40. Philadelphia: ASTM.

TOMLINSON, C., ED.

1852 *Cyclopaedia of Useful Arts, Mechanical and Chemical, Manufactures, Mining, and Engineering,* 2 vols. London and New York: George Virtue & Co.

TURGAN, [JULIEN]

1865 Mines et fonderie de zinc de la Vieille-Montagne. In *Les grandes usines. Études industrielles en France et a l'etranger,* vol. 3. Paris: Michel Levy Freres.

WAITE, J. G.

1991 National Building Museum lecture, Washington, D.C.

WALKER, R.

1974 The use of benzotriazole as a corrosion inhibitor for immersed zinc. In *Proceedings of the 14th Sem. Electrochem., Central Electrochemical Research Institute, Karaikudi, India, November 1973*, 394–400. Karaikudi: Central Electrochemical Research Institute.

WEIL, P. D.

1987 Problems in the conservation of zinc sculpture in outdoor exposure. In *Conservation of Metal Statuary and Architectural Decoration in Open-air Exposure*, 157–68. Rome: ICCROM.

WESTERN WHITE BRONZE COMPANY

1893 *White Bronze Monuments and Statuary*. Advertising pamphlet distributed at the World's Columbian Exposition.

WYPISKI, M.

1990 Technical examination of a 19th c. zinc outdoor sculpture. In *Papers Presented by Trainees at the Fifteenth Annual Art Conservation Training Programs Conference, April 28*, 93–109. Cambridge: Center for Conservation and Technical Studies, Harvard University Art Museums.

Biography

Carol A. Grissom studied art history at Wellesley College and the University of Chicago, earned her master's degree in art conservation from Oberlin College and received postgraduate training at the Institut Royal du Patrimoine Artistique in Brussels and at the Istituto Centrale per il Restauro in Rome. She served as assistant conservator at the Center for Archaeometry at Washington University in St. Louis from 1977 to 1982, where she worked on outdoor sculpture and developed particular interests in the treatment of stone and zinc. From 1982 to 1983 she worked as Special Exhibitions Conservator at the National Gallery of Art in Washington, D.C., and since 1984 she has been chief objects conservator at the Conservation Analytical Laboratory of the Smithsonian Institution.